PATHOLOGICAL SPECIMENS AND GENOMIC MEDICINE

Emerging Issues

PATHOLOGICAL SPECIMENS AND GENOMIC MEDICINE

Emerging Issues

editors

Michael A Tangrea
Sinai Hospital, USA

Avi Z Rosenberg
Johns Hopkins University, USA

Michael R Emmert-Buck
Avoneaux Medical Institute, USA

World Scientific

NEW JERSEY • LONDON • SINGAPORE • BEIJING • SHANGHAI • HONG KONG • TAIPEI • CHENNAI • TOKYO

Published by

World Scientific Publishing Co. Pte. Ltd.

5 Toh Tuck Link, Singapore 596224

USA office: 27 Warren Street, Suite 401-402, Hackensack, NJ 07601

UK office: 57 Shelton Street, Covent Garden, London WC2H 9HE

Library of Congress Cataloging-in-Publication Data
Names: Tangrea, Michael A., editor. | Rosenberg, Avi Z., editor. | Emmert-Buck, Michael R., editor.
Title: Pathological specimens and genomic medicine : emerging issues / edited by
Michael A Tangrea, Avi Z Rosenberg, Michael R Emmert-Buck.
Description: New Jersey : World Scientific, 2019. | Includes bibliographical references and index.
Identifiers: LCCN 2018049397 | ISBN 9789813276222 (hardcover : alk. paper)
Subjects: | MESH: Pathology, Molecular--methods | Genomics--methods
Classification: LCC QH447 | NLM QZ 50 | DDC 572.8/6--dc23
LC record available at https://lccn.loc.gov/2018049397

British Library Cataloguing-in-Publication Data
A catalogue record for this book is available from the British Library.

For any available supplementary material, please visit
https://www.worldscientific.com/worldscibooks/10.1142/11154#t=suppl

Typeset by Stallion Press
Email: enquiries@stallionpress.com

Foreword

The present book describes some of the major emerging issues in the field of molecular pathology and personalized diagnostics, both the successes to date and the challenges that affect the implementation and practice of the new field of genomic medicine. None of the chapters are intended to be a comprehensive treatise of the subject matter as many of these topics could produce an entire book unto themselves, but rather each chapter provides an overview and then highlights some of the most exciting and vexing issues that the field is addressing now or will need to confront in the future.

The chapter by Dr. Emmert-Buck is an introduction to molecular pathology and genomic medicine that describes the current state of the field and some of the key issues moving forward. Drs. Williams and Moskaluk provide an excellent overview of the history of the practice of molecular pathology, presenting a historical context and foreshadowing many of the developments of today. Chapters on sources of specimens and front-end tissue processing follow the Introduction section, which is a critical aspect of the genomic medicine workflow since what happens at this early stage influences everything that follows. The chapter by Dr. Chuaqui addresses the use of biorepositories in genomic medicine and the chapter by Dr. Tangrea discusses the topic of sample heterogeneity and the role of tissue microdissection in properly assessing samples in the research laboratory and clinic.

The next section of the book deals with fascinating and topical subject matter — the place of digital imaging in the practice of molecular pathology and genomic medicine. The four chapters cover multiple areas within this space, including the potential of the technology to augment or replace standard review of glass slides under the microscope, and importantly the implications of digitizing the information contained within histopathological images, such that it becomes amenable to a wide variety of machine learning and artificial intelligence-related computational methods and advances. The four sets of authors touch on and emphasize different topics in the chapters, yet one sees overlap in their views of this fast-moving area and a number of common themes emerge that deserve careful watching over the next several years. The implications for genomic medicine likely will be profound.

The following section of the book contains chapters that cover molecular analysis of clinical specimens in the laboratory. This section is not a comprehensive description of the use genomic assays to study tissue samples, nor does it provide a lengthy list of protocols or techniques as this sort of information is covered in many books and review articles. Rather, the authors select and discuss key issues they believe are most impactful on the genomic medicine workflow, today and looking forward to the future.

To provide a flavor for how all of this comes together in the real world, the final section of the book covers informatics and implementation. The chapter by Dr. Johann describes computational issues around high-end genomic analysis and DNA sequencing, and the chapter by Dr. Delsante describes molecular diagnostic principles and practice applied to renal disease. In order to view the role and evolution of genomic medicine in a specific clinical scenario, the chapter by Dr. Hill discusses the pathologic and molecular features of a tumor syndrome, including the role of molecular diagnostics in understanding the pathogenesis as well as the clinical management of children and families. The final chapter by Dr. O'Leary is a superb review of the establishment and use of large genomic cohorts. These resources are critical in understanding DNA variants and diseases *writ large*, including how genomic information can be used successfully for individual patients. Along with digital imaging and artificial intelligence applied to pathology samples, the development and utilization of genomic cohorts is a field well worth watching over the next several years.

The hope is the book is an interesting and useful overview of emerging issues in pathological specimen processing and genomic medicine that is accessible to all groups who are interested in this topic. Certainly, it is an exciting time for the field of molecular pathology, the many challenges ahead notwithstanding.

Contents

Introduction

1 Overview of Molecular Pathology and Genomic Medicine

Michael R. Emmert-Buck

Avoneaux Medical Institute, Baltimore, MD 21215, USA

Contents

I. Introduction

A convergence of new molecular and bioinformatics capabilities for DNA sequencing and functional genomics are advancing the field of medicine, offering great promise to improve patient-based research and clinical practice.[1–14] In their broadest scope, these developments likely will impact a large segment of the healthcare sector, starting with oncology but then expanding into drug metabolism, cardiovascular disease, and other applications.

Often referred to as personalized, precision, or genomic medicine, the strategy of utilizing data generated from individuals as a guide to prognosis or therapy is not new *per se* as physicians have long employed patient-specific information for clinical care. However, the field is evolving rapidly now due to the success of the Human Genome Project, development of new molecular analysis technologies, and creation of novel bioinformatics methods.[15–23] Moreover, measurements are advancing from single-gene mutation assays into more complex, often multiplex tests that include targeted gene panels, exomes, whole genome sequencing, and functional genomic analyses that evaluate transcripts, proteins, and metabolic processes.[24–34] (For additional information, see chapter by Williams and Moskaluk.)

In cancer, the new capabilities in nucleic acid sequencing and expression measurements make it increasingly possible to uncover actionable alterations that can extend the quantity and quality of life for patients.[35] For example, molecularly targeted treatment of lung tumors (see Section II) has shown a positive clinical benefit and is championed as one of the early success stories of precision medicine.

When one stands back and views the history of molecular pathology, oncology, and the development and use of new medicines, these advances in genome-based precision medicine are nothing short of breathtaking, representing a paradigm shift from a one-size-fits-all strategy into a science-driven, industrial-scale enterprise that identifies specific molecular alterations in individual patient tumors to optimize clinical care. Thus, it would appear that the dream of turning the accomplishments of many decades of basic research into improved outcomes for patients, the goal of a large cadre of investigators, clinicians, funding agencies, and patrons, is close at hand.

Or is it?

In opposition to proponents of this technology-intensive strategy for genomic medicine are skeptics who claim the promise of personalized medicine is over-hyped and the strategy has yet to produce significant benefit.[36–39] They argue the grand vision being portrayed today and promised for tomorrow may never materialize, or may not materialize without significant alterations in strategy, and that the health care field would be better served by pursuing alternative approaches that are more certain to produce value for society.[37]

Both sides of the argument make good points. Certainly, the early genomic medicine successes have enlivened the oncology field and provide a pathway for the future.[40] Yet successful proof-in-principle accomplishments and early successes are not the same as creating a robust and reliable system that significantly improves both the quantity and quality of patients' lives on a comprehensive scale, across many years and diseases, and does so in a relatively non-intrusive and cost-effective manner.[41] The scientific and clinical concerns of the skeptics are real and significant.

From a high-level view, current and future challenges for the field of genomic medicine include the cost of new diagnostic technologies and companion drugs, regulatory concerns, an inadequately trained workforce, a need for new bioinformatics tools, improved data interoperability and storage, and a stable and predictable reimbursement landscape.

Like many debates, the truth of the potential versus the over-promise of precision medicine probably lies somewhere in between; there is great possibility in a grand vision of genomic medicine and there are ordeals and limitations ahead, too.[42] Both statements are true to some extent, and the debate is not yet settled. Certainly, though, what seems clear today is that an optimal molecular pathology ecosystem is needed at the ground, day-to-day level to help uncover whatever potential in genomic medicine does exist.

II. Lung Cancer — Past, Present, and Future

In order to crystallize some of the opportunities and challenges for precision medicine in the molecular pathology laboratory, we consider lung cancer as an example. Historically, treatment of pulmonary tumors, both small-cell (SCLC) and non-small cell (NSCLC) has been notoriously difficult, with dismal survival rates for patients. The introduction of chemotherapy and radiation therapy in the second half of the 20th century, along with improved surgical management techniques, was effective in improving progression-free survival (PFS), offering oncologists and patients hope and an arsenal of new tools to combat the disease. However, as of today the clinical benefit of treatment is generally measured in months, not years or decades, and the field has hit a plateau in producing efficacious therapies.

Today, molecularly targeted treatment of lung carcinoma using genomic guidance is slowly beginning to show a positive clinical benefit. For NSCLC, many of the molecular alterations that underlie tumor formation and progression are known and there are new drugs available against several genetic lesions in lung cancers. Approximately 60% of tumors contain mutations in the *KRAS, EGFR, MET, PIK3CA, ERBBS* or *BRAF* genes, while others have genomic fusions in *ALK, ROS,* or *RET.* Using DNA sequencing, FISH, and/or immunohistochemical (IHC) staining as companion diagnostic techniques, molecular pathologists can determine the specific etiology in a relatively high percentage of cases. This matters clinically. For example, several studies have shown that patients with EGFR mutation-positive tumors who are treated with tyrosine kinase inhibitors (TKIs, e.g., gefitinib) exhibit a stronger survival benefit than those without mutations. These data highlight the important need to correctly identify genomic status in the pathology laboratory, a requirement that will grow steadily more important as new drugs and additional actionable molecular changes, or simultaneous sets of changes, are identified in the responsible gene(s) in primary lung tumors.

Looking ahead, it may become equally important to detect and accurately measure DNA mutations and alterations in metastases and/or in individual primary tumor sub-clones, to account for the effects of tumor heterogeneity (see Section III). An on-going evolution in methodologies is likely to be necessary in the future, with algorithms on when, how, and what molecular assays to employ changing over time as technological capabilities advance and more evidence is generated on clinical effectiveness of new agents. The diagnostic laboratory will need to be agile and forward leaning as this process unfolds.

In addition to novel drugs that target actionable DNA, RNA, or protein changes in lung tumor cells, there is new excitement in the field of immuno-oncology, both for lung cancer and other tumor types.[43–47] Several clinical trials have shown a positive benefit from targeting the PD-1/PD-L1 axis, a regulatory control system in lymphocytes, offering patients a novel and valuable therapy. The optimal long-term molecular pathology strategy for determining biomarker status in the tumor microenvironment is yet to be determined, although early efforts using IHC staining have shown value and there are now four FDA-approved companion diagnostic tests available in the marketplace. There is a complex interplay among neoplastic and host cells in the tumor environment and a substantial body of basic science knowledge that the molecular pathology community can draw from to assist in development of the optimal technologies and markers to guide therapy in the future. Thus, the methods for measuring immuno biomarkers may eventually need to include more complex genomic, transcriptomic, and/or proteomic assays to assess both tumor and inflammatory cells.[48]

In parallel with tissue analysis, liquid biopsies are likely to play an increasingly important role in monitoring tumor progression in patients with lung cancer, as an adjunct and complementary tool.[49,50] Blood and other body fluids provide a non-invasive method for monitoring tumor status via their disseminated component, with potentially better capability in detecting mutations from a heterogeneous tumor cell population in which alterations are present in only a subset of tumor cells.[51] Liquid biopsies may become particularly useful for identifying tumor progression and drug resistance when it occurs. For example, third-generation TKIs are effective against T790M *EGFR* gene mutations that arise in patients receiving front-line therapy. Detection of drug-resistant tumor cells by monitoring DNA mutations in the circulation may provide earlier detection of tumor progression and a chance to alter treatment plans as needed. The main questions moving forward are around the sensitivity and specificity of liquid biopsies and how to properly integrate them within the framework of solid tumor testing.

Taken together, the new tools, strategies and potential for personalized treatment of lung cancer are impressive, building on a solid basic science foundation and setting the stage for increased benefit for patients. Again, though, critics will push back on the apparent success for treating pulmonary neoplasms. Just like with the advent of standard chemotherapy that extended PFS in patients by a relatively short time, months as opposed to years or decades, the new genomic medicine-based therapies also produce only modest improvements in PFS, and often at great financial expense and sometimes with significant toxicity.

Importantly, the critics argue the relatively small clinical benefit observed thus far is not a harbinger of more progress to come, but rather a dead end with a limited long-term upside, much like the trajectory that occurred with standard chemotherapy. In particular, they will point to the substantial genomic instability observed in clinically aggressive NSCLC and SCLC, and the rapid evolution of drug-resistant tumor genotypes and phenotypes that emerge under selection pressure as insurmountable long-term challenges to successful targeting of specific DNA lesions. In their view, a precision medicine strategy will never work well for clinically problematic and aggressive tumor types, which are the ones that produce the largest share of patient morbidity and mortality.

Although the critics make good points here, there is still reason for hope and optimism. Continued basic science investigations of lung cancer biology, in both the laboratory and clinic, will uncover new insights into disease etiology, including additional actionable changes, and R&D efforts in pharmaceutical and biotechnology companies will provide an ever-increasing armamentarium of therapeutic agents. But perhaps the most important development will be effective use of combinations of drugs and other modalities, for example, treating patients with multiple TKIs that are rationally applied to inhibit pathway nodes in lung cancer cells, along with concomitant use of immunotherapeutic agents to positively affect the lymphocyte-based response to lung cancer.[52–54] However, such an evolution in therapeutic strategy will further complicate matters in the molecular pathology laboratory, adding to the list of challenges and opportunities that will need to be addressed in the workflow described in Section III.

III. Specimen Workflow — Issues and Considerations

The effects of specimen processing must be considered throughout the entire clinical workflow. There are many issues to evaluate regarding how best to collect, fix, embed, study, and interpret data derived from patient samples. Moreover, it is important to view each component as part of a larger, interconnected whole, where the upstream steps significantly affect downstream assay performance and the quality of data that are produced. The broader context of each particular patient sample also needs to be factored into the interpretation of results, especially if molecular measurements are performed on labile species such as RNA, or if one is determining DNA methylation profiles and epigenetic status. Patient-specific elements range from exposure history to individual genotype, age, sex, race, health status, and the type and timing of the medical intervention used to procure a clinical sample. Clearly, the genomic medicine workflow is complex and extends well beyond just a single molecular assay.

Notably, there is one cultural issue around integration of histopathology with new genomic technologies that requires a brief editorial comment, since it can significantly impact current implementation efforts as well as future improvements in the field of precision medicine. Some in the scientific community state that the goal of precision medicine is to replace the practice of anatomic pathology with molecular analyses and thus, by extension, replace pathologists, in other words to replace the "art of histopathology" with the "science of molecular measurements". Conversely, many pathologists decry this attitude and point to the multitude of past failures by those who have attempted to perform "pathology via instrumentation or algorithm" as an inability of laboratory scientists to grasp the complexities (and opportunities) associated with real-world pathobiology and clinical care.

An alternative view is to recognize the need to employ histopathology as a platform on which to build genomic medicine. The field of anatomic pathology and its practitioners represent a rich source of current and historical information on human biology and disease. Importantly, a histological slide is a unique look directly into morphological features that arise from the biomechanical and molecular status of cells and structures, aided by two- and three-dimensional assessments of the geographical localization of normal and disease cells. This is a valuable glimpse into fundamental aspects of biology and how disease is manifested in complex tissue environments, and it is essential to integrate anatomic pathology with genomic measurements to synergize biological knowledge and clinical utility. In my view, histopathology sits prominently at the forefront of the clinical workflow, the bedrock platform from which genomic medicine best flows.

Of course, the specimen workflow is a work in progress, based on a historical underpinning of past successes and development, and today with many moving parts (see the Chapter by Williams and Moskaluk for more information). For cancer as an example, the biomedical research enterprise has steadily moved toward more sophisticated interrogations of neoplastic cells and the tumor microenvironment, evolving from a "black box" approach that regarded cancers as a single homogeneous entity, into an appreciation of the complexity of *in vivo* tumor biology and the cellular, clonal, and structural heterogeneity that comprises cancer. Technological innovations were key at many of the inflection points over the past several decades, including development of IHC staining, fluorescence-based methods, and manual and laser dissection techniques that enabled new scientific insight into the rich biological milieu of the tumor microenvironment.

Certainly, there is a need for yet more technological innovation to address vexing scientific and medical aspects of cancer. How do tumor stem cells and non-stem cells differ at the molecular level? What are the unique proteomic characteristics of the invasive tumor edge that might represent new therapeutic targets? Are tumor cells with nuclear atypia an especially aggressive subset with extensive genomic heterogeneity that is reflected by the spatial disruption of the nucleus? What is the variability in molecular profiles across tumors and how does this influence the emergence of drug resistance? How do reactive fibroblasts associated with a protective desmoplastic response differ from other reactive fibroblasts in the microenvironment? Is there a molecular difference between subsets of inflammatory cells that are closely associated with tumor versus those that are located distant from the cancer? Are there unique and important expression levels of key targets in tumor cells that are in a specific phase of the cell cycle and identifiable via immunostaining?

To address complex and heterogeneous disease microenvironments, the issue of sample and cell purity becomes paramount. Today, most samples that undergo genomic testing are processed as whole tissue curls or are macrodissected under the guidance of a pathologist. However, looking ahead, more precise laser microdissection methods may become necessary, such as the instrumentation shown in Figure 1. There is an accumulating body of evidence indicating laser dissection is useful, and in some scenarios essential, for accurate analysis of cells and pathological processes in tissue specimens.[55–64] However, its day-to-day use as a clinical

Figure 1. Laser microdissection instrumentation.

tool is largely unexplored territory and it is not known if the technology could improve patient-based assays in a meaningful way. This topic is presented and examined in more detail in the chapter by Tangrea.

Due to technological improvements in DNA sequencing and related functional genomics assays, there is now a strong interest in the research and clinical communities in applying new single-cell analysis technologies to archival specimens to answer a broad range of biological questions.[20,22,23,65,66] In cancer, these include examining the role of stem cells in tumor formation, the molecular basis of progression, tumor heterogeneity, the interplay between neoplastic cells and inflammatory cells for immunotherapy, and other areas. However, a major, unsolved technical problem looms ahead that will need to be addressed by the molecular pathology community. Analysis of single cells (or even small numbers of cells) from formalin-fixed, paraffin-embedded (FFPE) tissues generally does not work well. This is in contrast to the many studies generating useful data from large (e.g., whole slide) FFPE input amounts. Certainly, one can obtain information on mutations or on expression levels of a few genes from a small number of FFPE cells, but generation of deep and reproducible molecular profiling data sets is difficult, and these data are typically neither informative nor reliable.[21,67–70] Moreover, mutational analysis of small numbers of formalin-fixed cells is complicated by chemical modifications induced in DNA that can produce an outsized effect as template input decreases (discussed in detail in the chapter by Williams).

It does not appear that the use of formalin fixation will change anytime soon for reasons of cost and cultural inertia, and for retrospective studies of archival specimens the damage has already been done. These archival specimens in particular represent a crucial and invaluable resource for understanding the clinical behavior of tumors, including the development of biomarkers and therapeutics linked to patient outcome data. Looking ahead, it will be essential that the research community fully embraces, utilizes, and learns from this precious resource. An exciting and useful advance over the past decade was the decision by many academics and companies to "meet pathologists where they live" by developing innovative molecular methods adapted to fragmented biomolecules present in FFPE specimens, e.g., designing PCR primers sets that produce small amplicons, or sensitive mass spectrometry methods for measuring peptides.

FFPE-adapted protocols are now highly effective; however, they still require relatively large cell input amounts and are not readily applicable to single-cell analysis. The gee whiz factor that investigators experience when analyzing one or a few FFPE cells is typically replaced by disappointment as the data sets are found to be superficial and non-reproducible. This is not surprising for FFPE tissue since it is known from decades-old pathology studies that input amounts are critical. For example, one can perform a Northern blot analysis from formalin-fixed tissue by starting with a large number of whole microtome sections. Even though many transcripts are fragmented and not observable as a distinct band, there are enough intact mRNAs that survive tissue processing for a successful RNA blot. If fewer microtome sections (or a small number of dissected cells) are used, then Northern blots are no longer as possible. The same phenomenon exists for today's FFPE-adapted molecular assays, which require enough input material so that a sufficient number of relatively large mRNA fragments or intact peptides are present to be amplified by PCR or measured by mass spectrometry. For single cells or small numbers of cells, this is generally not the case.

Due to the challenges in studying FFPE-processed specimens, investigators may need to consider a "single-cell subtype" strategy instead of "single-cell" analysis — In other words, to optimize protocols to microdissect and analyze a sufficient number of identical target cells that share key morphological, molecular, and/or spatial characteristics in order to procure sufficient biomolecules such that sequencing (DNA, mRNA, miRNA) or mass spectrometry (proteins) assays will provide rigorous and sensitive measurements.

For interrogation of tumor heterogeneity as an example, a single-cell subtype strategy would allow collection and robust molecular analysis of tumor sub-clones based on criteria such as cell size, nuclear atypia (see Figure 2 right, top panel), high or low levels of a given biomarker, and/or location in a particular geographic region of a tissue section (e.g., invasive edge). From each tissue specimen, one tumor sub-population or several sub-populations could be studied. The presence and evolution of distinct tumor sub-clones within a neoplasm is currently a subject of intensive

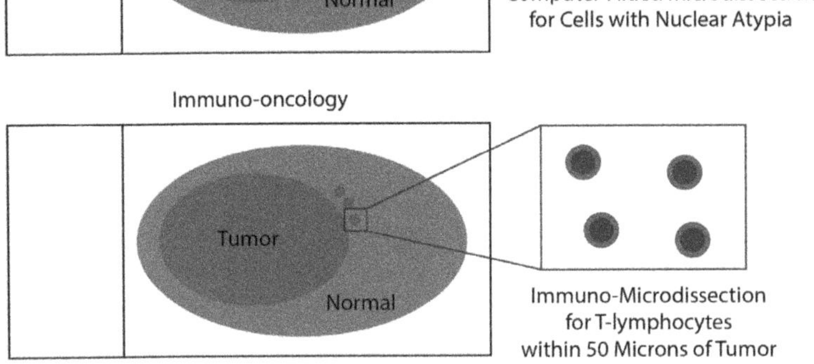

Figure 2. Strategies for single-cell analysis from FPPE samples, including phenotype-based targeting (upper panel) and immuno-based targeting (lower panel).

investigation in the cancer research community and a daunting clinical challenge due to the emergence of drug resistance. The problem is so significant for pharmaceutical development and for patient care that a growing number of thought leaders in oncology now question whether the implementation of personalized (precision) medicine strategies will ever have value. Additional laboratory and clinical research is needed to examine tumor heterogeneity in more detail.

As a second example, as discussed earlier in Section II, immunotherapy is showing promise in the clinic for treatment of many tumor types. However, the immune microenvironment is not yet well understood and current methods for discovering and utilizing companion diagnostic tests are imprecise. Best practices need to be developed so that investigators and clinicians can examine tumor cells and subtypes of tumor-infiltrating lymphocytes based on defined parameters such as: (a) antibody labeling (lineage, developmental stage, activation state); (b) location proximate to tumor (see Figure 2, bottom panel); and/or, (c) inducing cell death based on elevated apoptotic markers in an adjacent tumor cell.

Current thinking in the field is that long-term immunotherapy treatment success will depend on a multi-therapy strategy as opposed to a single agent, which will require improved measurement capabilities for panels of proteins or multiplex transcript patterns, including those that have relatively low expression levels. Today's predominant assay, IHC staining of a single biomarker, will not be sufficient. Newer, multiplex immunostaining technologies will likely provide some value and be effective for certain applications; however, both measurement sensitivity and antibody cross-reaction may limit its utility. Thus, a single-cell subtype dissection strategy might be useful for many future research and clinical applications in this space.

Development of the full suite of technologies necessary for future quantitative interrogation of patient tissue samples will require input and participation of a large number of investigators, sites, and scientific/medical disciplines. There are many diseases to study and many molecules and pathways to measure, and each will have unique challenges. Moreover, looking ahead, there will be a need for an evolving system that is capable of adapting to and taking advantage of new technical capabilities as they arise. As one example, an innovation in transcriptome amplification capability will change the calculus for the entire FFPE tissue analysis system by decreasing the number of input cells needed for mRNA measurements, which in turn will allow for new applications where the low number of target cells in histological sections had previously precluded analysis, which in turn will be enabled by improved automated dissection methods that allow for efficient collection of these rare cells. Since each component in the workflow will frequently change and since each component has a significant effect on the entire system, there is a need for protocols that are improved and updated over time.

Thus, it will be important moving forward that the molecular pathology and genomic medicine communities are integrated and collaborative. Patience and a sustained commitment will be important moving forward so that the value of genomic medicine, whatever it ultimately turns out to be, is maximized — boom, bust, or somewhere in between.

References

1. Brower V. NCI-MATCH pairs tumor mutations with matching drugs. *Nat Biotechnol* 2015; **33**: 790–1.
2. Cancer Genome Atlas N. Comprehensive molecular portraits of human breast tumours. *Nature* 2012; **490**: 61–70.
3. Cancer Genome Atlas Research N. Comprehensive genomic characterization defines human glioblastoma genes and core pathways. *Nature* 2008; **455**: 1061–8.
4. Cancer Genome Atlas Research N. Integrated genomic analyses of ovarian carcinoma. *Nature* 2011; **474**: 609–15.
5. Cancer Genome Atlas Research N. Comprehensive genomic characterization of squamous cell lung cancers. *Nature* 2012; **489**: 519–25.
6. Lih CJ, Sims DJ, Harrington RD, Polley EC, Zhao Y, Mehaffey MG, Forbes TD, Das B, Walsh WD, Datta V, Harper KN, Bouk CH, Rubinstein LV, Simon RM, Conley BA, Chen AP, Kummar S, Doroshow JH, Williams PM. Analytical validation and application of a targeted next-generation sequencing mutation-detection assay for use in treatment assignment in the NCI-MPACT trial. *J Mol Diagn* 2016; **18**: 51–67.
7. Newman AM, Bratman SV, To J, Wynne JF, Eclov NC, Modlin LA, Liu CL, Neal JW, Wakelee HA, Merritt RE, Shrager JB, Loo BW, Jr., Alizadeh AA, Diehn M. An ultrasensitive method for quantitating circulating tumor DNA with broad patient coverage. *Nat Med* 2014; **20**: 548–54.
8. Nielsen T, Wallden B, Schaper C, Ferree S, Liu S, Gao D, Barry G, Dowidar N, Maysuria M, Storhoff J. Analytical validation of the PAM50-based prosigna breast cancer prognostic gene signature assay and ncounter analysis system using formalin-fixed paraffin-embedded breast tumor specimens. *BMC Cancer* 2014; **14**: 177.

9. Paik S, Shak S, Tang G, Kim C, Baker J, Cronin M, Baehner FL, Walker MG, Watson D, Park T, Hiller W, Fisher ER, Wickerham DL, Bryant J, Wolmark N. A multigene assay to predict recurrence of tamoxifen-treated, node-negative breast cancer. *The New Engl J Med* 2004; **351**: 2817–26.
10. Reis PP, Waldron L, Goswami RS, Xu W, Xuan Y, Perez-Ordonez B, Gullane P, Irish J, Jurisica I, Kamel-Reid S. mRNA transcript quantification in archival samples using multiplexed, color-coded probes. *BMC Biotechnol* 2011; **11**: 46.
11. Schmitt MW, Kennedy SR, Salk JJ, Fox EJ, Hiatt JB, Loeb LA. Detection of ultra-rare mutations by next-generation sequencing. *Proc Nat Acad Sci USA* 2012; **109**: 14508–13.
12. Sparano JA, Gray RJ, Makower DF, Pritchard KI, Albain KS, Hayes DF, Geyer CE, Jr., Dees EC, Perez EA, Olson JA, Jr., Zujewski J, Lively T, Badve SS, Saphner TJ, Wagner LI, Whelan TJ, Ellis MJ, Paik S, Wood WC, Ravdin P, Keane MM, Gomez Moreno HL, Reddy PS, Goggins TF, Mayer IA, Brufsky AM, Toppmeyer DL, Kaklamani VG, Atkins JN, Berenberg JL, Sledge GW. Prospective validation of a 21-gene expression assay in breast cancer. *The New Eng J Med* 2015; **373**: 2005–14.
13. Thomas RK, Baker AC, Debiasi RM, Winckler W, Laframboise T, Lin WM, Wang M, Feng W, Zander T, MacConaill L, Lee JC, Nicoletti R, Hatton C, Goyette M, Girard L, Majmudar K, Ziaugra L, Wong KK, Gabriel S, Beroukhim R, Peyton M, Barretina J, Dutt A, Emery C, Greulich H, Shah K, Sasaki H, Gazdar A, Minna J, Armstrong SA, Mellinghoff IK, Hodi FS, Dranoff G, Mischel PS, Cloughesy TF, Nelson SF, Liau LM, Mertz K, Rubin MA, Moch H, Loda M, Catalona W, Fletcher J, Signoretti S, Kaye F, Anderson KC, Demetri GD, Dummer R, Wagner S, Herlyn M, Sellers WR, Meyerson M, Garraway LA High-throughput oncogene mutation profiling in human cancer. *Nat Genet* 2007; **39**: 347–51.
14. Zhao Y, Polley EC, Li MC, Lih CJ, Palmisano A, Sims DJ, Rubinstein LV, Conley BA, Chen AP, Williams PM, Kummar S, Doroshow JH, Simon RM. GeneMed: An informatics hub for the coordination of next-generation sequencing studies that support precision oncology clinical trials. *Cancer Inform* 2015; **14**: 45–55.
15. Dias-Santagata D, Akhavanfard S, David SS, Vernovsky K, Kuhlmann G, Boisvert SL, Stubbs H, McDermott U, Settleman J, Kwak EL, Clark JW, Isakoff SJ, Sequist LV, Engelman JA, Lynch TJ, Haber DA, Louis DN, Ellisen LW, Borger DR, Iafrate AJ. Rapid targeted mutational analysis of human tumours: A clinical platform to guide personalized cancer medicine. *EMBO Mol Med* 2010; **2**: 146–58.
16. Didelot A, Le Corre D, Luscan A, Cazes A, Pallier K, Emile JF, Laurent-Puig P, Blons H. Competitive allele specific TaqMan PCR for KRAS, BRAF and EGFR mutation detection in clinical formalin fixed paraffin embedded samples. *Exper Molecular Pathol* 2012; **92**: 275–80.
17. Emens LA, Butterfield LH, Hodi FS, Jr., Marincola FM, Kaufman HL. Cancer immunotherapy trials: leading a paradigm shift in drug development. *J Immunotherapy Cancer* 2016; **4**: 42.

18. Frampton GM, Fichtenholtz A, Otto GA, Wang K, Downing SR, He J, Schnall-Levin M, White J, Sanford EM, An P, Sun J, Juhn F, Brennan K, Iwanik K, Maillet A, Buell J, White E, Zhao M, Balasubramanian S, Terzic S, Richards T, Banning V, Garcia L, Mahoney K, Zwirko Z, Donahue A, Beltran H, Mosquera JM, Rubin MA, Dogan S, Hedvat CV, Berger MF, Pusztai L, Lechner M, Boshoff C, Jarosz M, Vietz C, Parker A, Miller VA, Ross JS, Curran J, Cronin MT, Stephens PJ, Lipson D, Yelensky R. Development and validation of a clinical cancer genomic profiling test based on massively parallel DNA sequencing. *Nat Biotechnol* 2013; **31**: 1023–31.
19. Garcia-Murillas I, Lambros M, Turner NC. Determination of HER2 amplification status on tumour DNA by digital PCR. *PloS One* 2013; **8**: e83409.
20. Gawad C, Koh W, Quake SR. Single-cell genome sequencing: current state of the science. *Nat Revi Genet* 2016; **17**: 175–88.
21. Greytak SR, Engel KB, Bass BP, Moore HM. Accuracy of molecular data generated with FFPE biospecimens: Lessons from the literature. *Cancer Rese* 2015; **75**: 1541–7.
22. Leung ML, Wang Y, Kim C, Gao R, Jiang J, Sei E, Navin NE. Highly multiplexed targeted DNA sequencing from single nuclei. *Nat Protoc* 2016; **11**: 214–35.
23. Picelli S, Faridani OR, Bjorklund AK, Winberg G, Sagasser S, Sandberg R. Full-length RNA-seq from single cells using Smart-seq2. *Nat Protoc* 2014; **9**: 171–81.
24. Geiss GK, Bumgarner RE, Birditt B, Dahl T, Dowidar N, Dunaway DL, Fell HP, Ferree S, George RD, Grogan T, James JJ, Maysuria M, Mitton JD, Oliveri P, Osborn JL, Peng T, Ratcliffe AL, Webster PJ, Davidson EH, Hood L, Dimitrov K. Direct multiplexed measurement of gene expression with color-coded probe pairs. *Nat Biotechnol* 2008; **26**: 317–25.
25. Gevensleben H, Garcia-Murillas I, Graeser MK, Schiavon G, Osin P, Parton M, Smith IE, Ashworth A, Turner NC. Noninvasive detection of HER2 amplification with plasma DNA digital PCR. *Clin Cancer Res* 2013; **19**: 3276–84.
26. Hedges DJ, Guettouche T, Yang S, Bademci G, Diaz A, Andersen A, Hulme WF, Linker S, Mehta A, Edwards YJ, Beecham GW, Martin ER, Pericak-Vance MA, Zuchner S, Vance JM, Gilbert JR. Comparison of three targeted enrichment strategies on the SOLiD sequencing platform. *PloS One* 2011; **6**: e18595.
27. Jumper N, Hodgkinson T, Paus R, Bayat A. Site-specific gene expression profiling as a novel strategy for unravelling keloid disease pathobiology. *PloS One* 2017; **12**: e0172955.
28. Kinde I, Wu J, Papadopoulos N, Kinzler KW, Vogelstein B. Detection and quantification of rare mutations with massively parallel sequencing. *Proc Nat Acad Sci USA* 2011; **108**: 9530–5.
29. Kyi C, Postow MA. Checkpoint blocking antibodies in cancer immunotherapy. *FEBS Lett* 2014; **588**: 368–76.
30. Li J, Wang L, Mamon H, Kulke MH, Berbeco R, Makrigiorgos GM. Replacing PCR with COLD-PCR enriches variant DNA sequences and redefines the sensitivity of genetic testing. *Nat Med* 2008; **14**: 579–84.

31. Ney JT, Froehner S, Roesler A, Buettner R, Merkelbach-Bruse S. High-resolution melting analysis as a sensitive prescreening diagnostic tool to detect KRAS, BRAF, PIK3CA, and AKT1 mutations in formalin-fixed, paraffin-embedded tissues. *Arch Pathol Lab Med* 2012; **136**: 983–92.
32. Su Z, Dias-Santagata D, Duke M, Hutchinson K, Lin YL, Borger DR, Chung CH, Massion PP, Vnencak-Jones CL, Iafrate AJ, Pao W. A platform for rapid detection of multiple oncogenic mutations with relevance to targeted therapy in non-small-cell lung cancer. *J Mol Diagn* 2011; **13**: 74–84.
33. Williams PM, Li R, Johnson NA, Wright G, Heath JD, Gascoyne RD. A novel method of amplification of FFPET-derived RNA enables accurate disease classification with microarrays. *J Mol Diagn* 2010; **12**: 680–6.
34. Zoon CK, Starker EQ, Wilson AM, Emmert-Buck MR, Libutti SK, Tangrea MA. Current molecular diagnostics of breast cancer and the potential incorporation of microRNA. *Expert Rev Mol Diagn* 2009; **9**: 455–67.
35. Kummar S, Williams PM, Lih CJ, Polley EC, Chen AP, Rubinstein LV, Zhao Y, Simon RM, Conley BA, Doroshow JH. Application of molecular profiling in clinical trials for advanced metastatic cancers. *J Natl Cancer Inst* 2015; 107.
36. Tannock IF, Hickman JA. Limits to precision cancer medicine. *The New Engl J Med* 2017; **376**: 96–7.
37. Letai A. Functional precision cancer medicine-moving beyond pure genomics. *Nat Med* 2017; **23**: 1028–35.
38. Fojo T. Precision oncology: A strategy we were not ready to deploy. *Seminars Oncol* 2016; **43**: 9–12.
39. Prasad V. Perspective: The precision-oncology illusion. *Nature* 2016; **537**: S63.
40. Abrahams E, Eck SL. Molecular medicine: Precision oncology is not an illusion. *Nature* 2016; **539**: 357.
41. West HJ. No solid evidence, only hollow argument for universal tumor sequencing: Show me the data. *JAMA Oncol* 2016; **2**: 717–8.
42. Swanton C, Soria JC, Bardelli A, Biankin A, Caldas C, Chandarlapaty S, de Koning L, Dive C, Feunteun J, Leung SY, Marais R, Mardis ER, McGranahan N, Middleton G, Quezada SA, Rodon J, Rosenfeld N, Sotiriou C, Andre F. Consensus on precision medicine for metastatic cancers: a report from the MAP conference. *Anna Oncol: Official J Euro Soc Med Oncol/ESMO* 2016; **27**: 1443–8.
43. Nakazawa Y, Suzuki S, Nishio N. Regulation of adoptive immunotherapy. [Rinsho ketsueki] *The Jap J Clini Hematol* 2016; **57**: 2373–80.
44. Pardoll DM. The blockade of immune checkpoints in cancer immunotherapy. *Nat Revi Cancer* 2012; **12**: 252–64.
45. Schumacher TN, Kesmir C, van Buuren MM. Biomarkers in cancer immunotherapy. *Cancer Cell* 2015; **27**: 12–4.
46. Shalapour S, Font-Burgada J, Di Caro G, Zhong Z, Sanchez-Lopez E, Dhar D, Willimsky G, Ammirante M, Strasner A, Hansel DE, Jamieson C, Kane CJ, Klatte T,

Birner P, Kenner L, Karin M. Immunosuppressive plasma cells impede T-cell-dependent immunogenic chemotherapy. *Nature* 2015; **521**: 94–8.

47. Wilson WH, Young RM, Schmitz R, Yang Y, Pittaluga S, Wright G, Lih CJ, Williams PM, Shaffer AL, Gerecitano J, de Vos S, Goy A, Kenkre VP, Barr PM, Blum KA, Shustov A, Advani R, Fowler NH, Vose JM, Elstrom RL, Habermann TM, Barrientos JC, McGreivy J, Fardis M, Chang BY, Clow F, Munneke B, Moussa D, Beaupre DM, Staudt LM. Targeting B cell receptor signaling with ibrutinib in diffuse large B cell lymphoma. *Nat Med* 2015; **21**: 922–6.
48. Yadav M, Jhunjhunwala S, Phung QT, Lupardus P, Tanguay J, Bumbaca S, Franci C, Cheung TK, Fritsche J, Weinschenk T, Modrusan Z, Mellman I, Lill JR, Delamarre L. Predicting immunogenic tumour mutations by combining mass spectrometry and exome sequencing. *Nature* 2014; **515**: 572–6.
49. http://www.pnas.org/content/early/2017/08/29/1704961114.short?rss=1.
50. Bettegowda C, Sausen M, Leary RJ, Kinde I, Wang Y, Agrawal N, Bartlett BR, Wang H, Luber B, Alani RM, Antonarakis ES, Azad NS, Bardelli A, Brem H, Cameron JL, Lee CC, Fecher LA, Gallia GL, Gibbs P, Le D, Giuntoli RL, Goggins M, Hogarty MD, Holdhoff M, Hong SM, Jiao Y, Juhl HH, Kim JJ, Siravegna G, Laheru DA, Lauricella C, Lim M, Lipson EJ, Marie SK, Netto GJ, Oliner KS, Olivi A, Olsson L, Riggins GJ, Sartore-Bianchi A, Schmidt K, Shih l M, Oba-Shinjo SM, Siena S, Theodorescu D, Tie J, Harkins TT, Veronese S, Wang TL, Weingart JD, Wolfgang CL, Wood LD, Xing D, Hruban RH, Wu J, Allen PJ, Schmidt CM, Choti MA, Velculescu VE, Kinzler KW, Vogelstein B, Papadopoulos N, Diaz LA, Jr.. Detection of circulating tumor DNA in early- and late-stage human malignancies. *Sci Transl Med* 2014; **6**: 224ra24.
51. Qazi MA, Vora P, Venugopal C, Sidhu SS, Moffat J, Swanton C, Singh SK. Intratumoral heterogeneity: Pathways to treatment resistance and relapse in human glioblastoma. *Ann Oncol:* Official *J Euro Soc Med Oncol*/ESMO 2017; **28**: 1448–56.
52. Mahoney KM, Rennert PD, Freeman GJ. Combination cancer immunotherapy and new immunomodulatory targets. *Nat Rev Drug Discovery* 2015; **14**: 561–84.
53. Melero I, Berman DM, Aznar MA, Korman AJ, Perez Gracia JL, Haanen J. Evolving synergistic combinations of targeted immunotherapies to combat cancer. *Nat Rev Cancer* 2015; **15**: 457–72.
54. Morrissey KM, Yuraszeck TM, Li CC, Zhang Y, Kasichayanula S. Immunotherapy and novel combinations in oncology: Current landscape, challenges, and opportunities. *Clin Transla Sci* 2016; **9**: 89–104.
55. Chowdhuri SR, Xi L, Pham TH, Hanson J, Rodriguez-Canales J, Berman A, Rajan A, Giaccone G, Emmert-Buck M, Raffeld M, Filie AC. EGFR and KRAS mutation analysis in cytologic samples of lung adenocarcinoma enabled by laser capture microdissection. *Mod Pathol* 2011; **25**: 548–55.
56. Cubi R, Vembar SS, Biton A, Franetich JF, Bordessoulles M, Sossau D, Zanghi G, Bosson-Vanga H, Benard M, Moreno A, Dereuddre-Bosquet N, Le Grand R, Scherf A.

Laser capture microdissection enables transcriptomic analysis of dividing and quiescent liver stages of Plasmodium relapsing species. 2017; **19**.

57. Eberle FC, Hanson JC, Killian JK, Wei L, Ylaya K, Hewitt SM, Jaffe ES, Emmert-Buck MR, Rodriguez-Canales J. Immunoguided laser assisted microdissection techniques for DNA methylation analysis of archival tissue specimens. *J Mol Diagn* 2010, **12**: 394–401.
58. Ebrahimi A, Jung MH, Dreyfuss JM, Pan H, Sgroi D, Bonner-Weir S, Weir GC. Evidence of stress in beta cells obtained with laser capture microdissection from pancreases of brain dead donors. *Islets* 2017; **9**: 19–29.
59. Erickson HS, Albert PS, Gillespie JW, Rodriguez-Canales J, Marston Linehan W, Pinto PA, Chuaqui RF, Emmert-Buck MR. Quantitative RT-PCR gene expression analysis of laser microdissected tissue samples. *Nat Protoc* 2009; **4**: 902–22.
60. Emmert-Buck MR, Bonner RF, Smith PD, Chuaqui RF, Zhuang Z, Goldstein SR, Weiss RA, Liotta LA. Laser capture microdissection. *Science* 1996; **274**: 998–1001.
61. Hanson JC, Tangrea MA, Kim S, Armani MD, Pohida TJ, Bonner RF, Rodriguez-Canales J, Emmert-Buck MR. Expression microdissection adapted to commercial laser dissection instruments. *Nat Protoc* 2011; **6**: 457–67.
62. Hipp J, Cheng J, Hanson JC, Yan W, Taylor P, Hu N, Rodriguez-Canales J, Tangrea MA, Emmert-Buck MR, Balis U. SIVQ-aided laser capture microdissection: A tool for high-throughput expression profiling. *J Pathol Inform* 2011; **2**: 19.
63. Ning X, Yin Z, Li Z, Xu J, Wang L, Shen W, Lu Y, Cai G, Zhang X, Chen X. Comparative proteomic analysis of urine and laser microdissected glomeruli in IgA nephropathy. *Clin Expe Pharmacol Physiol* 2017; **44**: 576–585.
64. Zhang L, Lanzoni G, Battarra M, Inverardi L, Zhang Q. Proteomic profiling of human islets collected from frozen pancreata using laser capture microdissection. *J Proteomics* 2017; **150**: 149–59.
65. Baslan T, Kendall J, Rodgers L, Cox H, Riggs M, Stepansky A, Troge J, Ravi K, Esposito D, Lakshmi B, Wigler M, Navin N, Hicks J. Genome-wide copy number analysis of single cells. *Nat Proto*c 2012; **7**: 1024–41.
66. Lin JR, Fallahi-Sichani M, Sorger PK. Highly multiplexed imaging of single cells using a high-throughput cyclic immunofluorescence method. *Nat Commun* 2015; **6**: 8390.
67. Carrick DM, Mehaffey MG, Sachs MC, Altekruse S, Camalier C, Chuaqui R, Cozen W, Das B, Hernandez BY, Lih CJ, Lynch CF, Makhlouf H, McGregor P, McShane LM, Phillips Rohan J, Walsh WD, Williams PM, Gillanders EM, Mechanic LE, Schully SD. Robustness of next generation sequencing on older formalin-fixed paraffin-embedded tissue. *PloS One* 2015; **10**: e0127353.
68. de Jong D, Verbeke S, Meijer D, Hogendoorn PC, Bovee JV, Szuhai K. Opening the archives for state of the art tumour genetic research: Sample processing for array-CGH using decalcified, formalin-fixed, paraffin-embedded tissue-derived DNA samples. *BMC Res Notes* 2011; **4**: 1.

69. Farragher SM, Tanney A, Kennedy RD, Paul Harkin D. RNA expression analysis from formalin fixed paraffin embedded tissues. *Histochem Cell Biol* 2008; **130**: 435–45.
70. Linton K, Howarth C, Wappett M, Newton G, Lachel C, Iqbal J, Pepper S, Byers R, Chan WJ, Radford J. Microarray gene expression analysis of fixed archival tissue permits molecular classification and identification of potential therapeutic targets in diffuse large B-cell lymphoma. *J Mol Diagn* 2012; **14**: 223–32.

2 Molecular Pathology — Past, Present, and Future

Eli S. Williams* and Christopher A. Moskaluk*,†

*Departments of Pathology, University of Virginia, Charlottesville, VA 22908, USA
†Biochemistry & Molecular Genetics, University of Virginia, Charlottesville, VA 22908, USA

Contents

I. Introduction

Advances in human genetics and genomics are impacting the way we diagnosis and treat disease. Sparked in part by the success of the Human Genome Project (HGP), the technology used for genetic testing has become more accurate and less expensive.[1] The decreased cost of generating DNA (and RNA) sequence data has made genomic information more accessible than ever before. Doctors are ordering genetic testing for a longer list of indications, and scientists are casting a wider net in the search for disease-causing genetic variants. Even the general public is ordering genomic testing, with over 1,000,000 customers purchasing their personal genomic profiles through direct-to-consumer laboratories.[2] The HGP has created a vision of a more efficient way to practice medicine and brought attention to the importance of one's genetic information. By basing treatment and lifestyle choices on one's genetic profile, medicine could be made more effective by not only identifying an individual optimized treatment regimen for disease ("personalized medicine") but also by preventing disease occurrence by personalized behavior modification regimens and preventative prescriptions ("chemoprevention").

One of the key technologies in the expansion of genomic medicine is next-generation sequencing (NGS). NGS is also known as massively parallel sequencing, as it performs millions of sequencing reactions at the same time. This huge number of sequencing reactions can be used to sequence the entire genome of one individual (whole genome sequencing, WGS), which represents approximately 3 billion base pairs. Coincidentally, the HGP spent approximately 3 billion dollars sequencing the first human genome,[3] approximately $1 per base pair. Today, high-volume genetic testing laboratories are approaching $1,000 per genome, a goal set in 2003 by the US National Human Genome Research Institute (NHGRI) aimed to make clinical genomic testing achievable.[4] NGS-based testing is being used in a variety of disciplines including oncology, cardiology, neurology, pediatrics, and others. Diagnoses of rare genetic diseases have been made using new genomic approaches, in some cases ending years of searching. In 2011, the reporting of just such a case won a Pulitzer Prize, reinforcing the promise of personalized medicine.[5,6]

Molecular diagnostic techniques are not new to medicine, of course, but the target size of NGS-based testing is much larger than traditional single gene testing. Whole exome sequencing (WES), testing the entire protein-coding region of a person's genome, has proven to be an effective approach to identify an underlying genetic etiology for diseases with non-specific symptoms, such as developmental delay or autism. Targeted gene panels have also been used successfully when the clinically relevant genes for a particular disease are well defined, such as in non-small cell lung cancer. While NGS can be a more efficient use of limited sample size, it produces perhaps more data than is needed to solve the clinical problem at hand. WES can reveal thousands of variants; each one must be categorized as a disease causing or not. An ever-growing number of genomic sequences available from both normal and affected populations have improved interpretation pipelines. However, we are still unable to classify a significant percentage of variants, designating these frustrating findings as variants of unknown significance (VUS). Another by-product of expanded gene testing is secondary genetic findings or incidental findings, genetic variants that predispose an individual for a disease unrelated to the current clinical concern. The reporting of incidental findings back to the patient can be challenging, especially when the patient is unaware of the additional information that NGS testing can reveal. Other challenges turning genetic information into medically actionable information have arisen. Questions have been raised regarding the regulation and oversight of clinical genomic testing, with the FDA and CMS reevaluating current standards.

The expansion of NGS-based testing in medicine depends on our ability to demonstrate improved outcomes as a result of genomic test results. The increased diagnostic yield of NGS versus the standard-of-care testing has been demonstrated in multiple patient populations.[7–10] However, there is still not clear evidence that this early diagnosis leads to better outcomes or that NGS is a more cost-effective approach compared to current practices. Nevertheless, the role of clinical genomics in medicine is currently increasing, driven in part by the growing number of genes associated with human disease. The application of any new technology in medicine often sparks a dramatic surge forward in the understanding of disease. In this chapter, we will look back at the applications of genetics in medicine, examining the impact of new genetic technologies. We will also look at the regulatory issues and quality assurance necessary to run the complex, clinical genomics laboratories of today.

There are also ethical, legal, and social issues that will undoubtedly help to shape the future of genomic medicine.

II. The History of Genetics in Medicine

Genetics represents a relatively new branch of medicine, especially when compared to fields such as anatomic pathology, which has its roots in the gross dissection and examination of human organs performed by Galen and the Ancient Greeks. It was only 150 years ago that Gregor Mendel famously described heredity using pea plants.[11] His discussion on inheritance was much more a mathematical approach, the establishment of rules and ratios, rather than a biological one. Exactly what underpinned the dominant and recessive inheritance patterns described by Mendel remained a mystery until the early 20th century. Two independent observations, by Walter Sutton[12] and Theodor Boveri,[13] established the chromosome theory of inheritance, which proposed that chromosomes are the vehicle on which the genetic information could be carried from generation to generation.

Around the same time as the chromosome theory of inheritance, which is now more commonly known as the Boveri–Sutton theory, German pathologists were speculating about the role of abnormal mitotic figures they were observing in human malignancies.[14] They theorized that it was abnormalities such as these that were responsible for cancer initiation. Boveri furthered that notion, postulating that mammalian tumors are initiated by mitotic abnormalities resulting in the change of the number of chromosomes in a cell.[15] He extrapolated the link between mitotic abnormalities, chromosomes, and cancer to other concepts including genetic instability of cancer cells and the idea of clonal evolution. Many of these theories, though initially controversial, came to be proved true through the experimentation of Thomas Morgan, who was initially skeptical of the idea.[16] While these discoveries provided a foundation of understanding of cancer biology, there was not yet a direct link between genetics and cancer, or any other human condition.

Chromosomes could be seen under the microscope, and the field of cytogenetics emerged in the early 20th century. Cytogenetics focuses on the structure of chromosomes, and by using cell culturing techniques cytogeneticists were able to enrich for cells in metaphase, when chromosomes are most easily visualized. These specialized techniques allowed cytogeneticists to begin to define the number of chromosomes in different organisms, as well as identify the underpinnings of sex determination and genetic recombination. In 1923, TS Painter established the XY method of sex determination in humans.[17] He also reported a significantly less helpful observation of human chromosomes, his determination that humans have 48 chromosomes. Unfortunately, this incorrect determination remained uncontested in the genetics world for over 30 years.[18] Eventually, a happy lab accident resulted in a hypotonic buffer being added to cultured cells, resulting in swollen cells. When these swollen cells were "dropped" onto microscope slides they burst like water balloons, spreading the chromosomes out widely on a slide, and facilitating chromosome counting.[19] Finally, in 1956, the human chromosome number was correctly established at 46.[20]

Clinical cytogeneticists subsequently flourished. Clear links between genetic abnormalities and human disease were made, the first by French cytogeneticist Jerome Lejeune, who described the presence of an additional chromosome in individuals with Down Syndrome in 1959.[21] Chromosome aneuploidy, the presence of too many or too few chromosomes, was firmly established as a recurrent mechanism of human disease in just a few short years. The genetic etiology of Patau syndrome (trisomy 13), Edwards syndrome (trisomy 18), Turner syndrome (45,X) and others was revealed in the months and years following Lejeune's important discovery.[22–24] Despite these clear examples of chromosomal change leading to disease, the importance of chromosomes in cancer was viewed with more skepticism. The discovery of the Philadelphia chromosome in CML,[25] although historically important, did not lead to the same flurry of activity in cancer cytogenetics that Lejeune's discovery did in congenital cytogenetics. It was not until the work of Janet Rowley in the early 1970s, which not only resolved the structure of the Philadelphia chromosome as a translocation between chromosomes 9 and 22, but also described the presence of other recurrent translocations in cancer, that cemented the role of chromosome aberrations and cancer.[26,27]

The study of the structure and function of genes was also taking hold in the early 1970s. Wally Gilbert and Fred Sanger simultaneously developed techniques for sequencing DNA.[28,29] Although it is Sanger's method that has stood the test of time, both earned the Nobel Prize in chemistry in 1980 for their pioneering work in genetics.[30] This new technology was quickly applied to a variety of model disease systems to try to identify the underlying genetic abnormality. Oncogenes were identified, as was the idea that these oncogenes existed in a non-activated state until they were altered by a mutation or a rearrangement. This search for genes expanded outside of the typical realm of academic investigators, with many private companies investing in the search for genes, with the aim to capitalize on these discoveries. It was tantalizing to imagine a drug designed specifically for the molecular defect of a particular disease. Unraveling the molecular mechanism of the Philadelphia chromosome seemed low-hanging fruit as the chromosomal nature of the translocation had been characterized and the presence of the Philadelphia chromosome is well correlated with leukemia, particularly chronic myelogenous leukemia (CML). However, it was not until the 1980s that researchers were able to demonstrate the critical molecular players in the Philadelphia chromosome.[31–35]

The determination that the abl-proto-oncogene located at 9q34 was juxtaposed with the bcr gene located at 22q11 provided a biological explanation into the oncogenesis of leukemia.[36,37] *ABL* is a tyrosine kinase involved in a variety of cellular pathways including cell differentiation and division. The chimeric protein produced by the *BCR–ABL* translocation lacked appropriate regulatory control, leading to unchecked tyrosine kinase activity and downstream activation.[38] The demonstration that the tyrosine kinase activity of the mutant BCR–ABL protein is critical for the transformative properties of the Philadelphia chromosome led to the targeting of specific inhibitors to the chimeric protein. In 1990, a phase I trial using a tyrosine kinase inhibitor with high specificity for BCR–ABL, imatinib, was found to have significant anticancer effect in Philadelphia chromosome-positive CML (Ph^+ CML) patients, as well as Ph^+ acute lymphoblastic leukemia (ALL) patients.[39,40] Thus, the era of precision medicine was born.

The year 1990 was significant in the field of genomics for another important event, the beginning of the HGP. The HGP was striking in its scope. Expected to take 15 years and cost 3 billion dollars, the goal of the

HGP was tripartite: (1) to develop technology to better analyze DNA; (2) to sequence and map the human genome, as well as other model organisms; and (3) to explore the ethical, legal, and social issues surrounding the broader implementation of genomic medicine and research.[41] The HGP was a rousing success, and the race to the finish between groups led by Francis Collins and Craig Venter captured the public interest. Just 10 years after the announcement of the HGP, Francis Collins and Craig Venter announced that they had both successfully completed sequencing the genome. The first draft of the human genome was published in the journal *Nature* by the Collins-led NIH group in February of 2001[42] and in the journal *Science* by the Venter-led Celera group one day later.[43]

III. Impact of the Human Genome Project and Next-Generation Sequencing on Medical Diagnostics

The impact of the HGP on the fields of clinical genetics and molecular diagnostics was profound. The HGP invigorated genetic research and sparked innovation in technology, which together have laid the groundwork for expanded medical genetics. The completion of the HGP provided researchers with a map not only of the human genome,[44] but of other model organisms including *C. elegans*,[45] *Drosophila*,[46] and *Arabidopsis*.[47] These reference genomes served as a platform for research, allowing insights into evolution, gene structure and function, and the mutational spectrum of human disease. Undoubtedly, the discovery of new disease-causing genes directly impacted molecular diagnostics. Physicians called upon laboratories to offer testing for these new genes when more routine evaluations failed to reach a diagnosis. As the number of genes associated with a particular disease grew, laboratories looked for ways to test all of the mutations and/or genes of interest with a single test. NGS technology was able to address this need for high-throughput genomics, and it was implemented in certain clinical situations. Although the upfront investment for instrumentation could be daunting, the NGS-based assays fit well within the existing framework of molecular pathology labs. There were already established workflows and safeguards in place to prevent contamination of the sensitive assays, and the techniques used in NGS were largely familiar to molecular genetics technologists.

Early clinical adoption of NGS was primarily focused on the diagnosis of rare disease, leveraging the ability of NGS to test a large part, if not all, of an individual's genome to try to identify a causative underlying genetic defect. The use of WES and WGS on these diagnostic dilemmas revealed a disease-causing mutation in a large percentage of cases, as high as 30% or 40% for some indications.[10,48,49] These early successes primed the system for further application of genomics. It quickly became clear, however, that the amount of data produced by NGS was something that would need to be considered by adopting laboratories. This data is on the order of gigabytes per patient, depending on the exact approach, and requires bioinformatics processing to be available for clinical analysis. The number of variants that were found in each patient was staggering, approximately 10,000 non-synonymous variants,[50] all of which must be annotated with evidence of contribution, or lack thereof, to the phenotype of interest. This data problem, colloquially referred to as "big data", continues to generate controversy in regards to storage and usage of patient data, interpretation and reporting of tests results, and other issues, as will be addressed in Section V.

There are particular attributes of NGS technology that are lending themselves to previously problematic aspects of molecular diagnostics. The first attribute is that NGS requires fragmentation of the genome into small molecules of DNA, no more than a few hundred base pairs (bp) in length.[51] This makes the technology ideal for utilizing specimens once thought to have DNA too degraded for whole genome sequencing. This is particularly the case for formalin-fixed paraffin-embedded (FFPE) tissue blocks, generated as part of routine histologic examination of tissue. Once regarded as a barrier to widespread implementation of molecular diagnostics, tissue processed via routine histologic technique is now regularly used in both targeted gene panels and WES/WGS.[52,53] The next attribute of NGS is its ability to perform genome-wide analysis on small amount of input DNA.[54] In addition to allowing the use of limiting tissue amounts in small biopsies of tumors, NGS has recently been extended to the analysis of circulating tumor cells and soluble cell-free tumor DNA found in cancer patients' bloodstreams.[55] This capability has also seen utilization in prenatal diagnostics, with cell-free fetal DNA being isolated from maternal blood, offering a source for fetal testing without the risk of miscarriage associated with amniocentesis and chorionic villus sampling.[56,57]

In parallel to the advent of genome-wide DNA analytic techniques, technologies that can survey other classes of biomolecules have been developed, including RNA transcripts (transcriptomics or gene expression profiling), proteins and phosphoproteins (proteomics), and various other small molecules (lipidomics, metabolomics, etc.). While it is outside the scope of this chapter to do justice to all of these technologies and their impact on clinical diagnostics, a generality of the "-omics" technological revolution is that single analyte assays will increasingly be supplanted by large-scale analyses of multiple analytes, impacting the classification of disease and prediction of therapeutic efficacy.

IV. Current Oncology Molecular Genetic Diagnostic Techniques and Emerging Trends

Currently, NGS is being adopted in routine diagnostics for many types of malignancies. Standard histopathology and cytopathology remain the bedrock of cancer diagnostics, but increasingly molecular attributes of cancer are required for the precise classification of neoplasms and to specify therapeutic options. These attributes are not only defined by the mutational profile of the cancer, but by other molecular biomarkers as well. The antibody-based detection of proteins or other antigenic expression on tumor cells by use of immunohistochemical techniques (tissue) or flow cytometry (blood, bone marrow, or other fluid specimens) is perhaps the predominant molecular analysis. These analyses provide discrete, phenotypic data points that are woven together into a gene expression signature that supports a specific tumor classification. These techniques are most commonly used in tumors in which histologic features alone can be insufficient or difficult to apply for robust classification (leukemias, lymphomas, and poorly differentiated carcinomas).

The genetic profile of a tumor provides independent evidence for a specific tumor classification or treatment modality. Cytogenetic analysis of viable cells may be used to detect characteristic translocations or other gross chromosomal alterations in leukemias and a small subset of solid tumors. Higher-resolution targeted assays must also be used to detect many of the important tumor-associated gene mutations. Fluorescence *in situ* hybridization (FISH) on tissue sections, cytologic, or cytogenetic preparations is commonly used to identify translocations or gene amplification/deletion using sets of DNA probes specific to a disease entity. Other targeted approaches include the testing of known tumor-associated genes or clinically relevant mutation hotspots by polymerase chain reaction (PCR)-mediated amplification of the desired areas from samples of tumor DNA, followed by Sanger sequencing or pyrosequencing. However, these PCR based methods generally interrogate small areas of DNA (typically less than 200 bp). For genes in which mutations can be found throughout almost the entire gene length (as is the case in most tumor suppressor genes such as *TP53*) or are found in widely separated exons (such as the oncogenes *KRAS* and *EGFR*), several separate PCR products must be individually sequenced to provide appropriate coverage. The increasing need to assay several genes or hotspots per tumor type is forcing labs to reevaluate the value equation of PCR-based sequencing assays versus NGS-based assays. NGS-based assays remain more expensive than standard PCR-based assays, but as with rare diseases, there exists a tipping point where NGS becomes more cost-effective. This exact tipping point may vary from lab to lab, but once the number of individual PCR products increases to over a handful (four or five separate PCR reactions), the costs of supplies, technical hours, and professional interpretative time of these individual tests make NGS-based targeted gene panels or WES/WGS more cost-effective for the testing of multiple targets.

One consequence of the increased mutational and molecular analysis of human cancers is the categorization of tumors based on their genetic profile. Cancers previously thought to be one entity arising in the same organ or cell type are now known to be composed of several subtypes of molecularly distinct cancers. The profiling of these cancers by gene expression analysis or genomic analysis has led to better understanding of the behavior of a particular cancer entity or to more effective therapy for previously refractory subtypes. One prominent example of this principle includes the subtyping of human breast cancer based on molecular classifiers. Traditional molecular markers in breast cancer such as hormone receptor status and HER2 status have provided predictive and prognostic information for many years.[58] The responsiveness of a particular patient to endocrine therapy can be largely predicted by hormone receptor status,[59] while the amplification of the *HER2* gene predicts response to anti-HER2 therapy.[60] Now, microarray-based gene expression profiling (GEP) incorporates these traditional molecular classifiers, along with a host of additional gene expression data, to classify breast cancer into molecularly distinct groups, e.g., basal, luminal, or triple negative.[61–63]

Mutational analysis of cancer provides similar insights, with the identification of discrete subgroups of lung adenocarcinomas serving as an important example. Mutations in well-defined oncogenes, such as those of the *RAS* family of genes, are frequently seen in non-small cell lung cancer (NSCLC). These mutations drive oncogenesis through dysregulation of critical signaling pathways leading to growth advantages, increased survival, and other hallmarks of cancer.[64] The ability to target these defective pathways with small-molecule inhibitors has been a boon to the treatment of many cancer types, including NSCLC. The use of these therapies is guided by companion diagnostics to identify the presence of the "druggable" mutation. The relatively high occurrence rate of NSCLC combined with availability of mutation-specific therapies for recurrent mutations in NSCLC has made it one of the early proving grounds for NGS.

The next consequence of increased genetic testing of cancer specimens is the realization that while cancers of a specific type may have a common set of gene mutations or similar GEPs, occasionally a tumor may have molecular profile not typical for that class. While the set of actionable gene mutations is still relatively small, for an individual patient, the identification of such an unanticipated therapeutic modality has a large impact. For this reason, tumors that are not responsive to proscribed therapeutic regimens are a growing indication for large NGS-based gene panels. The molecular profile generated by GEP has also been useful at identifying cancers previously thought to be insensitive to small-molecule inhibitors. The discovery of ALL patients exhibiting a GEP consistent with Ph^+ ALL, even in the absence of the *BCR–ABL* fusion, has led to the effective use of targeted tyrosine kinase inhibitors in this setting.[65,66] The expanded use of existing small-molecule inhibitors has led to additional impetus to perform the genome-wide screen "up front" in lieu of multiple gene-specific assays.

A third consequence of expanded molecular profiling is the understanding that cancer genomes can be extremely complex. Some cancers samples have yielded hundreds of genetic variations from the patients' normal genomes. Moreover, the variants are often not present in all cancer cells, with only certain populations of cells that contain only a portion of the variants. These findings have helped cement the concept of "driver" mutations, which underlie the oncogenic process and tend to be present in the vast majority of abnormal cells. In addition to driver mutations, cancers accumulate "passenger" mutations through the genomic instability inherent in cancer. Although passenger mutations are not thought of as oncogenic, they may influence the overall milieu of the tumor and potentially impact disease progression or treatment response.[67] We currently do not have a sophisticated enough understanding of cancer genomes to interpret the impact of all the detected variants. This plagues the process of turning genetic data into a clinically relevant molecular report. Nevertheless, the presence of a molecular marker with prognostic or predictive value can have profound implications for therapeutic options, and so we must be able to distinguish the clinically relevant variant (or variants) from those that are either benign or of unknown significance.

V. Impact of the Human Genome Project on Public Health Genomics

One of the aims of the HGP was to understand and begin to address the ethical, legal, and social issues (ELSI) that would arise with the increased use of genetic data.[41] The success of the HGP brought an enhanced public awareness to the field of genetics. People began to ask more questions about their genetics and seek information about their genetic makeup. This information became available from a variety of sources, from doctors seeking information to help treat disease, from researchers investigating the impact of variation in the human genome, and from private companies attempting to capitalize on a new direct-to-consumer market. The generation, use, and storage of genetic data in each of these settings raises ELSI that are both unique to each setting but share core concerns of privacy and safety.

Genetic information generated by medical professionals should be of high quality and well annotated. As will be discussed in the section "Regulation of Clinical Genomics," all medically actionable lab results must pass through a pipeline validated for accuracy and reproducibility of results. This ensures that the information the doctor is using to make medical decisions is dependable, allowing for safe and effective intervention, if indicated. Genetic data in this context tends to be used only for the reason it was ordered, a limitation in our current practices. The storage of medically generated genetic data should be done in a secure fashion, as it is considered protected health information (PHI) under the Health Insurance Portability and Accountability Act of 1996 (HIPAA). [ref 69]. However, this important health information often does not fit well within the confines of electronic medical records (EMRs). A survey of medical centers found that genetic information is most often stored under the heading of "Genetics," "Pathology," or "Laboratory," but this classification could depend on whether the report was normal or abnormal.[69] This lack of uniformity in the way genetic data is integrated into the EMR can lead to the oversight of potentially useful genetic data, and ultimately unnecessary genetic testing. Clear communication between the ordering physician and the lab professionals is needed to determine the clinical question and the best utilization of lab resources.

A primary concern in regards to ELSI for medically generated genetic information relates to incidental or secondary findings. These are findings unrelated to the indication for ordering the genetic testing, but may be of value to the individual. Mutations in cancer susceptibility genes often fall into this category, as mutations in genes such as *BRCA1* confer an elevated lifetime risk of cancer, which can be mitigated in many circumstances with increased surveillance or prophylactic intervention. Mutations in genes that predispose for hereditary cardiomyopathies or arrhythmias are also deemed of value to the patient, as cardiac evaluation may reveal previously unrecognized risks. Professional organizations have published guidelines on the interpretation and reporting of secondary findings.[70,71] The American College of Medical Genetics and Genomics have published a list of 56 genes in which mutations should be reported,[70] the so-called ACMG-56. Laboratories performing WES and WGS are beginning to adopt these guidelines into their reporting standards. These guidelines also emphasize the importance of proper genetic counseling when the results are being returned to the patient. In order to maximize the value of genomics in medicine, the individual tested must understand the implications of their test results. Licensed genetic counselors are trained to help patients understand the complicated results genomic testing can deliver. Increasingly, however, genome-wide tests are not ordered in the context of a genetic consult, but rather ordered by neurology, hematology, oncology, or other physicians. Proper handling of genomic test results is an essential part in relieving concerns of ELSI.

An interesting facet to ELSI has been the direct-to-consumer (DTC) genetic testing companies. These companies have sought to capitalize on increased public demand for genetic and genomic information. Consumers provide appropriate samples for genetic testing, typically buccal swabs or saliva, using kits provided by the DTC companies. The companies would perform genome-wide testing on these samples under the premise of either determining one's ancestry, or more problematically, of identifying genetic risk factors. Consumers were given their relative risk for a host of common diseases, including diabetes, hypertension, and others that may be mitigated with a lifestyle change or follow-up disease screening. People were empowered to use their own personalized genomics data to direct a healthy lifestyle. The actual application of this data, however, was fraught with issues[72] Most DTC testing was not performed in laboratories accredited to perform medical testing, hence the ability of medical professionals to assist individuals based on these analyses was problematic. Also, in a rush to stay current, variant alleles were reported to have disease associations based on limited data, sometimes single published studies or unverified disease associations found in public datasets. Most problematic, without guidance from licensed genetic counselors, lay individuals were left to grapple alone with difficult scenarios: predisposition for severe untreatable conditions (i.e., neurodegenerative disorders), implications for family members, disclosure to health insurers, etc.

The enhanced ability to identify genetic risk factors has larger societal implications of stigmatization, medical privacy, health insurance, and job discrimination. Legislation aimed to curb these issues has been passed, most notably the Genetic Information Nondiscrimation Act of 2008

(GINA),[73] however, not all issues can be foreseen. The field of genomics is in a phase of rapid growth, with increased involvement by the general public. Standards of practice are being written and rapidly rewritten as new guidelines are published, tests are being revalidated as new clinically relevant genes are discovered, and ELSI are being reviewed as new challenges arise.

VI. Regulation of Clinical Genomics

Ever-changing technology is undoubtedly a key force in shaping the field of genetics and its application in medicine. A second key force in clinical genetics and genomics is the regulation and oversight of these applications. The wider deployment of genetic analyses in the 1980s, particularly in cancer diagnostics, raised questions concerning the quality and implications of test results. The primary legislation relating to clinical genetics and genomics is the Clinical Laboratory Improvements Amendments (CLIA).[74] These amendments were first legislated in 1988, leading to the moniker CLIA '88, which is still used today. CLIA '88 continues to be the primary regulatory oversight for clinical genetic and genomic testing. One primary mandate of CLIA '88 is that only CLIA-certified laboratories can perform human testing that results in clinical-decision making. This mandate ensures that the quality of testing and integrity of patient safety remain relatively uniform across laboratories. These standards are ensured through laboratory inspections performed by qualified organizations, such as the College of American Pathologists (CAP).

A challenge in the regulation of clinical genetic and genomic tests is that many tests are developed by individual laboratories, for use in their facility only. In other words, relatively few molecular diagnostic tests are FDA-approved compared to other areas of laboratory medicine that rely more heavily on commercial kits and/or instrument manufacturers. These laboratory-developed tests (LDTs) offer individual labs more flexibility to develop a suitable molecular assay for their institutions' needs or areas of emphasis. However, as these LDTs are not passed through the stringent approval process required of FDA-approved tests, CLIA demands a more stringent test validation process than it does for FDA-approved tests. FDA-approved tests have been validated by the test manufacturer and the performance characteristics of the test have been established. Any laboratory implementing such a test without any modifications need only to verify the performance specifications. The implementation of an LDT (or a modified FDA-approved test) requires that the laboratory establish and confirm all performance metrics including accuracy, precision, reportable range, and other relevant metrics. Additionally, the analytical sensitivity (limit of detection) and analytical specificity must be determined through validation.

While CLIA defines what test specifications must be established through the validation of an LDT, it does not provide much guidance as to exactly how these specifications should be established. The validation protocol is left to the discretion of the laboratory director, and the determination of what constitutes an appropriate validation is often left to the discretion of the inspector of the laboratory. This lack of uniformity in validation is amplified as the complexity of the testing increases. Significant concerns have been raised regarding the ability of CLIA to continue to adequately oversee clinical genetic and genomic testing. In 2014, the FDA offered a proposal to dramatically increase the regulation of LDTs, essentially requiring each LDT to undergo a modified FDA-approval process.[75] The aim of this proposal is to help unify clinical genomic testing from laboratory to laboratory, but many professional organizations have opposed this measure, arguing that the exorbitant cost and time required to undergo such approval would bring innovation to a halt.

VII. The Future of Molecular Pathology

We are encountering genetics more now than ever before. Commercials for genetic testing services promise insights about your health, your traits, and your ancestry.[76] Doctors' appointments are more often including discussions of family history and decisions involving genetic testing. Investigations into the meaning of the human genome are discovering new genetic causes for disease and unraveling mysteries about human evolution. High-throughput genomic testing is becoming commonplace in labs across the world. Despite this rapid expansion, the field of genetics and genomics is not yet mature. If the turn of the century was the birth of personalized genomics, we are currently in the awkward teenage years.

The field is being influenced by a variety of factors, and it is unclear what the outcome will be. There is ample demand for genetic information, but the information that is delivered must meet certain expectations. The clinicians expect that the genomic tests that they are ordering will provide accurate and clinically useful information. The public expects that genomics will provide insight into their personal health goals. There are also expectations from regulatory agencies and insurers. These expectations include well-validated, high-quality tests and appropriate test ordering patterns. It is important that clinical laboratories do not become handcuffed by these expectations, but instead look for opportunities where genomics can improve clinical outcomes.

Hurdles do exist to the broader application of genomics in medicine. While the technology has become much cheaper, genetic tests are still expensive. While compassionate pricing is increasingly available, the price of an WES is ranges from about $1,100 to nearly $10,000.[77] Coverage of genetic and genomic tests by insurers varies widely and depends heavily on the clinical circumstance. In many cases, genomic testing is considered experimental and therefore not reimbursable by insurance. This can place an unnecessary burden on the patient, in some cases leading to large, unexpected medical bills. Another hurdle for genomic testing is the turnaround time. The complexity of both the current genomic procedures and the interpretation of test results leads to turnaround times of one to two weeks (at best) to months (at worst). This makes the implementation of genomic medicine impractical in many types of acute settings. However, as the volume of genomic tests increase and as the analytic platforms become more standardized, it is anticipated that both the costs and turnaround times will decrease to the point that return of results for known pathogenic variants may occur in one to two days, for fees comparable to those of standard radiologic analysis today (<$1,000).

Ultimately, to maximize the medical benefit of genomics, WES/WGS will be offered in the prenatal and/or neonatal setting. It is then that genomics offers the most benefit, a benefit that will pay out over the course of a lifetime. From this single test, we can learn if a baby has a mutation for a host of life-threatening newborn illnesses, we can identify children at increased risk for early-onset cancers and monitor them more carefully, we can provide information about drug sensitivities to the care team before prescriptions are written, and we can improve outcomes. The challenges that this type of scenario creates is the storage and accessibility of the raw genomic data. This data must be available throughout the individual's lifetime to allow for the periodic revisiting and reanalysis of the data in light of new knowledge of genetic disease associations We also must be able to make this data available in a concise and pertinent manner to health care providers at the right time and in the right context of the individual's life, disease course, and therapeutic regimen. Unlike a typical laboratory test where the value is generally static in regard to time and quickly becomes irrelevant to the patient's therapeutic decision-making, genetic data is not only relevant at all times, but its relevance changes according to circumstances (e.g., pharmacogenomics data only becomes relevant when a particular class of drugs is prescribed to a patient). WES/WGS will therefore need to be matched with a new type of medical informatics systems, in which genomic data is constantly revisited, genetic variants are constantly associated with new attributes, and this data is brought to the attention of the practitioner by specific triggers in the electronic medical record (i.e., patient age, blood pressure value, drug prescription, etc.).

In the special case of cancer genomics, it is likely that WES/WGS technology will not be implemented just once, but several times during the patient's course. The future of cancer diagnostics may not totally supplant traditional tissue diagnostics (core biopsies, fine needle aspiration, etc.) but may do so in certain scenarios. For instance, the radiographic finding of a small lung or pancreatic mass may elicit a genomic assessment of cell-free circulating DNA instead of an organ biopsy, and if specific criteria for DNA abundance and mutational spectrum are met, may obviate the need for routine tissue diagnostics. Certainly, as more and more targeted therapeutics are developed for specific types of genetic alterations in cancers, genomic tumor analyses will become standard to set initial therapy. Genomic analyses are also being used to monitor patient blood samples for the burden of disease and for recurrence during the disease course. If tumors recur in an individual, it is likely due to the selection of distinct genetic sub-clones of resistant cancer cells that will need to undergo secondary genomic analysis to assign further therapy. It is anticipated that advances in therapy linked to serial genomic analyses of the patient's tumor will relegate cancer to the status of chronic disease, if not totally curable, for a large number of cancer patients.

References

1. Wetterstrand K. DNA sequencing costs: Data from the NHGRI Genome Sequencing Program (GSP). [cited 2016 03/01/2016].
2. 23andMe, 23andMe Genotypes One Millionth Customer. 2015, 23andMe: http://mediacenter.23andme.com/blog/23andme-1million/.
3. The Human Genome Project completion: Frequently asked questions. 2010 [cited 2016 03/01/2016].
4. Spencer G. New grants support development of faster, cheaper DNA Sequencing. 2010: https://www.genome.gov/12513210/2004-release-nhgri-seeks-next-generation-of-sequencing-technologies/.
5. Worthey EA. *et al.* Making a definitive diagnosis: successful clinical application of whole exome sequencing in a child with intractable inflammatory bowel disease. *Genet Med* 2011; **13**(3): 255–62.
6. The 2011 Pulitzer Prize Winner in Explanatory Reporting. The Pulitzer Prizes 2011 [cited 2016 03/01/2016].
7. Shashi V. *et al.* The utility of the traditional medical genetics diagnostic evaluation in the context of next-generation sequencing for undiagnosed genetic disorders. *Genet Med* 2014; **16**(2): 176–82.
8. Stoddard JL. *et al.* Targeted NGS: A cost-effective approach to molecular diagnosis of PIDs. *Front Immunol* 2014; **5**: 531.
9. Yang Y. *et al.* Clinical whole-exome sequencing for the diagnosis of mendelian disorders. *N Engl J Med* 2013; **369**(16): 1502–11.
10. Yang Y. *et al.* Molecular findings among patients referred for clinical whole-exome sequencing. *JAMA* 2014; **312**(18): 1870–9.
11. Mendel GJ. *Versuche uber Pflanzen-Hybriden [Experiments Concerning Plant Hybrids]*. In *Verhandlungen des naturforschenden Vereines in Brünn [Proceedings of the Natural History Society of Brünn]*. Brunn, Germany (1865).
12. Sutton WS. The chromosomes in heredity. *The Biol Bullet* 1903; **4**(5): 231–250.
13. Boveri T. *Ergebnisse über die Konstitution der chromatischen Substanz des Zelkerns*. Jena, Germany: Verlag von Gustav Fischer, 1904.
14. von Hansemann D. Ueber asymmetrische Zellheilteilung in epithelkrebsen und deren biologische bedeutung. *Virschows Arch Pathol Anat* 1890; **119**: 299–326.
15. Boveri T. Uber mehrpolige Mitosen als Mittle ur Analyse des Zellkerns. *Verh Phys-med Ges Wurzburg NF* 1902; **35**: 67–90.
16. Morgan TH. Sex limited inheritance in Drosophila. *Science* 1910; **32**(812): 120–2.
17. Painter TS. Further observations on the sex chromosomes of mammals. *Science* 1923; **58**(1500): 247–8.

18. Hsu TC. Mammalian chromosomes *in vitro:* I. The Karyotype of Man. *J Heredity* 1952; **43**(4): 167–172.
19. Hsu TC. Pomerat C.M. Mammalian chromosomes *in vitro:* II. A Method for Spreading the Chromosomes of Cells in Tissue culture. *J Heredity* 1953;. **44**(1): 23–30.
20. Tjio JH. Levan A. The chromosome number of man. *Hereditas* 1956; **42**(1–2): 1–6.
21. Lejeune J. Gautier M. Turpin R. Study of somatic chromosomes from 9 mongoloid children. *C R Hebd Seances Acad Sci* 1959; **248**(11): 1721–2.
22. Edwards JH. *et al.* A new trisomic syndrome. *Lancet* 1960; **1**(7128): 787–90.
23. Ford CE. *et al.* A sex-chromosome anomaly in a case of gonadal dysgenesis (Turner's syndrome). *Lancet* 1959; **1**(7075): 711–3.
24. Patau K. *et al.* Multiple congenital anomaly caused by an extra autosome. *Lancet* 1960; **1**(7128): 790–3.
25. Nowell P. Hungerford D. A minute chromosome in human chronic granulocytic leukemia. *Science* 1960; **132**: 1497.
26. Rowley JD. Letter: A new consistent chromosomal abnormality in chronic myelogenous leukaemia identified by quinacrine fluorescence and Giemsa staining. *Nature* 1973;. **243**(5405): 290–3.
27. Rowley JD. Identificaton of a translocation with quinacrine fluorescence in a patient with acute leukemia. *Ann Genet* 1973;. **16**(2): 109–12.
28. Maxam AM, Gilbert W. A new method for sequencing DNA. *Proc Natl Acad Sci USA* 1977; **74**(2): 560–4.
29. Sanger F, Coulson AR. A rapid method for determining sequences in DNA by primed synthesis with DNA polymerase. *J Mol Biol* 1975; **94**(3): 441–8.
30. Nobel Prize in Chemistry. 2014 [cited 2016 20 Jan 2016].
31. de Klein A. *et al.* A cellular oncogene is translocated to the Philadelphia chromosome in chronic myelocytic leukaemia. *Nature*, 1982; **300**(5894): 765–7.
32. Groffen J. *et al.* Philadelphia chromosomal breakpoints are clustered within a limited region, bcr, on chromosome 22. *Cell* 1984; **36**(1): 93–9.
33. Heisterkamp N. *et al.* Localization of the c-ab1 oncogene adjacent to a translocation break point in chronic myelocytic leukaemia. *Nature* 1983; **306**(5940): 239–42.
34. Shtivelman E. *et al.* Fused transcript of abl and bcr genes in chronic myelogenous leukaemia. *Nature* 1985; **315**(6020): 550–4.
35. Stam K. *et al.* Evidence of a new chimeric bcr/c-abl mRNA in patients with chronic myelocytic leukemia and the Philadelphia chromosome. *N Engl J Med* 1985; **313**(23): 1429–33.
36. Daley GQ. Van Etten RA. Baltimore D. Induction of chronic myelogenous leukemia in mice by the P210bcr/abl gene of the Philadelphia chromosome. *Science* 1990; **247**(4944): 824–30.
37. Heisterkamp N. *et al.* Acute leukaemia in bcr/abl transgenic mice. *Nature* 1990; **344**(6263): 251–3.

38. Konopka JB. Watanabe SM. Witte ON. An alteration of the human c-abl protein in K562 leukemia cells unmasks associated tyrosine kinase activity. *Cell* 1984;. **37**(3): 1035–42.
39. Druker BJ. *et al.* Activity of a specific inhibitor of the BCR-ABL tyrosine kinase in the blast crisis of chronic myeloid leukemia and acute lymphoblastic leukemia with the Philadelphia chromosome. *N Engl J Med* 2001; **344**(14): 1038–42.
40. Druker BJ. *et al.* Efficacy and safety of a specific inhibitor of the BCR-ABL tyrosine kinase in chronic myeloid leukemia. *N Engl J Med* 2001; **344**(14): 1031–7.
41. NHGRI. International Consortium Completes Human Genome. 2003 [cited 2015 Nov 11].
42. Lander ES. *et al.* Initial sequencing and analysis of the human genome. *Nature* 2001; **409**(6822): 860–921.
43. Venter JC. *et al.* The sequence of the human genome. *Science* 2001; **291**(5507): 1304–51.
44. International Human Genome Sequencing Consortium, Finishing the euchromatic sequence of the human genome. *Nature* 2004; **431**(7011): 931–45.
45. Genome sequence of the nematode C. elegans: A platform for investigating biology. *Science* 1998; **282**(5396): 2012–8.
46. Adams MD. *et al.* The genome sequence of *Drosophila melanogaster. Science* 2000; **287**(5461): 2185–95.
47. Arabidopsis Genome Initiative, Analysis of the genome sequence of the flowering plant *Arabidopsis thaliana*. Nature 2000; **408**(6814): 796–815.
48. Farwell K.D. *et al.* Enhanced utility of family-centered diagnostic exome sequencing with inheritance model-based analysis: Results from 500 unselected families with undiagnosed genetic conditions. *Genet Med* 2015; **17**(7): 578–86.
49. Lee H. *et al.* Clinical exome sequencing for genetic identification of rare Mendelian disorders. *JAMA* 2014; **312**(18): 1880–7.
50. A map of human genome variation from population-scale sequencing. *Nature* 2010; **467**(7319): 1061–1073.
51. Metzker ML. Sequencing technologies — the next generation. *Nat Rev Genet* 2010; **11**(1): 31–46.
52. Corless CL. Spellman PT. Tackling formalin-fixed, paraffin-embedded tumor tissue with next-generation sequencing. *Cancer Discov* 2012; **2**(1): 23–4.
53. Hedegaard J. *et al.* Next-generation sequencing of RNA and DNA isolated from paired fresh-frozen and formalin-fixed paraffin-embedded samples of human cancer and normal tissue. *PLoS One* 2014; **9**(5): e98187.
54. Goswami RS. *et al.* Identification of Factors Affecting the Success of Next-Generation Sequencing Testing in Solid Tumors. *Am J Clin Pathol* 2016; **145**(2): p. 222–37.
55. Kidess E. Jeffrey SS. Circulating tumor cells versus tumor-derived cell-free DNA: rivals or partners in cancer care in the era of single-cell analysis? *Genome Med* 2013; **5**(8): 70.

56. Norton ME. *et al.* Cell-free DNA analysis for noninvasive examination of trisomy. *N Engl J Med* 2015; **372**(17): 1589–97.
57. Porreco RP. *et al.* Noninvasive prenatal screening for fetal trisomies 21, 18, 13 and the common sex chromosome aneuploidies from maternal blood using massively parallel genomic sequencing of DNA. *Am J Obstet Gynecol* 2014; **211**(4): 365:e1–12.
58. Fitzgibbons PL. *et al.* Prognostic factors in breast cancer. College of American Pathologists Consensus Statement 1999. *Arch Pathol Lab Med* 2000; **124**(7): 966–78.
59. Early Breast Cancer Trialists' Collaborative Group, Effects of chemotherapy and hormonal therapy for early breast cancer on recurrence and 15-year survival: an overview of the randomised trials. *Lancet* 2005; **365**(9472): 1687–717.
60. Goldhirsch A. *et al.* Meeting highlights: international expert consensus on the primary therapy of early breast cancer 2005. *Ann Oncol*, 2005; **16**(10): 1569–83.
61. Foekens JA. *et al.* Multicenter validation of a gene expression-based prognostic signature in lymph node-negative primary breast cancer. *J Clin Oncol* 2006; **24**(11): 1665–71.
62. van 't Veer LJ. *et al.* Gene expression profiling predicts clinical outcome of breast cancer. *Nature* 2002; **415**(6871): 530–6.
63. Wang Y. *et al.* Gene-expression profiles to predict distant metastasis of lymph-node-negative primary breast cancer. *Lancet* 2005; **365**(9460): 671–9.
64. Hanahan D. Weinberg RA. Hallmarks of cancer: The next generation. *Cell* 2011; **144**(5): 646–74.
65. Roberts KG. *et al.* Targetable kinase-activating lesions in Ph-like acute lymphoblastic leukemia. *N Engl J Med* 2014; **371**(11): 1005–15.
66. Roberts KG. *et al.* Genetic alterations activating kinase and cytokine receptor signaling in high-risk acute lymphoblastic leukemia. *Cancer Cell* 2012; **22**(2): 153–66.
67. McFarland CD. *et al.* Impact of deleterious passenger mutations on cancer progression. *Proc Natl Acad Sci USA* 2013; **110**(8): 2910–5.
68. Services, U.S.D.o.H.H. Health Information Privacy. 2015 [cited 2016 3/3/16].
69. Plon SE. *et al.* Display of Genetic Information in the Electronic Health Record: From Varied Chaos to Everything in Its Pace. In 2015 ACMG *Annual Clinical Genetics Meeting*. Salt Lake City, Utah, 2015.
70. Green RC. *et al.* ACMG recommendations for reporting of incidental findings in clinical exome and genome sequencing. *Genet Med* 2013; **15**(7): 565–74.
71. Richards S. *et al.* Standards and guidelines for the interpretation of sequence variants: A joint consensus recommendation of the American College of Medical Genetics and Genomics and the Association for Molecular Pathology. *Genet Med* 2015; **17**(5): 405–24.
72. Wright CF. Hall A. Zimmern RL. Regulating direct-to-consumer genetic tests: what is all the fuss about? *Genet Med* 2011; **13**(4): 295–300.
73. Green RC, Lautenbach D. McGuire AL. GINA, genetic discrimination, and genomic medicine. *N Engl J Med* 2015; **372**(5): 397–9.

74. Services, C.f.M.M. Clinical Laboratory Improvement Amendments (CLIA). 2016 [cited 2016 2/12/2016].
75. Howard S. Framework for Regulatory Oversight of Laboratory Developed Tests (LDTs), D.o.H.H. Services, Editor. 2014, FDA: http://www.fda.gov/downloads/MedicalDevices/ProductsandMedicalProcedures/InVitroDiagnostics/UCM407409.pdf.
76. 23andMe. 23andme. 2016 [cited 2016 June 6].
77. NextGxDx. the Genetic Testing Resource. 2016 [cited 2016 5/21/2016].

Tissue Biorepositories, Sample Heterogeneity, and Microdissection

3 The Use of Biorepositories in Genomic Medicine

Rodrigo Chuaqui

Cancer Diagnosis Program (CDP)
Division of Cancer Treatment and Diagnosis (DCTD)
National Cancer Institute
Rockville, MD, USA

Contents

I. Introduction

The advent of powerful molecular techniques has allowed the development of "precision medicine" in oncology, an approach based on biological agents that target specific molecular changes in individual tumors.[1] The molecular analysis of clinical samples, in particular tissue samples including primary, recurrent, and metastatic tumors, is essential in this process.[2] Access to biospecimens with associated high-quality clinical, treatment, recurrence, and outcome data is critical to developing and validating the tests needed for diagnosis, target identification, and prediction of response to therapy. In addition, biospecimens are needed for studies including cancer discovery research with a focus on basic and clinical applications, cancer etiology and early markers (earlier detection) studies, population-based screening studies, and cancer epidemiology including molecular epidemiology and genetics studies for cancer risk and outcome.

Frozen tissue samples have traditionally been the "gold standard" for molecular analysis[3] due to the higher integrity of biomolecules. However, in the last decade, technology developments have allowed the application of most of the molecular techniques to formalin-fixed, paraffin-embedded (FFPE) tissues, including high-throughput analysis approaches of DNA and RNA[1,2,4,5] as well as proteins.[6–8] Buffered formalin fixation is the standard method of preservation of clinical tissue samples for histological diagnosis, due to the higher quality of morphological detail preservation using this fixative. Most of the biorepositories include paraffin blocks in their collections, an approach useful for retrospective studies. Molecular analysis applied to FFPE material has allowed access to wider range of samples, such as rare tumors, as well as high number of specimens. This is reflected in the number of publications that have used archival material: between 1986 and 1996, only 20 publications contained "FFPE" in a search of all fields; in 2010 only, 160 such publications were identified in a study.[5] These numbers continue to rise.

However, working with FFPE tissues presents a number of technical challenges to the study of biomolecules. Challenges are faced at all steps, including at the pre-analytic (collection, storage, processing) as well as the downstream analysis phases. Important challenges faced at the biorepository level include, among others, issues assessing the histologic quality of the material prior to the molecular analysis, as well as streamlining the access process to samples for researchers. At the downstream analysis level, the cross-linking and processing present problems in the extraction of nucleic acids and proteins, with lower yields and integrity, affecting their use for downstream protocols. In this chapter, the different types of biorepositories that store FFPE are discussed, including the challenges they face. An overview of the biorepositories supported by the National Cancer Institute (NCI), most of them providing FFPE tissues for the cancer research community, is presented. The role the different NCI repositories play in the use of samples for the development of clinical markers and assays for cancer patients is explained. Some of the challenges such as the histologic quality assessment prior to use in downstream analysis protocols as well as the molecular techniques that are currently being applied to FFPE will be included in a separate chapter.

II. Biorepository Challenges in the Use of FFPE Tissues

a. Pre-analytics

The NCI has established Best Practice recommendations for Biospecimen Resources (https://biospecimens.cancer.gov/bestpractices/2016-NCI BestPractices.pdf). The document, last updated in 2016, includes two main sections, one referring to technical and operational aspects, the second to ethical, legal, and policy issues. Technical aspects include, among others, biospecimen collection, processing, storage, retrieval, and distribution; quality management; biosafety; collecting and managing clinical data; and data management, inventory control, and tracking. A variety of factors may affect biospecimen quality, and therefore research results might not necessarily reflect biological changes. These factors are grouped into "pre-analytic variables" and "analytic variables". Among pre-analytic variables, physiology of the patient, specimen collection practices, as well as specimen handling practices are included.

The International Society for Biorepositories and Environmental Research (ISBER) also has a Best Practices document (https://www.ucdmc.ucdavis.edu/biorepositories/pdfs/bestpractices/ISBER_Best-practices-repositories_3rd-edition.pdf), supporting most of the NCI recommendations. One of the key goals is to obtain uniformity in biospecimen collection and preservation practices. Some of the most frequent pre-analytical factors considered include: time of clamping during surgery ("warm ischemia"), time from removal of the specimen to processing ("cold ischemia"), and time of fixation itself. The literature is vast regarding the effect of these variables in the molecular data of FFPE tissues.[9–10] The different biomolecules are not affected equally by these variables, DNA being the most stable. For example, a recent study of 59 FFPE high-grade ovarian serous adenocarcinomas analyzed the effect of time of storage on the sequencing results.[11] DNA was extracted and subjected to whole exome sequencing (WES) in samples stored between 3 and 32 years. 58 of 59 specimens yielded enough DNA for WES. When storage time was taken into account, older specimens tended to have lower DNA sequencing quality metrics; however, 90% of specimens provided successful WES data regardless of storage time.

A more significant effect is seen at the expression level. A list of proteins and transcripts most affected both by cold and warm ischemia times include is reported by David *et al.* Some transcripts are upregulated, others downregulated with longer ischemia times. Genes that were significantly upregulated comprised transcription factors (EGR1, FOS), signaling molecules (CYR61, RGS1, SGK1) of the extracellular matrix, and phosphatases (DUSP1). Genes that were downregulated included oxidases 2 (DUOX2), solute carrier family 6 (SLC6A14), and vanin (VNN1, involved in vitamin B5 recycling). Changes differ between normal and tumor tissue and do not occur in all patients. EGFR pathway protein expression for example showed changes in 60% of the cases. However, many changes occur only in a few cases. Proteins, in particular the phosphorylation status of proteins, might be affected most, in particular by ischemia times. Signaling molecules within the AKT and MAPK pathways are significantly affected, in particular in tumor tissue compared to normal. Over-phosphorylation, as well as de-phosphorylation occur. The most consistent change was a decrease in protein phosphorylation observed for AKT, MTOR, p70-S6K, GSK3B, and ERK1/2.

However, others have found less effect of pre-analytical variables on molecular data. Boenisch[12] finds consistent staining in 26/30 antigens regardless of length of FF, even up to eight days. Sprung *et al.*[13] report that fixation of tissue for up to two days in neutral buffered formalin did not adversely impact protein identifications. Chung *et al.*[14] report that decreased RNA fragment length is observed only after long fixation times, over 48 hours.

Erickson *et al.*[15] found no variations in genes not known to be involved in the cellular response to hypoxia with various warm ischemia times. Only small changes occur in genes related to hypoxia. Zhang *et al.*[16] report that 80% of the probe sets in oligonucleotide microarrays were not differentially regulated with warm ischemia. Also, changes belong to defined groups of genes (maintenance of cell structure, division, differentiation, tricarboxylic acid cycle, and tissue repair).

At this point, there is no doubt that the molecular, proteomic, and morphologic endpoints can be affected to a certain degree by ischemia times and suboptimal processing conditions in FFPE. The magnitude and direction of the changes varies according to the analyte (DNA being the most stable, phosphoproteins the most affected). A useful evaluation of the most influential pre-analytical factors obtained by reviewing the most recent literature is presented by Bass *et al.*[10] Thresholds for pre-analytical factors are presented, including ischemia times, specimen size, decalcification, tissue to fixative ratio, fixative conditions such as concentration, duration, and other processing steps (clearing reagents, paraffin reagents, embedding time, block size, mounting on slides, tissue storage). However, it is difficult to standardize these thresholds, since they will depend on several variables, such as analyte (DNA, RNA, protein), platform used, whether amplification is involved (PCR, RT-PCR) or not (FISH), whether the approach is uniplex vs. multiplex, quantitative vs. semi-quantitative, the protocol and kits used, and the tissue studied (tissues with high enzymatic content such as pancreas might degrade differently than other with lower enzymatic content).

Some variables might be easier to standardize by the use of Best Practices, such as the consistent use of 10% buffered formalin, keeping the appropriate ratios of sample size and fixative volume, as well as some of the automated processing steps. However, important factors such as ischemia times, which seem to have important effects, especially on the proteome, are the most difficult to control and therefore to standardize. Ischemia times, i.e. the clamping time during surgery, is dictated by the clinical practice, and not by the potential effect on the quality of molecules for research. Even cold ischemia times are difficult to control, since operationally a coordinated effort is needed between the operating room staff and pathology personnel.

b. Tracking the pre-analytic features of a sample

Various approaches can be taken to the pre-analytic problem. Three groups can be described: (a) trying to capture the values for the pre-analytical variables of the sample, thus being aware of the "lifecycle" of the sample;(b) identifying molecular biomarkers of sample stability that can be standardized; and (c) improving/standardizing the methods of collection and handling of specimens.

(a) A concerted attempt to standardize the tracking of pre-analytical variables for biospecimens was described in 2010 by different members of the ISBER.[17] Recognizing the need for managing and tracking pre-analytical variations between Biobanks, the Standard PREanalytical Code (SPREC) was developed. A list of pre-analytical factors for clinical fluid and solid specimens was identified and each factor was given a code, thus standardizing and facilitating the implementation and integration of the system in Biobanks.

The SPREC uses the existing Laboratory Data Management System (LDMS) codes (https://www.fstrf.org/apps/cfmx/apps/ldms/manual/5.8/) for the sample types and the primary container types. LDMS is a widely used software-based laboratory management system with key features that support modern laboratory's operations. The LDMS system allows for data tracking and smart data exchanging among different interfaces. In SPREC, each sample is assigned a seven-element-long code that corresponds to seven pre-analytical variables, with a string of 11 letters (for fluids) or 13 (for solid samples) in a defined order, separated by six hyphens (Figure 1).

The SPREC elements applicable to solid tissue or tissue-derived cytologic biospecimens are: type of sample, type of collection, warm ischemia time, cold ischemia time, fixation type, fixation time and storage condition. Version 3.0 of the SPREC was recently published, including new technological developments and knowledge about the critical ranges of preanalytical times.[18]

Software tools were developed for SPREC labeling and effective exchange and search of stored biospecimens. These software tools that are used for generating, storing, managing, and exchanging SPREC-related information associated to specimens is open source.

Thus, adopting the SPREC facilitates and consolidates multi-center biomarker and biospecimen research in the clinical biobanking environment.

Other recommendations for reporting data have been published, such as the Biospecimen Reporting for Improved Study Quality.[19] Other data elements are included to be reported in studying biospecimens, and include clinical parameters such as disease status of the patient, other available medical information, clinical diagnosis, as well as histologic assessment of the quality of the sample (Figure 2). Again, this is an

Table 2. Preanalytical variables included in the SPREC (seven-element-long SPREC), version SPREC-01, applied to solid samples. Codes in bold come from the LDMS

Type of sample	
Fresh cells from nonblood specimen type	**CEN**
Cells from nonblood specimen type (e.g., disrupted tissue), viable	**CLN**
Cells from fine needle aspirate	FNA
Hair	**HAR**
Cells from laser capture microdissected tissue	LCM
Cells from nonblood specimen type (e.g., disrupted tissue), nonviable	**PEN**
Solid tissue	**TIS**
Cells from disrupted tissue	LCM
Other	ZZZ
Type of collection	
Autopsy <6 h postmortem	A06
Autopsy 6-12 h postmortem	A12
Autopsy 12-24 h postmortem	A24
Autopsy 24-48 h postmortem	A48
Autopsy 48-72 h postmortem	A72
Biopsy	**BPS**
Fine needle aspirate	FNA
Punction	PUN
Surgical excision	SRG
Swab	**SWB**
Other	ZZZ
Warm ischemia time	
<2 min	A
2-10 min	B
10-20 min	C
20-30 min	D
30-60 min	E
>60 min	F
Unknown	X
Not applicable (e.g., biopsy)	N
Other	Z
Cold ischemia time	
<2 min	A
2-10 min	B
10-20 min	C
20-30 min	D
30-60 min	E
>60 min	F
Unknown	X
Not applicable (e.g., autopsy)	N
Other	Z
Fixation type	
Nonaldehyde with acetic acid	ACA
Aldehyde based	ALD
Alcohol based	ETH
Nonbuffered formalin	FOR
Snap freezing	**SNP**
Nonaldehyde without acetic acid	**NAA**
Neutral buffered formalin	**NBF**

Table 2. Preanalytical variables included in the SPREC (seven-element-long SPREC), version SPREC-01, applied to solid samples. Codes in bold come from the LDMS (Cont'd)

Fixation type		
Optimum cutting temperature medium		**OCT**
RNA Later		**RNL**
Unknown		XXX
Other		ZZZ
Fixation time		
<15 min		A
15 min to 1 h		B
1-4 h		C
4-8 h		D
8-24 h		E
24-48 h		F
48-72 h		G
Unknown		X
Other		Z
Long-term storage		
PP tube 0.5-2 mL*	−85°C to −60°C	A
PP tube 0.5-2 mL	−35°C to −18)°C	B
Cryotube 1-2 mL	Liquid nitrogen†	C
Cryotube 1-2 mL	−85°C to −60°C	D
Cryotube 1-2 mL	Programmable freezing to <−135°C	E
Straw	Liquid nitrogen	F
Straw	−85°C to −60°C	G
Straw	−35°C to −18°C	H
Straw	Programmable freezing to <−135°C	I
PP tube ≥5 mL	−85°C to −60°C	J
PP tube ≥5 mL	−35°C to −18°C	K
Microplate	−85°C to −60°C	L
Microplate	−35°C to −18°C	M
Paraffin block	RT	P
Unknown		X
Other		Z

NOTE: RT, room temperature: 18°C to 25°C.
*PP, polypropylene.
†Liquid nitrogen refers to either vapor or liquid phase.

Figure 1. Pre-analytical variables included in the SPREC (from Ref. [17]).

TABLE 1. QUICK-REFERENCE BIOSPECIMEN REPORTING FOR IMPROVED STUDY QUALITY SUMMARY/CHECKLIST: TIER 1 ITEMS TO REPORT IF KNOWN AND APPLICABLE

Data elements	*Examples*
☐ Biospecimen type Solid tissue, whole blood, or another product derived from a human being	Serum, urine
☐ Anatomical site Organ of origin or site of blood draw	Liver, antecubital area of the arm
☐ Disease status of patients Controls or individuals with the disease of interest	Diabetic, healthy control
☐ Clinical characteristics of patients Available medical information known or believed to be pertinent to the condition of the biospecimens	Premenopausal breast cancer patients
☐ Vital state of patients Alive or deceased patient when biospecimens were obtained	Postmortem
☐ Clinical diagnosis of patients Patient clinical diagnoses (determined by medical history, physical examination, and analyses of the biospecimen) pertinent to the study	Breast cancer
☐ Pathology diagnosis Patient pathology diagnoses (determined by macro and/or microscopic evaluation of the biospecimen at the time of diagnosis and/or prior to research use) pertinent to the study	Her2-negative intraductal carcinoma
☐ Collection mechanism How the biospecimens were obtained	Fine-needle aspiration, preoperative blood draw
☐ Type of stabilization The initial process by which biospecimens were stabilized during collection	Heparin, on ice
☐ Type of long-term preservation The process by which the biospecimens were sustained after collection	Formalin fixation, freezing
☐ Constitution of preservative The make-up of any formulation used to maintain the biospecimens in a nonreactive state	10% neutral-buffered formalin, 10 United States Pharmacopeia heparin units/mL
☐ Storage temperature The temperature or range thereof at which the biospecimens were kept until distribution/analysis.	−80°C, 20°C–25°C
☐ Storage duration The time or range thereof between biospecimen acquisition and distribution or analysis	8 days, 5–7 years
☐ Shipping temperature The temperature or range thereof at which biospecimens were kept during shipment or relocation	−170°C to −190°C
☐ Composition assessment and selection Parameters used to choose biospecimens for the study	Minimum 80% tumor nuclei and maximum 50% necrosis

Figure 2. Data elements included in the biospecimen reporting for improved study recommendations (from Ref. [26]).

attempt to standardize the reporting of any information that might influence the endpoint results of studies using biospecimens.

(b) A second tool to overcome bias due to pre-analytical variation is to identify consistent molecular markers of sample quality. For example, selecting the most consistent RNA transcript changes (those occurring in all cases) can serve as a "biomarker set" for tissue quality regarding ischemia times.[9] Regarding proteins, a similar attempt, identifying a subset of proteins, including their phosphorylation status that can serve as biomarkers for the quality of the sample was reported.[20] Thus, the state of protein preservation and specimen quality can be assessed prior to any downstream analysis by identification of this panel of natural proteins, referred to as "surrogate stability markers".

(c) Finally, modifications of handling and especially processing protocols have been attempted to improve the molecular quality of tissues.[3,7,20,21] Espina[20] used multi-purpose fixative solutions designed to stabilize, preserve, and maintain protein, nucleic acid and tissue architecture. Use of alcohol fixatives[3] have been attempted to improve yield and integrity of RNA from paraffin-embedded material. A novel method for recovering proteins from archival tissue for proteomic analysis was developed by Fowler *et al.*[7] Up to 4-fold increase in protein extraction efficiency was achieved using a pressure-assisted protein extraction method. Variations in temperature of formalin fixation may also increase the yield of biomolecules from FFPE material.[21] These attempts could be useful for research purposes, especially in the hands of individual laboratories, but are more difficult to standardize among biobanks and pathology departments. None of these novel methods are included as standard practices by the NCI or ISBER.

c. Investigator access to specimens and utilization of the samples

One of the keys to the success of a biorepository is how well the samples are utilized. Important repository metrics regarding productivity include, among others, "samples collected and distributed", "investigators served", as well as the corresponding scientific output derived from the usage of the samples. Utilization of the samples might be a significant issue in repositories that are built mainly with "leftover samples" from previous projects that have been completed. Standard operating procedures (SOPs) have to be established regarding the access process to specimens. It is essential that the access process to the samples is streamlined and efficient. The SOPs of each repository have to include priorities in case they have competing requests for samples. Clinical care is always first priority, research needs coming next. Most repositories prioritize internal institutional investigators, such as cancer center researchers over outside investigator requests. Some repositories require collaboration in the projects that use their samples. As we will see, prospective collection of samples initiated by a scientific question is the most efficient biorepository model in terms of utilization rate of their samples.

In order to facilitate researchers interested in finding biorepositories that might have samples appropriate for their research, the CDP at the

NCI initiated the Specimen Resource Locator (SRL; https://specimens.cancer.gov/). The SRL is a searchable website that helps investigators to locate human specimens for research. The database includes resources such as tissue banks and tissue procurement services with access to normal/uninvolved, benign, pre-cancerous and/or cancerous human tissue. The SRL provides investigators access to the NCI Tissue Expediter (https://specimens.cancer.gov/contact/). The Tissue Expediter is a scientist who can assist investigators in locating specimen resources in the event of a failed search. The Expediter also has knowledge of other potential resources and collaborators that may be able to assist and corresponds directly with the inquiring investigator to offer potential leads. As a search is performed, the web interface results filter as the specimen criteria (e.g., tumor type, specimen type, etc.) are selected. The resource contact information is promptly displayed, making it easy to further inquire about the biospecimens needed. The search criteria include: Organ type, Tumor Type, Specimen Type, Method of Preservation, specimen annotations, and the number of specimens required.

III. Biorepository Types and Models

a. General features of biorepositories

Biorepositories have a range of functions and are diverse, complex organizations, with variable levels of specialization.[22] Some collections might have been generated as leftover samples from a specific scientific project. Other repositories collect with open-ended goals, such as the Tumor Cancer Genomic Atlas (TCGA) for deep molecular analysis of nucleic acids and proteins. Other repositories collect samples only with associated clinical or epidemiological data, this data significantly increasing the value of the samples, especially for clinical research. Most repositories collect samples and store them as they are procured for future utilization, while other repositories collect only prospective samples with a scientific question or general goal already in mind and for immediate usage. The goals might range from basic to translation to clinical research, may be focused on individuals, or epidemiology, or population-based issues.[23]

They might collect only fresh/frozen material, fixed material, or both, although most repositories collect and store FFPE to some extent. Some collect only tissue, some also matched fluids (blood, serum, plasma, urine, saliva, other bodily fluid). Some are specialized either in pediatric samples,[24] in one organ, such as the National Institutes of Health (NIH) NeuroBiobank (NBB) (https://neurobiobank.nih.gov/), or in one main disease such as The AIDS and Cancer Specimen Resource (http://acsr.ucsf.edu/). Normal tissue might be collected in some repositories, sometimes matched with diseased tissue from the same patients.

There are various ways of classifying Biobanks. Grizzle *et al.*[25] discuss 3 basic models: the traditional model of tissue banking for future usage, a prospective collection model for immediate utilization, and a combination of these two models, in detail. Standard operating procedures and functions vary in each model. The amount of annotation is also different in each model. Samples collected prospectively for immediate usage commonly are associated with little annotation, usually basic demographic data included in the pathology report and, obviously, no follow-up or response to treatment data. Retrospective collections might include more detailed annotation, especially clinically relevant data. The key is that the types of tissues and material match the tissue needs of the researchers.

b. The NCI-supported biorepositories

Several biospecimen repositories are supported by the various divisions and offices at NCI (see Table 1). From the organizational perspective, we can identify three main biorepository models:

(a) Traditional tissue bank model. Here, the collection and storage of specimens and associated data occurs at a single central facility, such as in the CCR-PI collections.
(b) A virtual bank model, where the specimens are collected and stored locally at various places, with the associated specimen data centralized. Requests for specimens are handled through a central coordinating office. Most NCI biospecimen resources belong in this category.

Table 1. NCI-supported Biorepositories. OD: Office of the Director. DCTD: Division of Cancer Treatment and Diagnosis. DCB: Division of Cancer Biology. DCCPS: Division of Cancer Control and Population Science. DCEG: Division of Cancer Epidemiology and Genetics. CCR: Clinical Cancer Center.

NCI Division/Program	Name of Biospecimen Resource/Funding Mechanism
OD/Office of Cancer Centers	50 Cancer Centers Shared Resources
OD/Center for Cancer Genomics	The Cancer Genome Atlas (TCGA)
OD/Office of HIV & AIDS Malignancy	AIDS & Cancer Specimen Resource (ACSR)
DCTD/Cancer Diagnosis Program	Cooperative Human Tissue Network (CHTN)
DCTD/Cancer Diagnosis Program	Cooperative Group Banks (CGB)
DCTD/Translational Research Program	SPORE Biospecimen Cores
DCTD/Developmental Therapeutics Program	Melanoma and Pediatric Astrocytoma Repositories
DCTD/Cancer Diagnosis Program	Biospecimen Research Network (BRN); Cancer Human Biobank (caHUB)
DCB	Chernobyl Tissue Bank
DCB	International Registry of Werner Syndrome
DCB	Barrett's Esophagus Translational Research Coordinating Center
DCCPS	Multiple Biospecimen Registries & Studies
DCEG (Intramural) & DCP	PLCO Biorepository
DCEG (Intramural)	DCEG Principal Investigator Collections
DCEG (Intramural)	U.S. Radiologic Technology Study
DCEG (Intramural)	Alpha-Tocopherol, Beta-Carotene Prevention Study
CCR/Office of Clinical Director (Intramural)	Blood Processing Core

(c) Prospective tissue procurement service. The collection of specimens is prospective and specific to meet the investigator requests. Most specimens (except for rare ones) are stored only for a limited period until they can be distributed to the investigators who requested them. Examples are the Cooperative Human Tissue Network and the TCGA.

Overall, the following five categories of Biospecimen Resources are supported by NCI:

(1) The resource is established for public use and funds are allocated for its infrastructure including collection, storage, and distribution of specimens. The CHTN falls in this category.
(2) The resource is established for use by the members of a scientific network of one or several institutions studying a variety of cancers in different organ sites. Cancer center biorepositories and SPOREs are examples.
(3) The resource is initially established for biospecimen collection for a specific research project which has been completed, and the specimens are now available for sharing with other investigators. Examples are repositories at the two NCI divisions: DCCPS and DCEG.
(4) The resource is established for biospecimen collection of a specific research project which is not yet completed. The specimens are only available to investigators in the project, such as the TCGA project.
(5) The resource is established by an intramural principal investigator and the specimens are used for the investigator's research and are only available to other researchers for collaboration. Examples are CCR (intramural) PI collections, and DCEG (intramural) PI collections.

Even though these are historically diverse, and are not centralized, two features make this set of biorepositories unique:

(a) The samples are available to the scientific community for their use. Even though repositories vary in their access process, most of them make the samples available to the entire scientific community, rather than to segments of it (such as members of specific networks). This point is key, since access to samples and utilization is one of the major challenges biorepositories face.
(b) They fulfill a different function within the "pipeline" of development for assays or markers for patients (Figure 3). This development usually covers from basic discoveries to "bedside", ready to be implemented in the clinic. Samples are needed throughout the process. Initially, feasibility of the technology has to be assessed as well as marker prevalence. After determining the reproducibility,

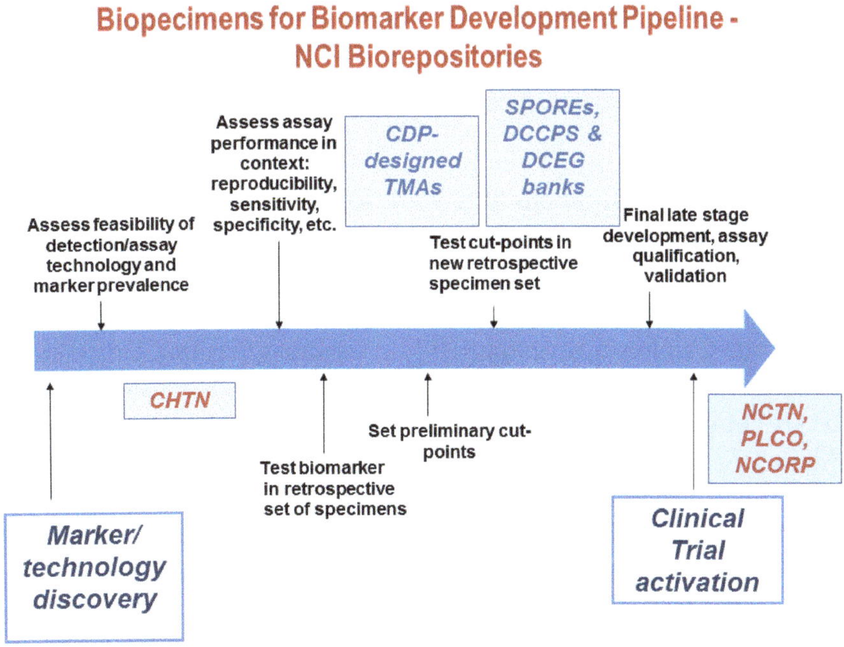

Figure 3. NCI-supported biorepositories and the assay/marker development pipeline. DCCPS: Division of Cancer Control and Population Science. DCEG: Division of Cancer Epidemiology and Genetics. SPOREs: Specialized Programs of Research Excellence.

sensitivity, and specificity of the assay, several tests in retrospective set of samples are usually performed. Preliminary cut-points are set, which might be repeated in new sets of samples. The final stage of development is called assay qualification and includes validation both of the assay protocol per se (technical validation), as well as its clinical utility (clinical validation). The types of samples and especially the degree of clinical annotation varies at different stages. Initially, highly annotated ones are usually not needed, and even clinical samples might not be required. Cell lines or animal models might be useful at this level. Later, during the different stages of testing and cut-point determinations, some annotation or associated data might become necessary, especially at the final stages of clinical qualification.

i. *Examples of NCI-supported biorepositories providing specimens suited for early-phase discoveries*

1. *The Cooperative Human Tissue Network (CHTN)*

The Cooperative Human Tissue Network (CHTN; http://www.chtn.org) is a network of six institutions that work together to provide prospective investigator-defined procurement of malignant, benign, diseased, and uninvolved (normal adjacent) tissues and body fluids to researchers throughout North America. The six participating institutions are: Nationwide Children's Hospital (CHTN Pediatric Division), University of Alabama at Birmingham (CHTN Southern Division), Ohio State University (CHTN Midwestern Division), University of Pennsylvania (CHTN Eastern Division), Vanderbilt University (CHTN Western Division), and University of Virginia (CHTN Mid Atlantic Division).

Several functionalities make the CHTN a unique biorepository. The CHTN is not a repository that stores tissues for unspecified future studies and research. It prospectively collects and distributes biospecimens upon investigators' requests.[26] Most of the biospecimens are distributed upon procurement, reaching almost 100% of utilization rates. CHTN only performs limited banking as needed to meet specific requests and longer-term banking of targeted rare and more difficult to obtain specimens and biomolecules.

Even though the majority of the specimens distributed are fresh or frozen samples, FFPE constitutes a significant proportion of the samples. The CHTN works prospectively with investigators to tailor biospecimen acquisition and processing to meet their specific project requirements. These may include various media to prepare fresh biospecimens, freezing method, preparation of slides, and various staining and fixation methods.

The collection of follow-up and clinical data is not part of the current mission of CHTN. Biospecimens are routinely de-identified; they are accompanied by basic demographic and diagnostic data including a copy of the redacted pathology report. High-quality, de-identified biospecimens fulfill a critical need in early-phase discoveries, translational research and analytical and clinical assay validation studies.[27] While the majority of the specimen requests are for neoplastic tissue with or without matching normal tissue, the CHTN is also unusual in that essentially all types of tissue and disease requests are considered, and many non-neoplastic disease specimens are provided including uninvolved (normal) tissue to the research community.

Access is open to the entire scientific community. CHTN Divisions coordinate investigator applications/requests based upon the investigator's geographic location. The Pediatric Division serves all investigators within North America. Requests for rare specimens are immediately networked to all divisions for prompt response. An informatics system facilitates the networking and tracking and is critical for the network to function as a seamless unit.

ii. *Examples of NCI-supported biorepositories/collections suited for translational research and middle-phase assay/marker development*

1. *Cancer Diagnosis Program (CDP)-developed NCI TMAs*

Most of the NCI-supported biospecimen resources provide samples with some annotations that can be useful for the middle phases of the assay development pipeline, but lack detailed clinical follow up or treatment data. A typical collection resource in the middle of the assay development pipeline corresponds to the CDP-designed tissue microarrays (TMAs) (now under the Mid-Atlantic Division of the CHTN, http://www.medicine.virginia.edu/research/cores/bio-tissue-core/TMA/TMAhome-page). These TMAs were carefully designed by an NCI biostatistician to ensure high statistical power, but lack follow up data. These TMAs include breast and colon TMAs. The breast cancer tissue microarrays (TMAs) are suited for investigations of marker prevalence associated with progression or prognosis of disease. All of the invasive cases are primary breast cancers with a principal histology of ductal cancer. Breast cancer progression TMAs are designed to investigate differences in prevalence of markers in three stages of invasive breast cancer: node-negative, node-positive, and metastatic disease. The breast cancer prognostic TMAs are designed to examine potential prognostic markers in non-metastatic breast cancer. Separate TMAs were constructed for each of the three non-metastatic TNM stages, I–III (as defined by AJCC Manual for Staging of Cancer, 5th edition).

The colon cancer prognostic TMAs are designed to investigate differences in prevalence of markers in THM stages I-IV, and include diverticulitis cores, normal colon tissue matched to colon cancer patient, as well as adenomatous colon polyps from patients without cancer.

2. *Specialized Program of Research Excellence (SPORE) biospecimen banks*

There are 64 SPOREs located at academic centers distributed over 20 states in the USA, and each SPORE is focused on a specific organ site. There are 18 organ sites represented in the SPORE. They provide sufficient, well-characterized, and high-quality tissues, including a wide range of neoplastic and non-neoplastic biospecimens from both fresh frozen and FFPE blocks, and other samples.

The Core also tracks biospecimens to facilitate sharing by the research laboratories according to a priority plan established with all SPORE investigators and approved by the Tissue and Blood Bank Core Pathologists. Biospecimen processing, inventory, distribution data, and histopathological analyses are maintained in the Core's computerized database with links to the Informatics component of the Biostatistics Core for storage and archiving to facilitate a web-based retrieval system. This database is designed for use by all SPORE investigators.

3. *DCCPS biospecimen banks*

The DCCPS biospecimen banks (http://epi.grants.cancer.gov/biospecimens.html) are established for specific research projects as part of population studies including molecular epidemiology studies to collect biospecimens or derivatives.

4. *DCEG biospecimen banks*

The DCEG has epidemiology resources including, Surveillance, Epidemiology, and End Results Program (SEER; http://seer.cancer.gov/biospecimen/). Currently, the banks apply electronic network for pathology and medical record reporting. FFPE specimens are associated with data such as demographic data, as well as the stage of the disease and survival in the majority of the cases. However, the resource has limited annotation of clinical data including treatment history and recurrence.[28]

iii. *Examples of NCI-supported biorepositories providing specimens suited for later stage clinical validation of markers and assays*

1. *Biospecimen banks to support NCI-clinical trials network (NCTN)*

These are the most highly annotated sample collections, coming from NCI-supported trial patients and therefore fall at the clinical end of the assay development pipeline. Cooperative Group Banks (NCTN) (https://research.usc.edu/nci-national-clinical-trials-network-nctn-program/) were established to collect, store, and provide to researchers well-annotated human specimens from cancer patients, who participate in NCI-funded phases III and II treatment trials. Specimen collections from such patients offer several advantages, such as (a) uniform information on medical history and treatment details; (b) consistency in standard operating procedures (SOPs) and biospecimen quality control; and (c) reasonable costs for biospecimens compared to other sources. These specimen collections are unique because they have detailed treatment histories and cancer recurrence data based on careful follow-ups of patients over long periods. NCTN biospecimens are distributed only to qualified projects in which this type of samples linked to high-quality clinical data (including treatment and outcome data) are critical for developing and validating biomarkers for cancer diagnosis, prognosis, and prediction of responses to therapy.

The NCTN banks have ability to receive and store FFPE blocks and histological slides from diagnostic biopsies and surgical resections from patients with cancer. Some of the banks have abilities to receive and store fresh frozen specimens collected in at least some of the clinical trials. Other specimen types collected and stored include but are not limited to: serum, plasma, whole blood, white blood cells, bone marrow, urine, skin, and sputum/buccal samples as defined in the trial protocol. In addition, these banks have capacities to prepare relevant biomolecules (DNA/RNA) from the collected biospecimens. These banks receive biospecimens collected by NCTN

Group member hospitals, and affiliated member institutions, including NCI Community Oncology Research Program (NCORP) hospitals, using special collection kits and SOPs provided by the banks. The biospecimens are then sent to the designated NCTN Bank of the lead group conducting the trial or to the bank designated in a specific protocol. The clinical diagnostic data as well as treatment and outcome data linked to the biospecimens are stored in Statistical Data Centers of the NCTN groups.

A comprehensive effort by the NCI has transformed the nine CGBs into five NCTN biospecimen banks aligned with the five NCI-funded NCTN Groups. The new structure of the NCTN Biospecimen Banking Resource comprises five NCTN Banks serving one pediatric (COG) and four adult NCTN Groups (ALLIANCE, ECOG-ACRIN, NRG, and SWOG).

A common informatics (IT) infrastructure to support biospecimen banking and integration of data management, statistical, and banking operation systems has been developed. Each bank has an established local IT system that connects the bank with the Group Statistical and Data Center, NCI enterprise, and other NCTN banks. In addition, each NCTN Bank uses a common biospecimen inventory IT system and search engine to connect to the Group Statistical, Data and Operations Centers and common Biospecimen IT Navigator system, and connect, biospecimen inventory to allow assessment of biospecimen availability based on trial design and clinical data endpoints. The biospecimen banks have the ability to perform a feasibility assessment on availability of banked specimens for research.

2. *DCP biospecimen banks*

Two biospecimen banking initiatives at DCP are Prostate, Lung Colorectal, and Ovarian (PLCO; http://epi.grants.cancer.gov/Consortia/members/plco.html), and NCI Community Oncology Research Program (NCORP; http://prevention.cancer.gov/major-programs/nci-community-oncology).

The PLCO cancer screening trial is a large population-based randomized trial designed to determine the effects of screening on cancer-related mortality and secondary endpoints in men and women between age 55 and 74.

The NCORP designs and conducts cancer prevention, control, screening, and post-treatment surveillance 21 clinical trials; conducts cancer care delivery research, including comparative effectiveness research; and integrates disparities research questions into clinical trials and cancer care delivery research. NCORP supports NCTN banks for storage and distribution of NCORP biospecimens prospectively collected from cancer screening, prevention, and control clinical trials across all NCORP sites.

Both these DCP resources are associated with trials and samples are clinically annotated, making them suitable for later validation studies in the marker/assay developmental pipeline.

IV. Future Directions

Biobanks face various challenges at all stages of their operations. With the advent of personalized medicine, these challenges continue to grow.[29] Most of the current molecular techniques can be applied to FFPE archival samples. Biopsy material has become less abundant and as sensitivity improves with new techniques, i.e. more information can be obtained from less material.[5] High-throughput techniques such as NanoString or microarray for expression analysis, or next-generation sequencing can be performed with just few hundred nanograms of nucleic acids. Clinical biopsy-type specimens (liquid material or small biopsies from recurrent or metastatic tumors) are collected at different time points and in a prespecified clinical context of treatment. These "next-generation biospecimens" are made available using multidimensional high-throughput techniques.[2] They differ from traditional biobank specimens, which usually contain large sections of primary tumors. Sample procurement and processing of these new specimens have become more complex and sophisticated, requiring a multidisciplinary therapeutic team, including interventional radiologists, molecular pathologists, and bioinformaticians.[2] As histological variation might be considerable in these small biopsies, histologic evaluation by pathologists is essential in confirming the diagnosis and tumor content. New ethical challenges are involved in the collection of these samples, including obtaining patients' consent at the initial steps of the disease. It is predicted that these types of samples are going to be the prevalent material coming from the clinic for molecular analysis.[2] The molecular analysis of these samples will be the basic guide for cancer treatment. Greater integration between treating physicians, nurse, study coordinators, radiologists, and pathologists will be required.

References

1. Dubbinka JZ, Deans ZC, Tops BBJ, van Kemenade FJ, Koljenovic S, van Krieken HJM, Blokx WAM, Dinjens WNM, Groenen PJTA. Next generation diagnostic molecular pathology: Critical appraisal of quality assurance in Europe. *Mol Oncol* 2014; **8**: 830–839.
2. Basik M, Aguilar-Mahecha A, Rousseau C, Diaz Z, Tejpar S, Spatz A, Greenwood CMT, Batist G. Biopsies: next-generation biospecimens for tailoring therapy. *Nat Rev Clin Oncol* 2013; **10**: 437–450.
3. Perlmutter MA, Best CJM, Gillespie JW, Gathright Y, González S, Velasco A, Linehan WM, Emmert-Buck MR and Chuaqui RF. Comparison of snap freezing versus ethanol fixation for gene expression profiling of tissue specimens. *J Mol Diag* 2004; **6**: 371–377.
4. Berg D, Hipp S, Malinowsky K, Böllner C, Becker KF. Molecular profiling of signalling pathways in formalin-fixed and paraffin-embedded cancer tissues. *Eur J Cancer* 2009; **46**: 47–55.
5. Fairley JA, Gilmour K, Walsh K. Making the most of pathological specimens: Molecular diagnosis in formalin-fixed, paraffin embedded tissue. *Curr Drug Targets* 2012; **13**: 1475–1487.
6. Shi SR, Liu C, Balgley BM, Lee CL, Taylor CR. Protein Extraction from Formalin-fixed, Paraffin-embedded Tissue Sections: Quality Evaluation by Mass Spectrometry. *J Histochem Cytochen* 2006; **54**: 739–743.
7. Fowler CB, Waybright TJ, Veenstra TD, O'Leary TJ, Mason JT. Pressure-Assisted Protein Extraction: A Novel Method for Recovering Proteins from Archival Tissue for Proteomic Analysis. *J Proteome Res* 2012; **11**: 2602–2608.
8. Gustafsson OJR, Arentz G, Hoffmann P. Proteomic developments in the analysis of formalin-fixed tissue. *Biochim Biophys Acta* 2014; **1854**: 559–580.
9. David KA, Unger FT, Uhlig P, Juhl H, Moore HM, Compton C, Nashan B, Dörner A, de Weerth A, Zornig C. Surgical procedures and postsurgical tissue processing significantly affect expression of genes and EGFR-pathway proteins in colorectal cancer tissue. *Oncotarget*, 2014. www.impactjournals.com/oncotarget.
10. Bass BP, Engel KB, Greytak SR, Moore, HM. Review of preanalytical factors affecting molecular, protein, and morphological analysis of formalin-fixed, paraffin-embedded (FFPE) tissue. How well do you know your FFPE specimen? *Arch Pathol Lab Med.* 2014; **138**: 1520–1530.
11. Carrick D, Mehaffey M, Sachs M, Altekruse S, Camalier C, Chuaqui R, Cozen W, Das B, Hernandez BY, Lih ZJ, Lynch CF, Makhlouf H, McGregor P, McShane L,

Phillips Rohan JA, Walsh W, Williams PM, Gillanders EM, Mechanic LE, Schully SD. Robustness of next generation sequencing on older formalin-fixed paraffin-embedded tissue. Plos One (accepted).

12. Boenish T. Effect of heat-induced antigen retrieval following inconsistent formalin fixation. *Appl Immunohistochem Mol Morphol* 2005; **13**: 283–286.
13. Sprung RW Jr, Brock JW, Tanksley JP, Li M, Washington MK, Slebos RJ, Liebler DC. Equivalence of protein inventories obtained from formalin-fixed paraffin-embedded and frozen tissue in multidimensional liquid chromatography-tandem mass spectrometry shotgun proteomic analysis. *Mol Cell Proteomics* 2009; **8**: 1988–1998.
14. Chung JY1, Braunschweig T, Williams R, Guerrero N, Hoffmann KM, Kwon M, Song YK, Libutti SK, Hewitt SM. Factors in tissue handling and processing that impact RNA obtained from FFPE tissue. *J Histochem Cytochem* 2008; **56**: 1033–42.
15. Erickson HS, Josephson JW, Vira M, Albert PS, Gillespie JW, Rodriguez-Canales J, Pinto PA, Linehan WM, Chuaqui RF, Emmert-Buck MR, Coleman JA. Influence of hypoxia induced by minimally invasive prostatectomy on quantitative rt-pcr gene expression analysis: Implications for molecular bio-markers of diagnosis, progression and treatment. *Am J Transl Res* 2010; **2**: 210–222.
16. Zhang S, He XS, Zhu XF, Wang DP, Ma Y, Ju WQ, Wu LW, Zhang J, Huang JF. An oligonucleotide microarray analysis of differential gene expressions of liver cells after warm ischemic injury. *Zhonghua Gan Zang Bing Za Zhi* 2007; **15**: 50–54.
17. Betsou F, Lehmann S, Ashton G, Barnes M, Benson EE, Coppola D, DeSouza Y, Eliason J, Glazer B, Guadagni F, Harding K, Horsfall DJ, Kleeberger C, Nanni U, Prasad A, Shea K, Skubitz A, Somiari S, Gunter E. Standard preanalytical coding for biospecimens: defining the sample PREanalytical code. *Cancer Epidemiol Biomarkers Prev* 2010; **19**: 1004–11.
18. Betsou F, Bilbao R, Case J, Chuaqui R, Clements JA, Souza Y, De Wilde A, Geiger J, Grizzle W, Guadagni F, Gunter E, Heil S, Kiehntopf M, Koppandi I, Lehmann S, Linsen L, Mackenzie-Dodds J, Aguilar Quesada R, Tebbakha R, Selander T, Shea K, Sobel M, Somiari S, Spyropoulos D, Stone M, Tybring G, Valyi-Nagy K, Wadhwa L. Standard PREanalytical Code version 3.0. Biopreserv Biobank. 2018 Jan 29. doi: 10.1089/bio.2017.0109.
19. Moore HM1, Kelly AB, Jewell SD, McShane LM, Clark DP, Greenspan R, Hayes DF, Hainaut P, Kim P, Mansfield E, Potapova O, Riegman P, Rubinstein Y, Seijo E, Somiari S, Watson P, Weier HU, Zhu C, Vaught J. Biospecimen reporting for improved study quality (BRISQ). *J Proteome Res* 2011; **10**: 3429–38.
20. Espina V, Mueller C, Edmiston K, SciroM, Petricoin EF, Liotta LA. Tissue is alive: New technologies are needed to address the problems of protein biomarker pre-analytical variability. *Proteomics Clin Appl* 2009; **3**: 874–882.
21. Gündisch S, Annaratone L, Beese C, Drecol E, Marchiò C, Quaglino E, Sapino A, Becker KF, Bussolati G. Critical roles of specimen type and temperature before and during fixation in the detection of phosphoproteins in breast cancer tissues. *Lab Invest* 2015; **95**: 561–71.

22. Henderson GE, Cadigan RJ, Edwards TP, Conlon I, Nelson AG, Evans JP, Davis AM, Zimmer C, Weiner BJ. Characterizing biobank organizations in the U.S.: results from a national survey. *Genome Med* 2013, **5**: 3, doi:10.1186/gm407.
23. Pukkala E. Biobanks and registers in epidemiologic research on cancer. *Methods Mol Biol* 2011; **675**: 127–64.
24. Brisson AR, Matsui D, Rieder MJ, Fraser DD. Translational research in pediatrics: Tissue sampling and biobanking. *Pediatrics* 2012; **129**: 153–.
25. Grizzle WE, Bella WC, Sexton KC. Issues in collecting, processing and storing human tissues and associated information to support biomedical research. *Cancer Biomarkers* 2011; **9**: 531–549.
26. Grizzle WE, Sexton KC, Bell WC. Quality assurance in tissue resources supporting biomedical research. *Cell Preserv Technol* 2008; **6**: 113–118.
27. Jewell SD, Srinivasan M, McCart LM, Williams N, Grizzle WE, MD, LiVolsi V, MacLennan G, Sedmak DD. Analysis of the molecular quality of human tissues: An experience from the cooperative human tissue network. *Am J Clin Pathol* 2002; **118**: 733–741.
28. Altekruse SF, Rosenfeld GE, Carrick DM, Pressman EJ, Schull SD, Mechanic LE, Cronin KA, Hernandez BY, Lynch CF, Cozen W, Khoury MJ, Penberthy LT. SEER cancer registry biospecimen research: yesterday to tomorrow. *Cancer Epidemiol Biomarkers Prev* 2014; **23**: 2681–2687.
29. Olson JE, Bielinski SJ, Ryu E, Winkler EM, Takahashi PY, Pathak J, Cerhan JR. Biobanks and personalized medicine. *Clin Genet* 2014: **86**: 50–55.

4 Specimen Heterogeneity and Tissue Microdissection

Kristen D. Noyes, Mary Kamos, Kevin B. Delfino and Michael A. Tangrea

Sinai Hospital of Baltimore, LifeBridge Health, Baltimore, MD, USA

Contents

I. Introduction

The field of genomic medicine has advanced rapidly over the past decade. This unprecedented growth has been driven primarily by the dramatic decrease in the cost of next-generation sequencing (NGS), which has reduced from costing millions of dollars per sample in the early 2000s to less than $1,000 today.[1] The launch of the Illumina and Ion Torrent technologies in the mid-2000s, with their distinct parallel sequencing methodologies, has fueled this field and democratized sequencing.[2] As a result, the use of genomic testing is becoming commonplace in community hospitals to better diagnose patients and pair them with the appropriate drug treatments, reducing side effects and increasing efficacy, i.e., precision medicine (Figure 1).

However, the ideals of Genomic or Precision Medicine are not yet fulfilled due to a number of factors, including those related to the complexity of specimens. In oncology, the tumor and surrounding microenvironment contain a wide variety of cell types, making molecular analysis of the tumor cells alone complex and challenging. The rapid

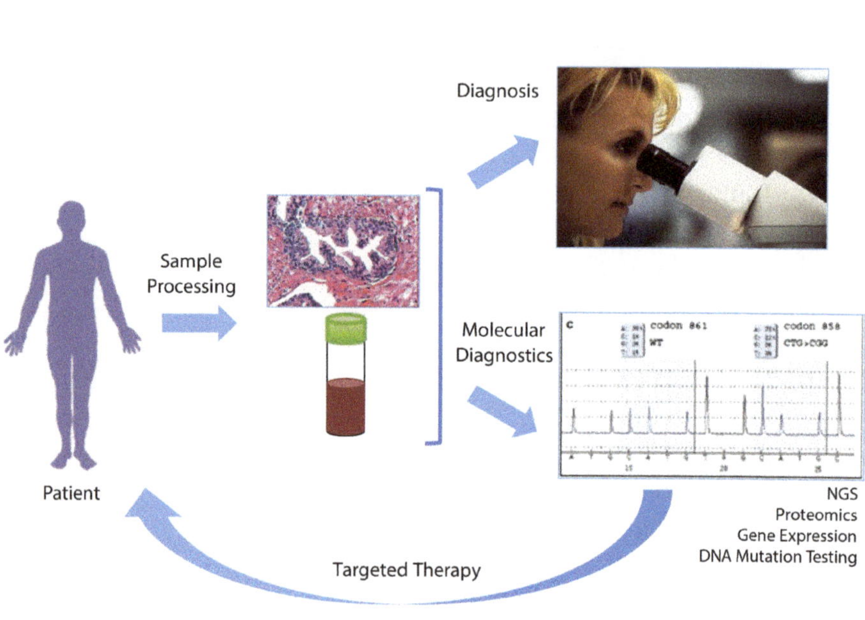

Figure 1. Precison medicine workflow.

evolution of NGS technologies have provided insights into the biology and roles of these specialized cells, yet we do not fully understand the interaction within the tumor microenvironment and the impact on disease progression. In addition, it is well known that the tumor itself is heterogeneous, displaying a variety of clones within a single tumor focus.[3–8] In this chapter, we will define and address the importance of specimen heterogeneity in solid tumors and discuss various tissue dissection strategies, including laser capture microdissection (LCM), expression microdissection (xMD), and the use of image analysis tools to drive the microdissection process.

II. Sample Heterogeneity

Today, cancer is defined by the tumor site, i.e., breast cancer, colon cancer, etc.; however, there is interest in tailoring treatment regimens to specific molecular markers regardless of tumor site.[9] Unfortunately, there is currently no universal therapy that works across patient populations within the same tumor type. This lack of a "magic bullet" for cancer is due in part to heterogeneity, heterogeneity between patients with the same disease type, heterogeneity among tumor foci within the same patient, and intra-tumor heterogeneity.[10] This level of complexity greatly impacts our ability to treat patients efficiently.

a. Tumor microenvironment

Tumors consist of neoplastic cells proper as well as the surrounding tumor microenvironment, which includes fibroblasts, inflammatory cells, vasculature, and other cells (Figure 2). This variance within and surrounding the tumor is also reflected at the molecular level.[11] Knowledge about the tumor microenvironment field has grown tremendously over the past few years, especially in how the microenvironment both supports and fights off tumor growth.[12] Although, these normal host cells are not considered part of the tumor *per se*, they have been shown to play a large role in tumor progression and development.[12,13] The role of the microenvironment was proposed as early as 1889 by Stephen Paget, who proposed the "seed and soil" concept for cancer metastasis.[14] Yet, the role

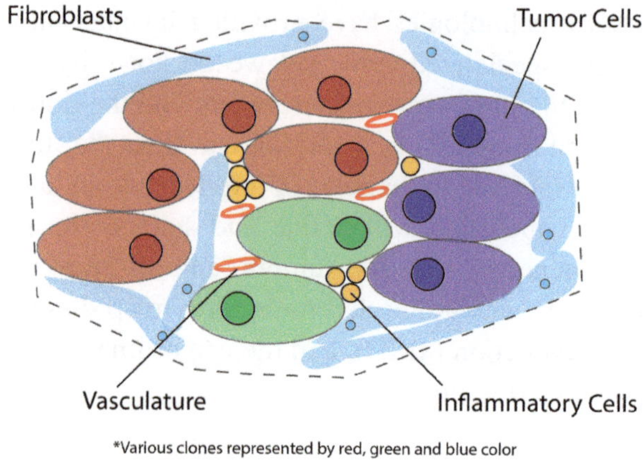

Figure 2. Tumor heterogeneity and microenviroment.

of the tumor microenvironment in promoting tumor growth and progression is still not completely understood today.

Immune cells within the microenvironment that invade and surround the tumor have become the focus on a number of new treatments.[11,15] Hanahan and Weinberg describe how tumor evasion of the immune system is a new hallmark of cancer.[16] The role of immunotherapy in the clinic has dramatically increased over the past few years with the development of nivolumab, pembrolizumab, and ipilimumab, among other therapeutics. In addition to immunotherapies, novel treatments are being developed that target the tumor microenvironment and not the tumor cells themselves to improve outcomes.[17–20] For example, Blansfield *et al.* developed a three-drug combination anti-angiogenic strategy to disrupt endothelial cell growth in an animal model study.[17] These multi-pronged approaches aim to provide a more durable response for cancer patients.[21]

b. Tumor heterogeneity

Today's NGS technologies allow researchers to sequence genomes and gene panels more rapidly than ever before. In contrast to Sanger sequencing, which was developed in the late 1970s,[22] today's technologies are based on parallel methodologies, which allow researchers to sequence DNA libraries overnight versus months or years.[23] The ability to rapidly and affordably sequence a tumor specimen has led to an improved understanding of tumor biology and intra-tumor heterogeneity.[4,5,24] Among the many benefits of these advances is the ability to classify cancers into distinct molecular subgroups.

The high level of heterogeneity within many tumors raises concerns on the reliance of a single biopsy for diagnosis.[5] It is currently not possible for a surgeon to identify the "bad actors" within the tumor *a priori*. Thus, the amount of genetic variability within the tumor further complicates the ability to treat patients.

c. Liquid biopsies

As an aside to the discussion on tumor heterogeneity, a brief mention of liquid biopsies is in order since this approach strategically attacks the issue of inter- and intra-tumor heterogeneity in a completely different and important manner. Liquid biopsies can complement mutation detection in conventional needle biopsy specimens via the capture and molecular evaluation of circulating tumor cells (CTCs) or circulating tumor DNA (ctDNA).

As tumors grow, cancer cells or nucleic acids can be shed into the blood stream, which lead to distant metastases (Figure 3). The ability to identify and count the number of circulating tumor cells (CTCs) in a 7–10 mL

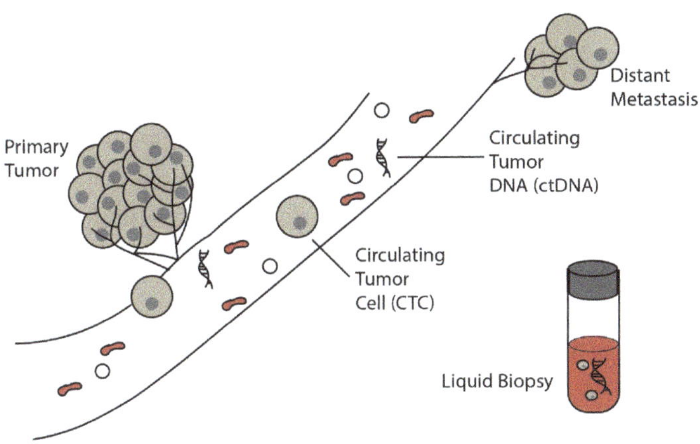

Figure 3. Circulating tumor cells and circulating tumor DNA.

vial of blood is attractive as a non-invasive way to track disease progression. However, the blood components greatly outnumber the CTCs, which is the greatest limitation of this method. A wide variety of CTC isolation strategies and analytical methods have been developed that utilize CTC data to predict prognosis, specifically in metastatic breast, colon, lung, and prostate cancers.[25–29] The clinical utility of liquid biopsies is still under debate, but the approach provides another precision medicine tool.[30] For a more thorough discussion of liquid biopsies and their clinical utility, please see the recent review by Hench *et al.*[31]

d. Molecular evaluation of tumors

Immunohistochemistry (IHC), a technique that detects antigens in a heterogeneous tissue section, is routinely used in pathology departments to determine the primary tumor site for tumors of unknown origin.[32] This concept, originally developed by Coons *et al.* in 1941, employs a labeled antibody targeted against an antigen of interest for visual detection.[33] IHC allows the user to assess the histological localization of the protein via either light or fluorescent microscopy and has become a workhorse technique for the evaluation of tissues. In their initial publication, Coons *et al.* utilized a fluorescent-tagged antibody for their detection system. Although a powerful technique, the need to directly label the antibody made the method cumbersome and difficult to employ. It was not until the mid-1960's that a more universal approach was established.[34–36] This iteration involved the incorporation of an enzyme attached to a secondary antibody, thus amplifying the detection signal and increasing specificity. This modification also allows for the preservation of staining for extended periods of time, since the standard diaminobenzidine (DAB) stain is one that can be coverslipped and stored. The technique went through additional changes in the 1980s, incorporating different labeling enzymes and adopting the biotin–streptavidin reaction.[37]

III. Microdissection Methods

As discussed above, there is a wide variety of cell types within the tumor and surrounding microenvironment, as well as clonal heterogeneity within the tumor itself. Despite the knowledge of this complexity, most clinical NGS tests today combine multiple formalin-fixed paraffin-embedded (FFPE) tissue sections, typically 10–15, for downstream molecular analysis with little to no effort to enrich for the tumor cell population.[38] For most DNA assays this may be sufficient. However, as the precision medicine field moves towards transcriptomic and proteomic assays, it becomes more crucial to evaluate a pure population of tumor cells for analysis.[39]

a. Macrodissection methods

Today, manual macrodissection is the dominant method to isolate tumor cells from a histological tissue section in the clinical setting. This is due to a number of factors, mainly cost and ease of use. However, macrodissection lacks the resolution necessary for sophisticated downstream molecular analysis.[40,41] For macrodissection, a pathologist reviews a hematoxylin & eosin (H&E)-stained tissue section under a microscope and highlights the tumor by marking the area. Next, a technician takes this "guide slide" and carefully scrapes away the "same" area of interest on a subsequent unstained serial recut. This approach has the advantage that it is rapid and inexpensive. However, now that we have molecular tools that can rapidly and precisely provide molecular analysis of a single cell, there is a need for more precise upfront tissue processing methods. By combining multiple cell populations together, such as those described earlier as part of the tumor microenvironment, the tumor cell's specific molecular profile may be masked.

b. Laser capture microdissection (LCM)

LCM is a technique for the isolation of specific cells from a microscopic region of a tissue sample. The technique was initially developed in the mid 1990s by a team of researchers at the National Institutes of Health (NIH) in the United States; however, many academic groups and companies have contributed to the growth of the dissection field and there is now a whole range of LCM instruments available commercially from a number of different vendors (Figure 4)[42].

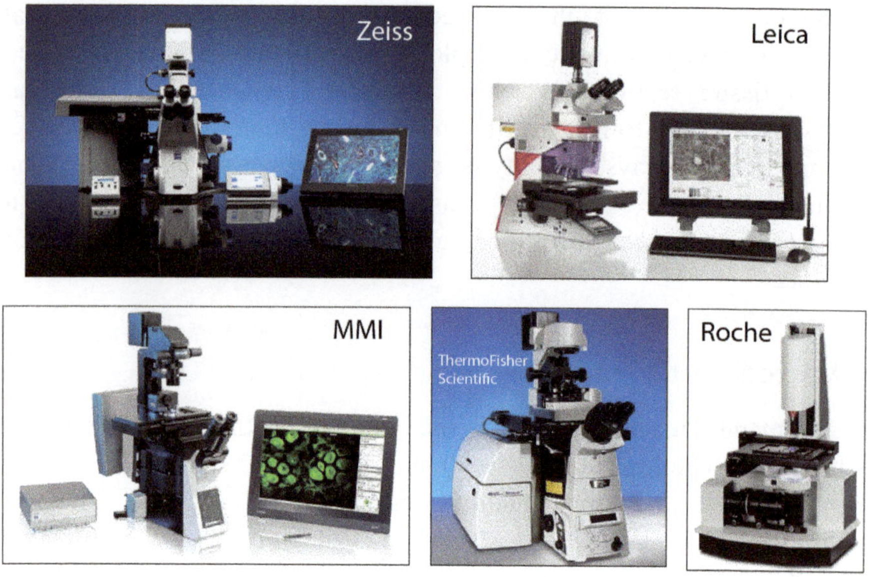

Figure 4. Microdissection technologies.

LCM allows for the rapid procurement of target cells from a heterogeneous tissue section. The isolation of tissue subpopulations via LCM provides the opportunity for downstream testing of cells, such as DNA genotyping and loss-of-heterozygosity (LOH) analysis, RNA transcript profiling, cDNA library generation, proteomics discovery, and signal-pathway profiling.[40,43–46]

The initial LCM method employed the use of a thin transparent ethylene vinyl acetate (EVA) film and an infrared laser. The EVA film is placed on top of a tissue section and the sample is viewed microscopically. The target cells are located, and the infrared laser is fired in a focused pulse. This laser pulse causes the EVA film to melt and bind the cells of interest underneath the film. The EVA film is then removed from the tissue sample with the target cells remaining adhered to the film. The cells can then be lysed with the appropriate buffer for downstream molecular analysis.

The technology was originally commercialized by Arcturus Engineering (now part of ThermoFisher Scientific) and has improved substantially over time. The LCM technology specifically has many advantages over manual macrodissection methods. First, it has a resolution of approximately seven microns, which allows for the procurement of single cells. The current Arcturus XT LCM device also has the ability to perform UV cutting to isolate large areas of cells from tissue. For this feature, the tissue must be sectioned onto a specialized membrane slide, whereas standard IR LCM can be performed on tissue mounted on standard glass slides, an important advantage.

The field of laser-based dissection technologies has advanced the biology and molecular pathology fields, with over 4,000 publications to date, and LCM is utilized to study molecular changes in a wide range of normal functions and pathologies, including neoplastic progression, cellular physiology, embryonic development, growth and differentiation, aging, and other processes.[47–50]

c. Image analysis LCM

The morphological assessment step in the LCM procedure can be time consuming and rate limiting when many cells and/or many samples need to be dissected. Thus, several groups have evaluated the use of computer-aided cell identification methods to improve the process. A more detailed description of image-analysis technologies can be found in several chapters in the book subsection on Digital Imaging, Machine Learning, and Artificial Intelligence.

One method for image segmentation and feature extraction of histopathological images is known as spatially invariant vector quantization (SIVQ), which fundamentally differs from other established image analysis algorithms.[51–53] SIVQ exhibits significant freedom from context-specific performance restrictions and at the same time provides for a rapid and interactive discovery workflow model. Its potential use as a tool for computational-assisted LCM is further enhanced by a simple user interface and associated modest training requirements. SIVQ-LCM improves the workflow by shifting the role of the pathologist from performing the entire microdissection process to a limited-contact supervisor-enabling large-scale extraction of tissue via an expedited process.

In the example shown in Figure 5, target cells with a specific nuclear morphology (prominent nucleoli and a glassy peri-nucleolar space; see inset, panel (a) were dissected from prostate tissue. This nuclear feature

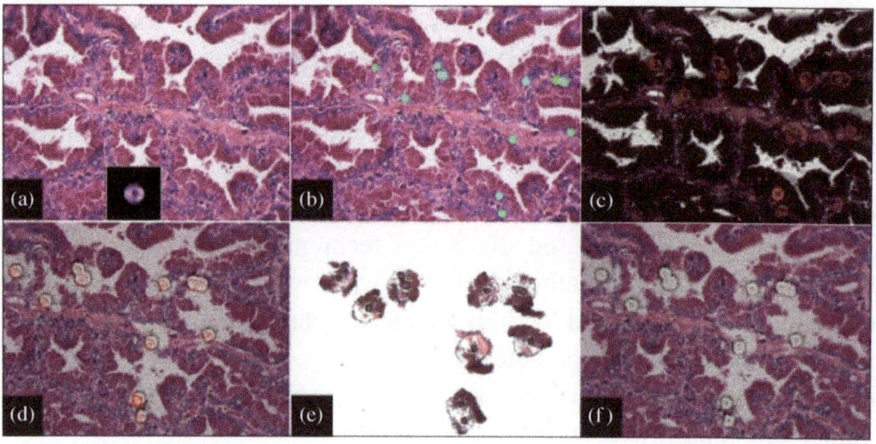

Figure 5. Isolation of single cells via SIVQ-LCM. From Ref. 51.

was then queried against the entire field of view, including luminal, basal, and stromal cells. SIVQ identified the particular nuclear size and morphology in cells (b), the processed image was loaded into an LCM instrument (c), and the cells were auto-dissected (d). The bottom-middle (e) and bottom-right (f) panels show the specific microdissection of the target cells, consisting of their nuclei and corresponding luminal cytoplasm.

To perform SIVQ analysis, a pathology slide is photographed using a microscope camera adapter. The SIVQ engine is compatible with a majority of the commonly used digital image file formats, so the image can be formatted in numerous ways. The technician identifies a distinct image within the file that exhibits the morphology of interest. The selected feature is used to search the entirety of the file (and therefore the entirety of the tissue sample) for regions that are morphologically similar to the selected region. The engine provides a heatmap of all statistically probable regions of similarity that then guide the dissection process.

d. Immuno-LCM

In contrast to computer-based dissection, immuno-LCM is a method developed to allow investigators to target specific cell types by visually aiming the dissection laser at IHC or immunofluorescence (IF)-stained cells. The technique has been used in numerous studies over the past several years, including epigenetic analysis, tumor microenvironment studies, and neuroscience research.[54–56] There are now several variations of immuno-LCM, including xMD and expression microlabeling (xML) that are more automated advances based on direct molecular targeting.

In addition to employing an individual dissection tool for a specific application, investigators can combine them together (e.g., computer-aided and immuno-guided) for certain applications. For example, SIVQ analysis can be applied to an immunostained slide by setting the targeting parameters to recognize all stained cells as well as cells with different labeling intensities (see Figure 6). Panel (a) show initial immuno-based dissection. Next, further targeting with SIVQ in panel (b) shows identification of only the dark brown chromogen, lighter staining is seen in (c), and medium staining in (d). After SIVQ targeting, the image was transferred to an LCM instrument and the IHC stained cells dissected (e, f).

e. Expression microdissection (xMD)

xMD is an adaptation of the LCM technique that optimizes the transfer of cells of interest from a tissue sample to a film. xMD is a more efficient method than LCM for some applications because it uses commercially available tools and can be done rapidly without an operator-based, cell-by-cell selection process. This is possible because xMD utilizes antibody probes and IHC in place of a pathologist or histologist to visually identify target cell populations.[57]

Rather than selectively adhering cell subpopulations to an EVA film via user-driven guidance and an infrared laser, xMD instead relies on IHC and broad heat activation. A sample of tissue is stained for diagnostic markers and chromogen using IHC. Then, a clear EVA film is placed over the tissue sample and the entire sample is exposed to a light source. The chromogen bound to the cells of interest absorbs the light energy, which causes focal heat activation of the dissection film and adheres the target cells to the film. Since the tissue sample had been stained to mark the cells of interest, and the light source was used to flash the entire sample, removal of the EVA film yields the transfer of *all cell subpopulations* of interest. This large transfer of target cells yields relatively large amounts of DNA, RNA, and protein, which can be utilized for downstream testing.

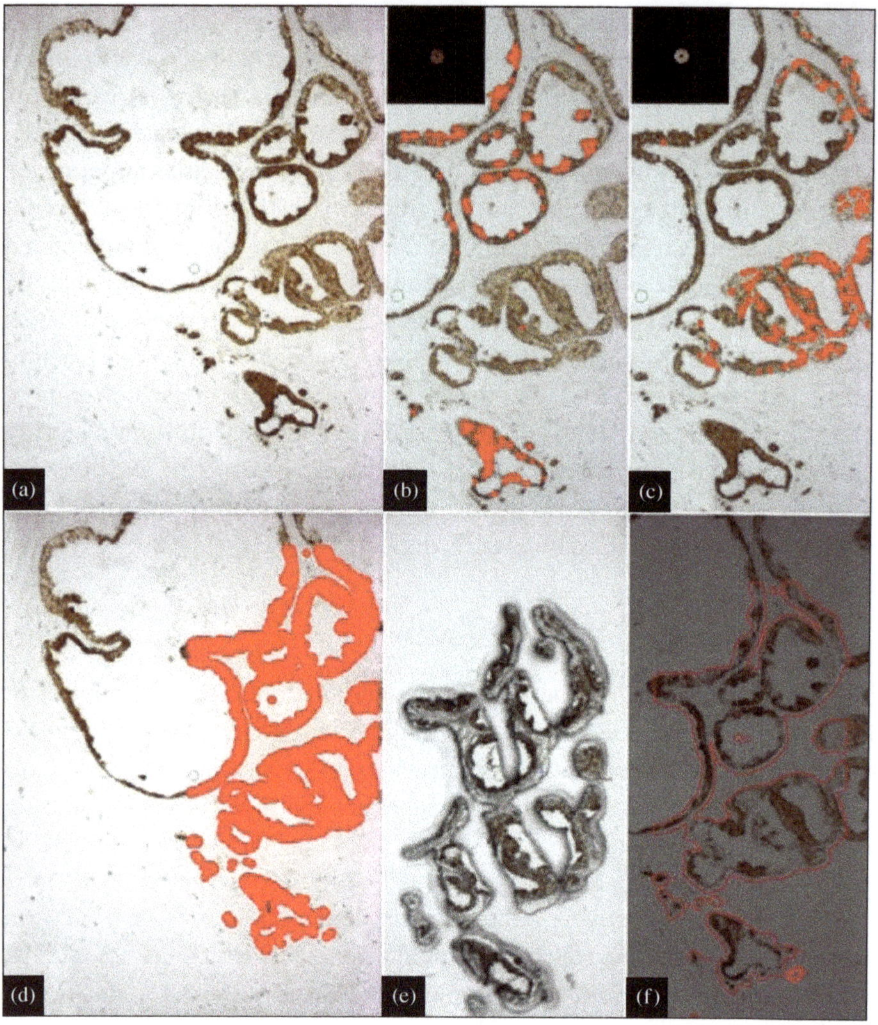

Figure 6. Isolation of immuno-stained cells via SIVQ-LCM. From Ref. 51.

xMD has been applied to the study of the tumor microenvironment in prostate, esophageal, lung, and brain tissue.[58–63]

f. Mesodissection

A more recent technology, mesodissection, developed by Parry *et al.* in 2013, employs a novel strategy to isolate specific cell populations from a tissue section.[64,65] This simple approach utilizes a milling technology rather than lasers to perform the dissection process. A spinning mill tip on a syringe scratches off the tissue of interest while a plunger pulls up the content into the capture tube with lysis buffer. Krizman *et al.* recently published a manuscript comparing mesodissection, LCM, and manual microdissection showing that they are complementary tools.[65] Mesodissection is another technology in the clinical or laboratory investigator's toolbox that can be used to isolate specific tissue regions of interest.

IV. Conclusion

Cancer is a complex disease, and both the tumor microenvironment and heterogeneity are challenging issues to address. The variety of different cell types in and around the tumor complicate the molecular understanding of the disease process. The use of microdissection tools aims to isolate specific cells, based on computer algorithms, IHC, or other means to provide a more specific evaluation of molecular profiles, especially with the explosion of single-cell genomic tools that were developed over the past 5–10 years, such as whole genome amplification[66,67] and advances in NGS-based bioinformatics.[68–70] All these technologies can be leveraged together to enhance an understanding of tumor development and progression.

References

1. Hayden EC. Technology: The $1,000 genome. *Nature* 2014; **507**: 294–5.
2. Pettersson E, Lundeberg J, Ahmadian A. Generations of sequencing technologies. *Genomics* 2009; **93**:105–11.
3. Burrell RA, McGranahan N, Bartek J, Swanton C. The causes and consequences of genetic heterogeneity in cancer evolution. *Nature* 2013; **501**: 338–45.
4. Fisher R, Pusztai L, Swanton C. Cancer heterogeneity: Implications for targeted therapeutics. *Br J Cancer* 2013; **108**: 479–85.
5. Gerlinger M, Rowan AJ, Horswell S, Larkin J, Endesfelder D, Gronroos E, Martinez P, Matthews N, Stewart A, Tarpey P, Varela I, Phillimore B, Begum S, McDonald NQ, Butler A, Jones D, Raine K, Latimer C, Santos CR, Nohadani M, Eklund AC, Spencer-Dene B, Clark G, Pickering L, Stamp G, Gore M, Szallasi Z, Downward J, Futreal PA, Swanton C. Intratumor heterogeneity and branched evolution revealed by multiregion sequencing. *The New England Journal of Medicine* 2012; **366**: 883–92.
6. Jamal-Hanjani M, Quezada SA, Larkin J, Swanton C. Translational implications of tumor heterogeneity. *Clin Cancer* Res 2015; **21**: 1258–66.

7. Swanton C, Burrell RA, Futreal PA. Breast cancer genome heterogeneity: A challenge to personalised medicine? *Breast Cancer* Res 2011; **13**: 104.
8. Yap TA, Gerlinger M, Futreal PA, Pusztai L, Swanton C. Intratumor heterogeneity. Seeing the wood for the trees. *Sci Transl Med* 2012; **4**: 127ps10.
9. Coyne GO, Takebe N, Chen AP. Defining precision: The precision medicine initiative trials NCI-MPACT and NCI-MATCH. *Curr Probl Cancer* 2017; **41**: 182–93.
10. Allison KH, Sledge GW. Heterogeneity and cancer. *Oncology* (Williston Park) 2014; **28**: 772–8.
11. Gorris MAJ, Halilovic A, Rabold K, van Duffelen A, Wickramasinghe IN, Verweij D, Wortel IMN, Textor JC, de Vries IJM, Figdor CG. Eight-color multiplex immunohistochemistry for simultaneous detection of multiple immune checkpoint molecules within the tumor microenvironment. *J Immunol* 2018; **200**: 347–54.
12. Hu M, Polyak K. Microenvironmental regulation of cancer development. *Curr Opin Genet Dev* 2008; **18**: 27–34.
13. Paterlini-Bréchot P. About seed and soil. *Cancer Microenviron* 2014; **7**: 91–3.
14. Paget S. The distribution of secondary growths in cancer of the breast. 1889. *Cancer Metastasis Rev* 1989; **8**: 98–101.
15. Gajewski TF, Schreiber H, Fu YX. Innate and adaptive immune cells in the tumor microenvironment. *Nat Immunol* 2013; **14**: 1014–22.
16. Hanahan D, Weinberg RA. Hallmarks of cancer. the next generation. *Cell* 2011; **144**: 646–74.
17. Blansfield JA, Caragacianu D, Alexander HR, 3rd, Tangrea MA, Morita SY, Lorang D, Schafer P, Muller G, Stirling D, Royal RE, Libutti SK. Combining agents that target the tumor microenvironment improves the efficacy of anticancer therapy. *Clin Cancer Res* 2008; **14**: 270–80.
18. Ji T, Zhao Y, Ding Y, Nie G. Using functional nanomaterials to target and regulate the tumor microenvironment: Diagnostic and therapeutic applications. *Adv Mater* 2013; **25**: 3508–25.
19. Li M, Zhang F, Su Y, Zhou J, Wang W. Nanoparticles designed to regulate tumor microenvironment for cancer therapy. *Life Sci.* 2018 May 15; **201**: 37–44.
20. Sau S, Tatiparti K, Alsaab HO, Kashaw SK, Iyer AK. A tumor multicomponent targeting chemoimmune drug delivery system for reprograming the tumor microenvironment and personalized cancer therapy. *Drug Discov Today*. 2018 Jul; **23**(7): 1344–1356.
21. Sharma P, Allison JP. Immune checkpoint targeting in cancer therapy. Toward combination strategies with curative potential. *Cell* 2015; **161**: 205–14.
22. Sanger F, Nicklen S, Coulson AR. DNA sequencing with chain-terminating inhibitors. *Proc Natl Acad Sci USA* 1977; **74**: 5463–7.
23. Head SR, Komori HK, LaMere SA, Whisenant T, Van Nieuwerburgh F, Salomon DR, Ordoukhanian P. Library construction for next-generation sequencing: Overviews and challenges. *BioTechniques* 2014; **56**: 61–4, 6, 8, passim.
24. Martelotto LG, Ng CK, Piscuoglio S, Weigelt B, Reis-Filho JS. Breast cancer intratumor heterogeneity. *Breast Cancer Res* 2014; **16**: 210.

25. Bidard FC, Peeters DJ, Fehm T, Nole F, Gisbert-Criado R, Mavroudis D, Grisanti S, Generali D, Garcia-Saenz JA, Stebbing J, Caldas C, Gazzaniga P, Manso L, Zamarchi R, de Lascoiti AF, De Mattos-Arruda L, Ignatiadis M, Lebofsky R, van Laere SJ, Meier-Stiegen F, Sandri MT, Vidal-Martinez J, Politaki E, Consoli F, Bottini A, Diaz-Rubio E, Krell J, Dawson SJ, Raimondi C, Rutten A, Janni W, Munzone E, Caranana V, Agelaki S, Almici C, Dirix L, Solomayer EF, Zorzino L, Johannes H, Reis-Filho JS, Pantel K, Pierga JY, Michiels S. Clinical validity of circulating tumour cells in patients with metastatic breast cancer: A pooled analysis of individual patient data. *The lancet oncology* 2014; **15**: 406–14.
26. Cohen SJ, Punt CJ, Iannotti N, Saidman BH, Sabbath KD, Gabrail NY, Picus J, Morse MA, Mitchell E, Miller MC, Doyle GV, Tissing H, Terstappen LW, Meropol NJ. Prognostic significance of circulating tumor cells in patients with metastatic colorectal cancer. *Ann Oncol: Off J Euro Soc Med Oncol/ESMO* 2009; **20**: 1223–9.
27. Hou JM, Krebs MG, Lancashire L, Sloane R, Backen A, Swain RK, Priest LJ, Greystoke A, Zhou C, Morris K, Ward T, Blackhall FH, Dive C. Clinical significance and molecular characteristics of circulating tumor cells and circulating tumor microemboli in patients with small-cell lung cancer. *J Clin Oncol: Off J Am soc Clin Oncol* 2012; **30**: 525–32.
28. Krebs MG, Sloane R, Priest L, Lancashire L, Hou JM, Greystoke A, Ward TH, Ferraldeschi R, Hughes A, Clack G, Ranson M, Dive C, Blackhall FH. Evaluation and prognostic significance of circulating tumor cells in patients with non-small-cell lung cancer. *J Clin Oncol: Off J Am Soc Clin Oncol* 2011; **29**: 1556–63.
29. de Bono JS, Scher HI, Montgomery RB, Parker C, Miller MC, Tissing H, Doyle GV, Terstappen LW, Pienta KJ, Raghavan D. Circulating tumor cells predict survival benefit from treatment in metastatic castration-resistant prostate cancer. *Clini Cancer Rese: An Off J Am Assoc Cancer Rese* 2008; **14**: 6302–9.
30. Gold B, Cankovic M, Furtado LV, Meier F, Gocke CD. Do circulating tumor cells, exosomes, and circulating tumor nucleic acids have clinical utility? A report of the association for molecular pathology. *J Mol Diagn* 2015; **17**: 209–24.
31. Hench IB, Hench J, Tolnay M. Liquid biopsy in clinical management of breast, lung, and colorectal cancer. *Front Med* (Lausanne) 2018; **5**: 9.
32. Kulkarni A, Pillai R, Ezekiel AM, Henner WD, Handorf CR. Comparison of histopathology to gene expression profiling for the diagnosis of metastatic cancer. *Diagn Pathol* 2012; **7**: 110.
33. Coons AH, Creech HJ, Jones RN. immunological properties of an antibody containing a fluorescent group. *Proceedings of the Society for Experimental Biology and Medicine* 1941; **47**: 200–2.
34. Nakane PK, Pierce GB, Jr.. Enzyme-labeled antibodies. Preparation and application for the localization of antigens. *J Histochem Cytochem* 1966; **14**: 929–31.
35. Nakane PK. Simultaneous localization of multiple tissue antigens using the peroxidase-labeled antibody method. A study on pituitary glands of the rat. *J Histochem Cytochem* 1968; **16**: 557–60.

36. Avrameas S, Uriel J. [Method of antigen and antibody labelling with enzymes and its immunodiffusion application]. *C R Acad Sci Hebd Seances Acad Sci* D 1966; **262**: 2543–5.
37. Burnett R, Guichard Y, Barale E. Immunohistochemistry for light microscopy in safety evaluation of therapeutic agents. An overview. *Toxicology* 1997; **119**: 83–93.
38. Foundation Medicine.
39. Mueller C, deCarvalho AC, Mikkelsen T, Lehman NL, Calvert V, Espina V, Liotta LA, Petricoin EF, 3rd. Glioblastoma cell enrichment is critical for analysis of phosphorylated drug targets and proteomic-genomic correlations. *Cancer Res* 2014; **74**: 818–28.
40. Kim HK, Kim J, Korolevich S, Choi IJ, Kim CH, Munroe DJ, Green JE. Distinctions in gastric cancer gene expression signatures derived from laser capture microdissection versus histologic macrodissection. *BMC Med Genomics* 2011; **4**: 48.
41. Klee EW, Erdogan S, Tillmans L, Kosari F, Sun Z, Wigle DA, Yang P, Aubry MC, Vasmatzis G. Impact of sample acquisition and linear amplification on gene expression profiling of lung adenocarcinoma: Laser capture micro-dissection cell-sampling versus bulk tissue-sampling. *BMC Med Genomics* 2009; **2**: 13.
42. Emmert-Buck MR, Bonner RF, Smith PD, Chuaqui RF, Zhuang Z, Goldstein SR, Weiss RA, Liotta LA. Laser capture microdissection. *Science* 1996; **274**: 998–1001.
43. Harrell JC, Dye WW, Harvell DM, Sartorius CA, Horwitz KB. Contaminating cells alter gene signatures in whole organ versus laser capture microdissected tumors. A comparison of experimental breast cancers and their lymph node metastases. *Clin Exp Metastasis* 2008; **25**: 81–8.
44. Andres SA, Wittliff JL. Relationships of ESR1 and XBP1 expression in human breast carcinoma and stromal cells isolated by laser capture microdissection compared to intact breast cancer tissue. *Endocrine* 2011; **40**: 212–21.
45. Baldelli E, Haura EB, Crino L, Cress DW, Ludovini V, Schabath MB, Liotta LA, Petricoin EF, Pierobon M. Impact of upfront cellular enrichment by laser capture microdissection on protein and phosphoprotein drug target signaling activation measurements in human lung cancer: Implications for personalized medicine. *Proteomics Clin Appl* 2015; **9**: 928–37.
46. El-Serag HB, Nurgalieva ZZ, Mistretta TA, Finegold MJ, Souza R, Hilsenbeck S, Shaw C, Darlington G. Gene expression in Barrett's esophagus: Laser capture versus whole tissue. *Scand J Gastroenterol* 2009; **44**: 787–95.
47. Dias MF, Blumenstein R, Russo J. Use of laser capture microdissection allows detection of loss of heterozygosity in chromosome 9p in breast cancer. *Oncol Lett* 2017; **13**: 3831–6.
48. Zhang Y, Garcia-Ibanez L, Brown G, Toellner KM. Detecting gene expression in lymphoid microenvironments by laser microdissection and quantitative RT-PCR. *Methods Mol Biol* 2017; **1623**: 21–36.

49. Zhang PGY, Yeung J, Gupta I, Ramirez M, Ha T, Swanson DJ, Nagao-Sato S, Itoh M, Kawaji H, Lassmann T, Daub CO, Arner E, de Hoon M, Carninci P, Forrest ARR, Hayashizaki Y, Goldowitz D, consortium F. Discovery of transcription factors novel to mouse cerebellar granule cell development through laser-capture microdissection. *Cerebellum.* 2018 Jun; **17**(3): 308–325.
50. Qin Z, Worthen CA, Quan T. Cell-size-dependent upregulation of HGF expression in dermal fibroblasts: Impact on human skin connective tissue aging. *J Dermatol Sci* 2017; **88**: 289–97.
51. Hipp J, Cheng J, Hanson JC, Yan W, Taylor P, Hu N, Rodriguez-Canales J, Hipp J, Tangrea MA, Emmert-Buck MR, Balis U. SIVQ-aided laser capture microdissection: A tool for high-throughput expression profiling. *J Pathol Inform* 2011; **2**: 19.
52. Hipp JD, Cheng JY, Toner M, Tompkins RG, Balis UJ. Spatially Invariant Vector Quantization. A pattern matching algorithm for multiple classes of image subject matter including pathology. *J Pathol Inform* 2011; **2**: 13.
53. Hipp JD, Cheng J, Hanson JC, Rosenberg AZ, Emmert-Buck MR, Tangrea MA, Balis UJ. SIVQ-LCM protocol for the ArcturusXT instrument. *J Vis Exp.* 2014 Jul 23; (89).
54. Demarest TG, Murugesan N, Shrestha B, Pachter JS. Rapid expression profiling of brain microvascular endothelial cells by immuno-laser capture microdissection coupled to TaqMan(®) low density array. *J Neurosci Methods* 2012; **206**: 200–4.
55. Waller R, Woodroofe MN, Francese S, Heath PR, Wharton SB, Ince PG, Sharrack B, Simpson JE. Isolation of enriched glial populations from post-mortem human CNS material by immuno-laser capture microdissection. *J Neurosci Methods* 2012; **208**: 108–13.
56. Simpson JE, Wharton SB, Heath PR. Immuno-laser-capture microdissection for the isolation of enriched glial populations from frozen post-mortem human brain. *Methods Mol Biol* 2018; **1723**: 273–84.
57. Tangrea MA, Chuaqui RF, Gillespie JW, Ahram M, Gannot G, Wallis BS, Best CJ, Linehan WM, Liotta LA, Pohida TJ, Bonner RF, Emmert-Buck MR. Expression microdissection. Operator-independent retrieval of cells for molecular profiling. *Diagn Mol Pathol* 2004; **13**: 207–12.
58. Appleby-Mallinder C, Wyles MD, Simpson JE, Wharton SB, Ince PG, Heath PR. Expression microdissection isolation of enriched cell populations from archival brain tissue. *J Neurosci Methods* 2016; **268**: 125–30.
59. Du Q, Yan W, Burton VH, Hewitt SM, Wang L, Hu N, Taylor PR, Armani MD, Mukherjee S, Emmert-Buck MR, Tangrea MA. Validation of esophageal squamous cell carcinoma candidate genes from high-throughput transcriptomic studies. *Am J Cancer Res* 2013; **3**: 402–10.
60. Grafen M, Hofmann TR, Scheel AH, Beck J, Emmert A, Kuffer S, Danner BC, Schutz E, Buttner R, Ostendorf A, Strobel P, Bohnenberger H. Optimized expression-based microdissection of formalin-fixed lung cancer tissue. *Lab Invest* 2017; **97**: 863–72.

61. Hanson JA, Gillespie JW, Grover A, Tangrea MA, Chuaqui RF, Emmert-Buck MR, Tangrea JA, Libutti SK, Linehan WM, Woodson KG. Gene promoter methylation in prostate tumor-associated stromal cells. *J Natl Cancer Inst* 2006; **98**: 255–61.
62. Hanson JC, Tangrea MA, Kim S, Armani MD, Pohida TJ, Bonner RF, Rodriguez-Canales J, Emmert-Buck MR. Expression microdissection adapted to commercial laser dissection instruments. *Nature Protocols* 2011; **6**: 457–67.
63. Rosenberg AZ, Armani MD, Fetsch PA, Xi L, Pham TT, Raffeld M, Chen Y, O'Flaherty N, Stussman R, Blackler AR, Du Q, Hanson JC, Roth MJ, Filie AC, Roh MH, Emmert-Buck MR, Hipp JD, Tangrea MA. High-throughput microdissection for next-generation sequencing. *PLoS One* 2016; **11**: e0151775.
64. Adey N, Emery D, Bosh D, Callahan S, Schreiner J, Chen Y, Greig A, Geiersbach K, Parry R. A mill based instrument and software system for dissecting slide-mounted tissue that provides digital guidance and documentation. *BMC Clin Pathol* 2013; **13**: 29.
65. Krizman D, Adey N, Parry R. Application of tissue mesodissection to molecular cancer diagnostics. *J Clin Pathol* 2015; **68**: 166–9.
66. Wang J, Fan HC, Behr B, Quake SR. Genome-wide single-cell analysis of recombination activity and *de novo* mutation rates in human sperm. *Cell* 2012; **150**: 402–12.
67. Gole J, Gore A, Richards A, Chiu YJ, Fung HL, Bushman D, Chiang HI, Chun J, Lo YH, Zhang K. Massively parallel polymerase cloning and genome sequencing of single cells using nanoliter microwells. *Nat Biotechnol* 2013; **31**: 1126–32.
68. Voet T, Kumar P, Van Loo P, Cooke SL, Marshall J, Lin ML, Zamani Esteki M, Van der Aa N, Mateiu L, McBride DJ, Bignell GR, McLaren S, Teague J, Butler A, Raine K, Stebbings LA, Quail MA, D'Hooghe T, Moreau Y, Futreal PA, Stratton MR, Vermeesch JR, Campbell PJ. Single-cell paired-end genome sequencing reveals structural variation per cell cycle. *Nucleic Acids Res* 2013; **41**: 6119–38.
69. Baslan T, Kendall J, Rodgers L, Cox H, Riggs M, Stepansky A, Troge J, Ravi K, Esposito D, Lakshmi B, Wigler M, Navin N, Hicks J. Genome-wide copy number analysis of single cells. *Nat Protoc* 2012; **7**: 1024–41.
70. Baslan T, Kendall J, Rodgers L, Cox H, Riggs M, Stepansky A, Troge J, Ravi K, Esposito D, Lakshmi B, Wigler M, Navin N, Hicks J. Corrigendum: Genome-wide copy number analysis of single cells. *Nat Protoc* 2016; **11**: 616.

Digital Imaging, Machine Learning, and Artificial Intelligence

5 Application and Practice of Digital Pathology

Joshua D. Webster and Jeff Eastham-Anderson

Department of Pathology, Genentech, Inc., South San Francisco, CA, USA

Contents

I. Introduction

Targeted therapies that are directed against specific proteins involved in disease pathogenesis have become the norm for drug development. Although the concept of personalized medicine was initially focused on oncology, this approach is being increasingly applied to other disease areas as well, most notably for the treatment of immunologic diseases. Aside from providing therapeutic breakthroughs, targeted therapies have also revealed disease heterogeneity in the context of responders vs. non-responders. Therefore, it has become increasingly important to identify predictive biomarkers that can identify patients who are more likely to respond to a specific molecularly based therapeutic regime.

A good predictive biomarker is defined by not just its ability to discriminate between responders and non-responders, but also by its validity, reproducibility, and ability to be applied in a clinical setting. Predictive biomarkers can take many forms including identification of chromosomal alterations by fluorescent *in situ* hybridization (FISH); identification of point mutations by next-generation sequencing (NGS); or identification of changes in protein expression, pathway activation, or characterization of cellular infiltration by immunohistochemistry (IHC). Pathology-based endpoints are generally measured by the subjective evaluation of a glass slide performed by a trained pathologist who evaluates the slide using an upright, brightfield microscope. For many qualitative diagnostic questions, this approach is tried and true. However, the increasing complexity of predictive biomarker assays and the need for reproducible, unbiased results suggests an increasing need and importance for objective quantitative assessment of pathology endpoints. In the last decade, there have been tremendous developments in the field of digital pathology and quantitative image analysis that allow for increasingly quantitative assessment of pathologic specimens. In the coming years, it is likely that there will be a greater reliance on these methods for assessment of pathology-based predictive biomarkers.

Digital pathology is the practice of pathology using digitized images of pathologic specimens.[1,2] In its broadest definition, this includes the capture, storage, and assessment of gross, histologic, cytologic, or ultrastructural specimens using digital file formats. The origin of digital pathology is rooted in static digital photomicrographs that could be transmitted between remote sites, allowing for second opinions or consultations. Robotic telepathology subsequently allowed a pathologist at a distant site to control an automated stage and view specimens in real time via a monitor. In the last decade, the development of whole-slide scanners that create high-resolution, digital images of entire histologic or cytologic sections has revolutionized the field of digital pathology and, as such, has touched all practical facets of pathology from teaching to diagnostics to investigative pathology.[3]

II. Whole-slide Imaging

Whole-slide imaging is the process by which high-resolution digital images are created of entire histologic or cytologic specimens. Several whole-slide scanners are commercially available, with each platform having its own unique set of attributes. Although the mechanics of slide loading and scanning vary between platforms, most platforms capture a series of high-resolution images from across the entire section to create the final whole-slide image. Individual images can consist of tiles or line scans that are individually collected and stitched together. Prior to collecting these images, the scanner needs to identify multiple focal points throughout the section to assure consistent auto-focusing during the scan. Although most scanners have gone through multiple development iterations to improve tissue detection and auto-focusing, the most critical variable for slide scanning is the quality of the tissue sample being captured.[4,5] Uneven sections, under-stained sections, or thick sections with multiple focal planes create challenges for even the best slide scanners. Additionally, leaking mounting media, and uneven or hanging slide labels can stall or potentially damage the scanner's autoloader. As such, similar to most elements of pathology, the quality of the input largely dictates the quality of the output.

Most slide scanners can acquire images equivalent to a 20× or 40× objective, although some scanners have higher magnification oil immersion objectives to facilitate imaging cytologic specimens. Since images will be viewed on variably sized monitors, the size of the objective used for scanning does not directly relate to the final image magnification.

However, the objective used for scanning and the detection system (camera) determine the final image resolution, which is defined by micrometers per pixel. For most scanners, 20× objectives produce images with approximately 0.5 μm/pixel resolution and 40× objectives produce 0.25 μm/pixel resolution images.[6,7] Images created with 20× objectives are suitable for most pathologic evaluations, although occasionally 40× scanned images are desired to evaluate infectious agents or detailed cytologic features. The final file size for brightfield images is dependent on the size of the specimen and the objective used for scanning. Typical 20× scanned images range from 200 MB to 1 GB.[1] Scanning with a 40× objective will quadruple both the scan time and the image size.[6]

Whole-slide images are viewed through slide viewing software. Stand-alone and web-based viewers are both available, and they each have their own distinct advantages and disadvantages. If digital images are being viewed from a hard drive and broadband access is limited or intermittent, then a stand-alone viewer is preferable. However, web-based viewers offer increased user flexibility, especially when reviewing images remotely, assuming internet access is not a limiting factor. In their most rudimentary form, slide viewing software allows users to move across slides at multiple magnifications, similar to a traditional microscope. Magnifications can either be fixed (2×, 5×, 10×, 20×, 40×) or be more granularly set with a slide bar depending upon the interface. Since image file sizes are large, software will retrieve data from only the requested viewing area at the displayed magnification at a given time.[8,9] This speeds up the image viewing time considerably, but still requires a reliable network that can consistently push and pull large amounts of digital image data at a given time. Many slide viewers have additional functionalities including camera tools to capture photomicrographs, scale bars, annotation tools, digital magnification, and thumbnails of the entire tissue to identify the current location within the tissue and to quickly move to different areas within the tissue (Figure 1).[5,8,10] If quantitative image analysis is being performed on whole-slide images, integration of image analysis masks that detail image segmentation is critical. Reviewing image masks enables optimization of analysis algorithms and quality control review of the final image segmentation during data review. There are no current standards for digital pathology digital image file formats,[1] and many vendors have developed

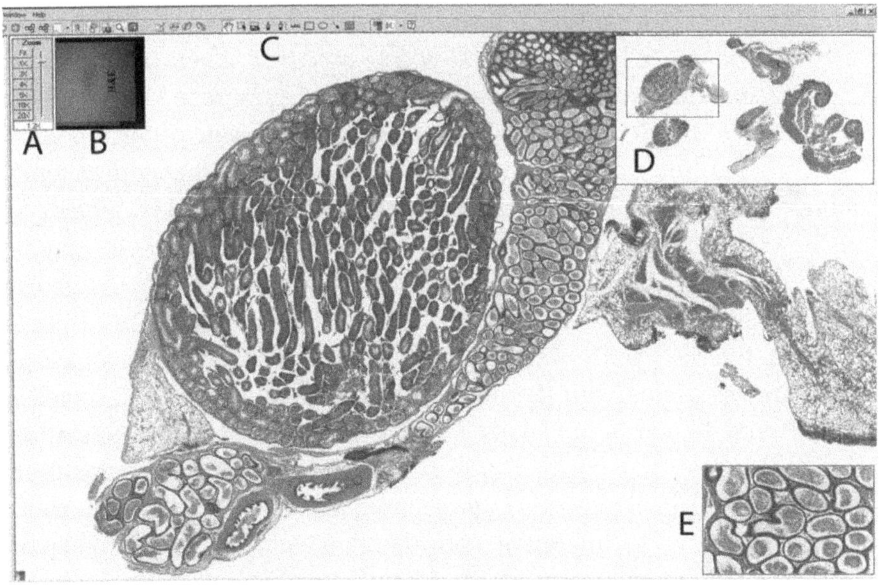

Figure 1. Digital pathology slide viewer, ImageScope (Leica Microsystems, Vista, CA). The slide viewer interface recapitulates the microscope interface with additional features including (A) a slide rule to change magnifications, (B) a thumbnail of the slide label, (C) annotation tools, (D) a thumbnail of whole-slide image to determine location in the sections and move quickly between locations, and (E) digital zoom.

proprietary image formats to accompany their scanners. Therefore, image file conversion may be necessary to view images from different scanning platforms on a single slide viewer. Most file formats permit at least some degree of compression,[8] the application of which, to even a modest degree, can dramatically reduce the file size of the digital image.

III. Images, Data, and Metadata

Whole-slide images produce large amounts of data that require a thoughtful IT infrastructure.[8] The three basic nodes of the whole-slide imaging system are the slide scanner, the data storage server, and the end-user interface/slide viewer. The complexity of the infrastructure will vary depending upon needs.[11,12] If digital images are used intermittently for teaching sets or used in rounds, a very simple workflow might be adequate. However, if whole-slide images are part of routine workflows and are being viewed by multiple end-users, used for image analysis, or viewed by external consultants or across physical sites, then IT considerations become much larger. Specifically, these considerations include image storage and backup plans, internet bandwidth, integration with Laboratory Information Management Systems (LIMS), linked metadata, and firewall and security, with an emphasis on the protection of patient data.[12–15] Additionally, computer processing power and distributed image processors are critical for large-scale image analysis (Figure 2).

Image file sizes are further increased if image analysis masks or annotations are associated with the image. Data storage has become increasingly affordable and is not likely to be a limiting factor in whole-slide imaging. Routine data backup, preferably at a secondary site, should be established at defined intervals to protect the valuable data. Internet bandwidth capacity is an important consideration, since large amounts of image data are continuously pushed and pulled during both image acquisition and viewing.[12]

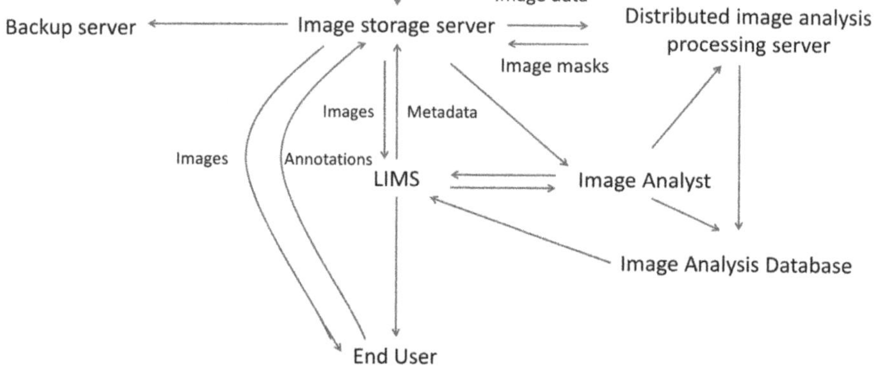

Figure 2. Diagram of the digital pathology data workflow from tissue processing to the end user.

Many whole-slide scanners include slide barcode readers that can directly link patient metadata with the image. This requires both a slide barcoding system and integration with the LIMS. Integration of whole-slide images with patient metadata offers several benefits. First, this serves as a permanent link between the slide/specimen and the patient, preventing annotation or dictation errors, and eliminating data duplication.[15] Second, the availability of associated metadata also enables for the specimen to be reviewed within the context of the patient's history. All data is provided simultaneously, which both facilitates the simultaneous review of the patient's history and decreases errors in cross-referencing. The ability to review patient data with the slide has been reported as a key factor in decreased digital slide review compared to traditional microscopic review.[16] Finally, this integration allows for refined search queries and reviewing of pathologic specimens based on specific patient or pathologic criteria. Although links between digital pathology images and patient data are extremely powerful, they do come with added concerns about security of personal data. Therefore, it is important to go about establishing an integrated system with consideration of HIPAA compliance in mind.[14]

IV. Digital Pathology in Archiving, Education, and Collaboration

Whole-slide imaging has impacted all aspects of digital pathology including archiving, teaching, collaboration and consultation, diagnostics, and quantitative image analysis in diagnostic and investigative pathology. Whereas glass slides can be broken, misplaced, or fade over time, whole-slide images can serve as a permanent and stable repository of pathologic data.[10] Additionally, digital image archives allow slides to be immediately retrievable and directly linked with patient metadata to further facilitate both searches and slide reviews. Digital image libraries also allow for the establishment of permanent and standardized teaching sets in medical and graduate education. Instead of requiring a defined amount of tissue with characteristic features to be available to make multiple slide sets, single representative digital sections can be provided to all students simultaneously. This standardizes the samples provided to all students and allows students to review images remotely at any time. Digital images can be shared among institutions, which can be important to provide examples of rare or regionally specific diseases. This also eliminates the risk of slides being lost, broken, or faded over time. Additionally, instructors can make annotations on whole-slide images to demonstrate specific pathologic features present in the sections and projected images can be reviewed together as a group.[9]

Whole-slide digital imaging platforms further enable the sharing of pathology data between multiple institutions for consensus conferences, tumor boards, or other educational or investigational interactions. Cloud-based platforms have been specifically developed for sharing whole-slide images. These platforms can move limited pathology data out from behind firewalls to enable multiple sites to review the slides simultaneously. Additionally, video conferencing software allows for multiple sites to review single pathologic images together, similar to a multi-headed microscope, during a teleconference via screen sharing. Digital slide conferences allow participants at different sites to not only review the identical pathology sections but also to look at and discuss specific features present within a tissue section in real time. This greatly enhances both the efficiency and quality of the communication of the slide review.[3,5,11]

V. Digital Pathology in Routine Diagnostics

In diagnostic medicine, digital pathology has been primarily utilized in intra-operative frozen section evaluation and in telepathology and teleconsultations where pathologists, especially those at distant sites with specializations, are consulted remotely.[3] The American Telemedicine Association has detailed guidelines for the clinical practice of telepathology.[17] Digital pathology has been used in primary diagnosis most frequently outside the United States. Regulatory considerations have been a significant concern for routine implementation of digital pathology in the United States, since the U.S. Food and Drug Administration (FDA) has classified whole-slide scanners as Class III medical devices, requiring premarket approval.[18] Recently, the FDA has published technical performance guidelines for whole-slide imaging devices.[19] The College of American Pathologists' (CAP) Pathology and Laboratory Quality Center has also provided guidelines for validating whole-slide imaging in diagnostic pathology.[14] The CAP guidelines recommend that each institution should validate their systems independently and that systems should be validated as a whole rather than as separate components. However, significant changes to the system would require revalidation of the entire system. In these guidelines, it is recommended that the validation should include 60 cases for one application (e.g., routine formalin-fixed paraffin-embedded histopathology) that reflects the real-world case population. Twenty cases should then be added for each additional application.

The goal of validation studies is to confirm that the diagnosis made on a digital image is comparable to a diagnosis made on the glass slide. Therefore, intra-observer concordance between the digital image and the glass slide is the most critical component of the validation process. Since pathologists may remember cases, especially challenging cases, it is important to have a washout period between reviewing the glass slide and digital image. The CAP guidelines recommend a two-week washout period. In prior validation studies, there has been approximately 86% concordance identified between glass slides and digital images; however, concordances and minor discordances were actually 98% in these studies with reportedly no statistically significant difference between whole-slide images and glass slides.[14]

In addition to regulatory concerns, digital pathology has faced many other hurdles in its implementation in routine diagnostic medicine. Fiscally, digital pathology systems are a sizable investment. Aside from the cost of slide scanners, there is continual instrument maintenance, cost of IT infrastructure including data storage and integration with LIMS, and personnel to perform the validation, scanning, and quality control. Slide scanning, including quality control, needs to be integrated into the laboratory's workflow. Within these considerations are several questions,[5,7,12,15] including: What proportion of cases will be digitized? How many slide scanners are needed to cover scanning demands? How will cases be batched to maximize efficiency? How does slide scanning affect the overall turnaround time? Issues of speed and the user interface are consistent concerns, especially among pathologists that are less familiar with digital pathology. Specific modifications can be made to enhance the user interface, such as larger monitors to increase the field of view displayed[20] and double displays that allow pathologists to view the clinical data and whole-slide image simultaneously.[16] In both cases, these interface enhancements have been suggested to decrease the diagnostic time. Additionally, novel methodologies including manual controls to pan across a digital slide and change magnifications can be integrated into the interface to better mimic a microscope and make the user more comfortable with navigating digital images.[4]

VI. Multiplexing

In recent years, there has been increasing emphasis on multiplexed immunolabeling. Multiplexed assays allow for protein co-localization, characterization of the relationship of multiple cell types within a single tissue and for the evaluation of protein expression within a specific cell type. Although two to three antibodies can be evaluated with chromogenic assays, immunofluorescence offers increased multiplexing, better discrimination of proteins localized to the same cellular or subcellular location and has a greater and more linear dynamic range than chromogenic assays. In addition to chromogenic and fluorescence immunolabeling, there have been recent strides in the development of ion-beam mass spectrometry assays in which antibodies are labeled with metal isotopes that are detected by mass spectrometry. Mass spectrometry approaches have the potential to allow for increased multiplexing, but the technology is still in early development.[21]

With the increased interest in multiplexed immunofluorescence, there have been great improvements in the development of fluorescent slide scanners. Fluorescent slide scanners generally require separate scans for each fluorophore, creating both increased acquisition times and file sizes. Most slide scanners allow for the detection of four to five fluors in a single tissue section, although up to seven fluors may be detected with multispectral imaging (discussed later).

Immunofluorescence in tissues sections offers unique challenges that need to be considered from the outset. Fluors should be chosen to minimize bleed–through in other channels, and antibody/ fluor combinations should be chosen to balance the signal strength of different antibodies. Selection of the appropriate exposure times is also essential to ensure high-quality images, and similar exposure times need to be used across all study samples to allow for quantitative comparisons in signal. Variable autofluorescence across tissue types needs to be accounted for as well.[22] Separate scans at wavelengths not used by the target fluors can be used to capture the autofluorescent signal within a tissue, which then can be digitally subtracted following slide scanning. Since fluorescent images do not provide intrinsic morphologic details, aside from autofluorescent signal, consideration must be given to labeling of tissue components beyond the primary markers of investigative interest. DAPI stain is commonly used to label nuclei, providing both a morphologic skeleton to the tissue and allowing for cell counting. However, additional labels, like cytokeratin, may be needed to define tumor regions or other specific tissues of interest.

Multispectral imaging allows for the separation of fluorescent signals within a tissue based on each signal's specific emission spectrum across a range of wavelengths. In multispectral imaging, a spectral library is created by measuring the signal in tissues labeled with a single fluor. Software can then separate out different signals within tissue sections based on these defined spectra. This approach should allow more robust separation of signals and compensation for autofluorescence.[22] The primary drawbacks of multispectral imaging are increased scanning times and the vast amounts of data and computing power needed to manage this data.

IHC performed on serial tissue sections serves as an alternate approach to multiplexed labeling on a single slide. Software can be used to align histologic images from serial sections and digitally construct a multiplexed image. This approach lowers the technical bar of immunolabeling by avoiding concerns with bleed-through of fluors across multiple channels and concerns about primary and secondary antibody combinations and cross-reactivity, but post-processing and alignment provide their own unique challenges. It should also be considered that this approach utilizes multiple tissue sections, which may be restrictive if tissues are limited, and there is not true co-localization of proteins in a single cell since it cannot be assured that the same cells are in each tissue section.

Data outputs for multiplexed assays can be incredibly complex. Determining how to visualize, analyze, and interpret data outputs serve as unconquered challenges. Despite the platform, all multiplexed systems will require the use of digital pathology tools for image viewing, segmentation, and analysis.

VII. Image Analysis

The simplest conceptualization of image analysis, whether it results in a readout generated by a human or a computer algorithm, is reducing vast amounts of image data to facilitate interpretation. This transformation enables a great deal of quantitative comparisons to be made with conclusions drawn regarding the statistical significance of those comparisons. Incorporating computer analysis has the potential to increase the accuracy, sensitivity, and speed of generating readouts. However, these benefits must be balanced against some very real costs, which must be put into the context of the number of samples subject to the assay.

Traditionally, analysis of pathologic data has been based on subjective and qualitative assessments of morphologic changes that are determined by the experience and expertise of a trained pathologist. These qualitative assessments have proven invaluable to understanding and characterizing disease processes and for providing diagnoses.[4] However, these assessments provide only descriptive or, at best, ordinal data. As such, this data lacks dynamic range and granularity and is less amenable to statistical analyses. Additionally, some evaluations, such as assessments of percent areas, labeling intensities, and discrete cellular counting are difficult to do manually.[23] Manual estimations of percentages are vulnerable to error, especially in the estimation of small percentages or estimates of percentages within heterogeneous tissues. Additionally, the human eye has limited capacity and dynamic range to distinguish intensities.[24]

Quantitative image analysis is one of the most powerful and transformative applications of digital pathology, and whole-slide scanning and automated image analysis can address and complement many of the limitations of manual assessments. Digital image analysis is based on the development of mathematical workflows to segment images based on specific, definable features such as color, intensity, texture, or object relationships, and to quantify these features within the image. Since image analysis algorithms have defined workflows and segmentation criteria, identical analytical criteria are applied to all samples. Therefore, the uniform application of rule sets to each specimen results in unparalleled reproducibility and eliminates the potential for intra- and inter-observer variability.[5,25] Increased automation included in the algorithm progressively removes bias from the analysis.[26] For example, additional bias can be removed if tumor areas to be analyzed can be defined by the algorithm rather than by manually drawn regions of interest.

Since digital image analysis directly measures pixel-based features of an image, it has the ability to directly quantify features within the image with an increased dynamic range. While human gray-level acuity is thought to distinguish 55–64 levels at most,[27] basic cameras are able to discriminate 256 levels, providing a greatly improved dynamic range. Similarly, direct feature measurements within an image provide data on a continuous, rather than ordinal, scale.[5,28] The ability to reproducibly measure pathologic features with an improved dynamic range allows for detailed statistical analyses, group comparisons, and definition of cut points that are not possible with other approaches.[28] Additionally, these measurements allow for unique queries, including evaluation of spatial relationships within tissues that could have unique biologic and/or prognostic ramifications. For example, one could evaluate cellular infiltrates or protein expression levels in the context of their tissue localization.[5]

Computers are particularly well suited to perform repetitive tasks, whether it is at the scale of quantifying multiple endpoints in a single sample or a single endpoint across many samples. Execution of an algorithm will vary depending upon its mathematical complexity, with some algorithms taking days of computer runtime to read out an entire slide, but others taking only a matter of minutes. Although significant upfront time may be needed to develop an algorithm, once developed the algorithm can be applied in a high-throughput manner. Application of distributed computing solutions (either cloud based or local server based), where the processing load is spread between many individual processors, can speed analysis significantly.[29]

a. Image analysis platforms

There are many software programs available to facilitate the development of automated algorithms for image analysis. Each has its own strengths and weaknesses, but not surprisingly there tends to be a fairly strong inverse relationship between functionality and ease of use.[30] Typically, the simpler a software platform is to use, the smaller the range of use cases will be and vice versa. Simple software programs will generally feature a graphical user interface (GUI) that enable a user to quickly customize a range of parameters to solve a narrowly defined problem, like counting labeled cells. A novice user is typically able to develop a solution in about an hour. More complex programs will allow a certain amount of free-form flexibility to mix and match individual processing steps while usually still adhering to a GUI interface. While they are generally limited in their application, novice users should be able to utilize these programs fairly easily. However, harnessing their full functionality requires weeks or months of applied learning. The most advanced software programs are basically programming languages that tend not to use GUI interfaces but are extremely versatile in the solutions they can provide. Indeed, all of the GUI-centered programs mentioned previously utilize a programming language to provide their underlying functionality; more advanced programs expose a language with exquisite control. As such, novice users must commit significant investment to reap even the basic rewards of image analysis using these programming languages, which are typically reserved for individuals whose sole responsibility is developing image analysis solutions.

b. Image analysis algorithm development

Image analysis outputs may include measurement of immunolabeled areas, cell counts, measures of proportion of labeled cells (Figure 3), co-localization, and assessments of relationships between objects such as cell counts within a defined area of a blood vessel. The first step in designing an image analysis algorithm is defining the desired measurements. In their most basic form, image analysis algorithms are composed of at least two measurements: (1) the feature of interest (the numerator, e.g., IHC labeled area, cell counts, immunofluorescent intensity) and (2) the reference measurement to normalize the data (the denominator, e.g., total cell count, total tissue area, tumor area). Both measurements are equally important and need to be well defined. The reference measurement can be more challenging to define, but is critical, since comparisons or cut points need to be defined based on a standardized unit.[4] For example, if variable amounts of tumor are present in

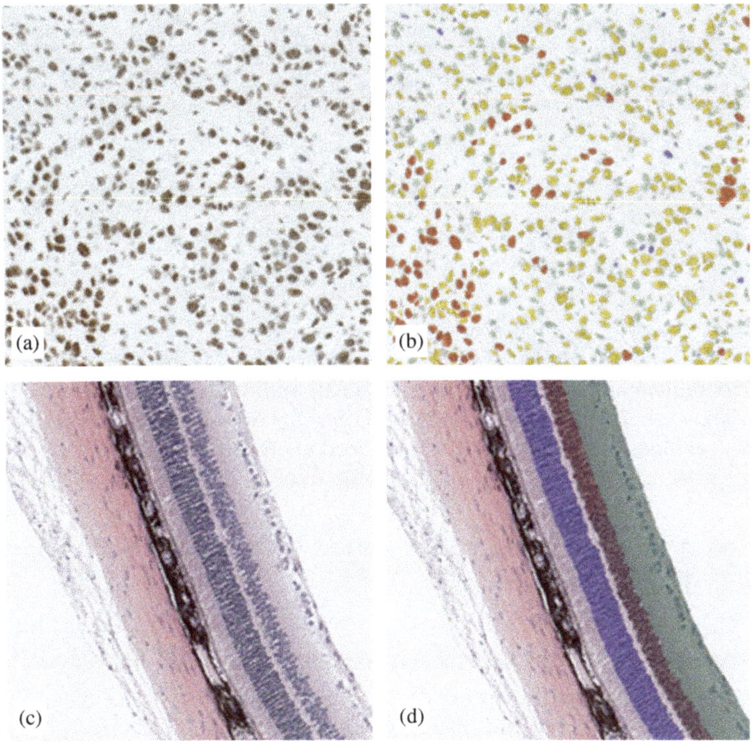

Figure 3. Image analysis masks allow reviewing image segmentation as a means of quality control. (a) Nuclear immunohistochemistry with DAB chromogen and hematoxylin counterstain. (b) Image mask demonstrating nuclear segmentation and color-coding of high (red), medium (yellow), low (green), and none (blue) DAB labeling intensity of the same region shown in (a). /(c) Hematoxylin and eosin-stained section of retina. (d) Segmentation overlay of the outer nuclear layer (blue), inner nuclear layer (red), and inner plexiform layer and ganglion cells (green) of the retina shown in (c).

each tissue section, it is important to consider if measurements of immunolabeled cells should be based on the total numbers of cells within the tissue or only the total number of cells within the tumor. This requires an understanding of both the disease and target biology. While counting the total number of cells within the tissue might be analytically simpler, it could result in severely biased results. Defining the desired measurements and performing quality control as the algorithm is developed

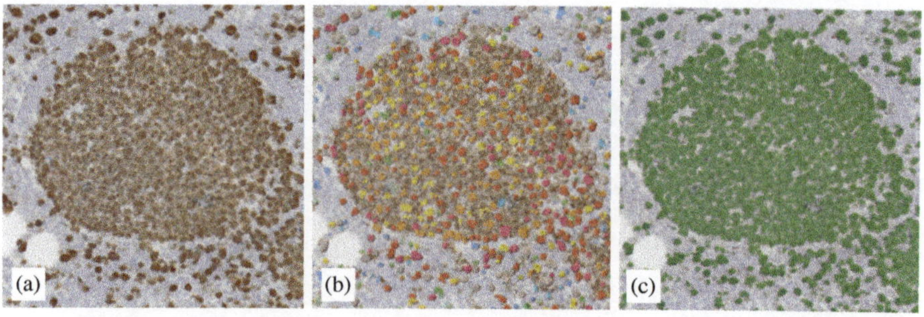

Figure 4. Segmentation challenges of closely packed cells. (a) Ki-67-immunolabeled lymphocytes are densely packed within a lymphoid follicle. (b) Many lymphocytes are not identified with a nuclear segmentation algorithm because that are too closely packed and lack definable cell–cell boundaries. (c) Area-based assessment of lymphocytes provides more accurate and unbiased analysis.

relies on a close working collaboration between the pathologist and image analyst.[5]

After a question is identified, the next step is to determine what form the answer should take. Area measurements can be very straightforward readouts for the quantification of complex structures. Simply thresholding on stain intensity introduces very few additional biases and is an elementary concept. Area measurements are also preferred if closely packed cells (e.g., lymphocytes) are difficult to individually discriminate (Figure 4). Intensity measurements are also easily produced, but some care must be taken in their interpretation. For example, a numerically small area of dense staining can be equivalent to a large area of sparse staining. Whether or not they should be interpreted as equivalent may require additional consideration. Furthermore, immunofluorescent images provide greater and more linear dynamic range compared to chromogenic intensity measurements, which should also be considered when setting reproducible and validated cut points. Higher-order identification of coherent objects, or segmentation, requires additional algorithmic complexity but allows for more complex measurements such as distribution patterns and inter-object relationships. A robust segmentation step is a prerequisite of many algorithms that proceed to measure features of the identified objects. In general, the use of cell or tissue markers to define the morphologic features to be quantified will greatly enhance the success and consistent performance of an algorithm.[31]

Once the desired readouts are identified, a typical next step involves selecting a subset of images or image fragments that will be used to develop an algorithm. This is necessitated by a key defining characteristic of digital pathology images, which is their immense size. A large section on a 1 × 3 slide scanned at 20× can be over a gigapixel (one billion pixels) in size.

The time it takes to process a given set of images primarily depends upon its total size, the complexity of the algorithm, and the amount of computing power dedicated to the task. Given an average image at 20× magnification, the task of reading the image data from disk, applying a modestly complex algorithm, and potentially writing segmentation images back out to disk can take anywhere from a few minutes to several hours. Algorithm development typically involves multiple iterations, and significant delays between iterations can be largely avoided by using a much smaller data set.

In addition to a distributed computing solution, there are strategies to mitigate the impact of image processing on project timelines and computer resources. One easily implemented strategy is to downsample the image data. For example, downsampling from a 20× image to a 5× image can be thought of as taking a 4 × 4 square grid of pixels and reducing that into a single pixel. This results in $1/16^{th}$ of the image data that must be processed, and generally results in a similar reduction in processing time. Some analysis algorithms, such as the detection of tissue area on a slide, are routinely run on highly downsampled image data. Once the tissue area is identified at low magnification, only those regions specific to the tissue are subjected to more complicated analyses at higher magnification. Another strategy to speed up processing is to randomly sample and analyze a fraction of the whole image.

The overall complexity of the algorithm will be influenced by the question being interrogated, the complexity of the tissue, and the degree of desired automation. For example, if only diseased tissues are included in the samples, then segmentation of diseased vs. normal adjacent tissue will not be necessary. Alternatively, manual annotation of regions of interest eliminates computational needs of automated tumor detection. Area measurements based on intensity thresholding will be simpler than cell segmentation and counting. However, in the end, the ease of algorithm development should be weighted based on the potential for bias, robustness of the assay, and a cost–benefit ratio of the time it takes to develop the algorithm vs. the quality of data output.

c. Data output and analysis

Data output is an important consideration for the application and interpretation of the final analysis results. While there is temptation to simplify algorithms by limiting a report to a few summary readouts, there is value to providing additional contextual data. For example, it is unlikely that a sample with a 50% prevalence of Ki-67 positive cells but only ten total positive cells will carry as much weight as a sample that has the same prevalence, but 10,000 positive cells. Apparently mundane measurements can be useful indicators of poor algorithm performance. As in the previous example, if the two samples showed approximately the same total tissue area, the first sample may be under-stained by the cellular marker.

Beyond simple reporting of tabular data, modern digital pathology solutions are providing additional functionality by creating a tighter link between the numbers and the images from which they are generated. Retaining the positional information allows data points at the slide, tissue, or cellular level to be filtered or gated, and then rendered back on the image itself. Even beyond a simple marking of positive or negative events, segmentation images can be used to indicate numerical readouts that would otherwise be difficult to interpret, for example, overlaying a heat map on an image to represent the signal intensity as it varies across a sample (Figure 5). For multiplexed assays, some have also transitioned to flow cytometry-based scatter plots that can gate and describe the cellular populations within the tissue.[32]

d. Validation and quality control

Validation and quality control of an image analysis algorithm must be considered in the context of the use case and gold standard in the field. The results of an image analysis algorithm should be compared against

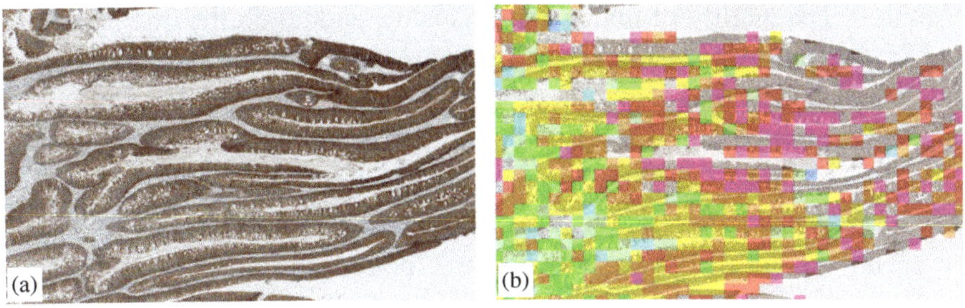

Figure 5. Heat map image mask demonstrating variation in labeling density. (a) DAB-immunolabeled mouse intestine. (b) Heat map image mask demonstrating the signal density of (a).

a gold standard, but this gold standard may vary for different analyses. If possible, the best gold standard will be comparisons to manual annotations.[33] This can be done in focused regions of interest in a subset of samples. Comparisons to manual annotations can provide measurements of the algorithm's sensitivity and specificity, which can be reported with the results. However, in many situations this is not practical, and the gold standard considered is the pathologist's semi-quantitative scores, H-scores,[33] or other qualitative manual assessment. In these situations, the correlation between the pathologist's score and the image analysis algorithm should be confirmed.[31] Alternatively, comparison of the pathology output to similar measures performed by other assays, such as ELISA, Western blots, or FACS, is an additional means of validating the assay. If the image analysis algorithm is to be applied as a clinical predictive marker, then validation should be focused on FDA requirements for biomarker validation and qualification, or determination of equivalency to an already validated manual pathology assay to undertake FDA 510(K) validation of equivalency.[35]

There are three critical quality control check points in the image analysis process. The first is immediately following immunolabeling. This check point should evaluate the consistency and quality of the immunolabeling and counterstaining. Variations or inconsistencies of either of these within or between studies can significantly affect the results if a standardized algorithm is to be applied. The second check point is following slide scanning. The quality of image analysis is dependent on the quality of the image. Therefore, it is important to assure that the entire tissue section was scanned, all areas are in focus, and the images are free from dust and debris. A quality control check point at the time of scanning assures that these issues are caught prior to analysis. The final check point is evaluation of the image analysis masks. Although most potential problems will be identified during the first two check points, unforeseen issues may become apparent only after the analysis is completed. For example, subtle variations in staining can result in dramatic segmentation inaccuracies if these variations were not captured during algorithm development. Therefore, the final check point is important to assure the images are segmented as expected.

e. Image analysis challenges and potential pitfalls

Digital image analysis algorithms contain specifically defined criteria for image segmentation and quantification. As such, these algorithms are susceptible to variability in the sample preparation, staining, and tissue quality (Figure 6). Algorithms can be developed to accommodate for some variability or noise between samples, but greater standardization in the sample collection, sectioning, and staining will increase the quality of the final results (Figure 7). Like most tissue-based assays, every step in sample preparation from the tissue collection to slide scanning can affect the quality of the sample. Diseased tissue is frequently heterogeneous. Inappropriate grossing of the specimen can result in evaluation of necrotic, sclerotic, or normal tissue. At the time of collection, both warm and cold ischemic times can result in protein and mRNA degradation, especially if the target is a phospho-epitope. Similarly, every step in tissue processing from fixation (considering both the time of fixation and the size of tissue being fixed), dehydration, and sectioning to staining can have an effect on the quality and standardization of the final stained slides and the quality of the final results. Since image analysis can rely on cut points from a continuous scale, inconsistencies in staining can result in inappropriate predictive classification. Therefore, standardization and quality control is essential for image analyses based assessments.[4,5]

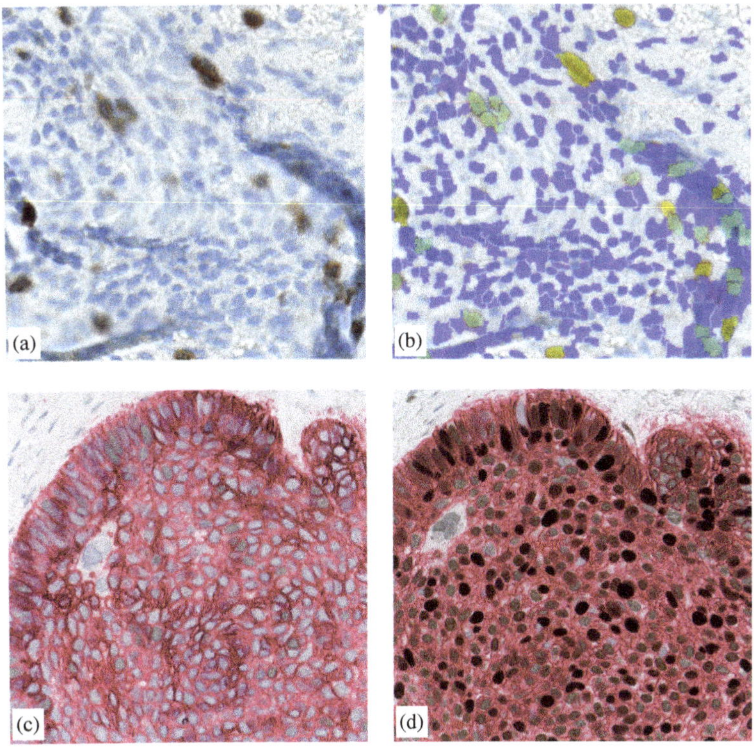

Figure 6. Tissue processing and staining artifacts complicate image analysis. (a) Brightfield image of nuclear marker (DAB, brown) and hematoxylin counterstain (blue) exhibiting multiple artifacts, including out of focus areas and folds with multiple distinct appearances. (b) Segmentation overlay of individual cells showing poor separation of cells in out of focus areas, in addition to false identification of positive cells (green and yellow) localized to folds. (c) Brightfield dual-IHC image of a skin section of a tumor marker (red) and proliferation marker (brown). Extremely weak immunolabeling of the proliferation marker in both the tumor and typically proliferative epidermis is indicative of poor immunolabeling. (d) Brightfield image from section adjacent to (c), re-labeled and showing the anticipated pattern.

VIII. Current and Future Outlooks

Many diseases have traditionally been classified based on their clinical symptoms, body systems affected, pathologic lesions, and classical disease progression. However, as the understanding of the molecular pathogeneses

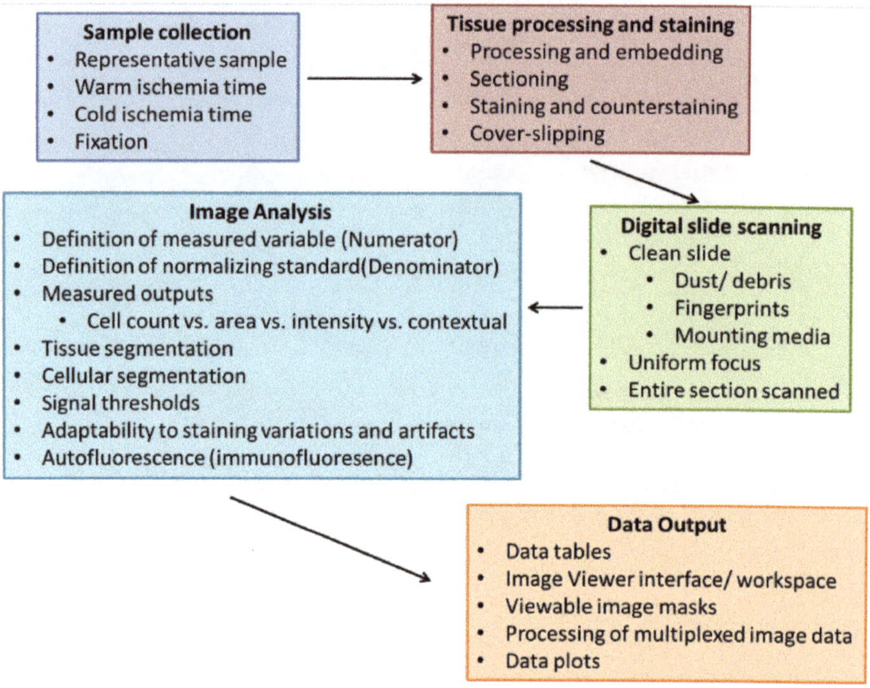

Figure 7. Digital pathology sample handling workflow highlighting considerations and potential factors that could impact final image analysis results.

of diseases grow, there is an increasing recognition that many clinical disease entities are heterogeneous in terms of their development, progression, and molecular pathogenesis. A discussion of sample heterogeneity can be found in the chapter by Noyes *et al.*

Molecularly targeted therapies have become a mainstay for treating neoplastic and immunologic diseases. But due to the molecular heterogeneity of many diseases, it is likely that very specific targeted therapies will only be effective in a subpopulation of patients and an additional subpopulation will only have a temporary response. Therefore, with the increased use of targeted therapies, there is an increased need for biomarkers that can stratify patients that will likely respond to a given therapeutic regimen. Especially in oncology, pathology-based biomarkers have played an important role in therapeutic stratification.

Few digital pathology image analysis assays, including image analysis solutions for estrogen receptor, progesterone receptor, Ki67, and Her2/neu, have been FDA approved through the 510(K) process.[35] In this process, the manufacturer must demonstrate equivalency to an existing device. These assays have relied on the translation of manual pathologist scores to an automated platform. There is a continuous push to develop new biomarkers that will differentiate responders from non-responders and to select treatment regimes. Although single biomarkers can be evaluated manually, there is always a risk of inter- and intra-observer variability that can invalidate the assay. Increased standardization and automation in tissue collection, processing, immunolabeling, and analysis is necessary to ensure the reproducibility and validity of an assay. Additionally, it is likely that biomarkers will increase in their complexity, incorporating multiple pathologic features or markers. Therefore, image analysis-based approaches will be increasingly needed to dissect and simplify these complex biological processes to elaborate an interpretable predictive endpoint. The use of digital pathology-based automated image analysis offers the potential to reproducibly quantify biomarkers in a high-throughput manner. This allows an additional layer of standardization, improves granularity of the analysis due to the continuous scale, and removes biases associated with manual observer assessments. As such, image analysis approaches will be essential to harvest the full potential of novel pathology-based biomarkers.

References

1. Ying X, Monticello TM. Modern imaging technologies in toxicological pathology: An overview. *Toxicol Pathol* 2006; **34**: 815–26.
2. Jara-Lazaro AR, Thamboo TP, The M, Tan PH. Digital pathology: Exploring its applications in diagnostic surgical pathology practice. *Pathol* 2010; **42**: 512–18.
3. Farahani N, Pantanowitz L. Overview of telepathology. *Clin Lab Med* 2016; **36**: 101–12.
4. Dunstan RW, Wharton KA Jr, Quigley C, Lowe A. The use of immunohistochemistry for biomarker assessment — can it compete with other technologies? *Toxicol Pathol* 2011; **39**: 988–1002.
5. Webster JD, Dunstan RW. Whole-slide imaging and automated image analysis: Considerations and opportunities in the practice of pathology. *Vet Pathol* 2014; **51**: 211–23.

6. Cornish TC, Swapp RE, Kaplan KJ. Whole-slide imaging: Routine pathologic diagnosis. *Adv Anat Pathol* 2012; **19**: 152–9.
7. Weinstein RS, Graham AR, Richter LC, Barker GP, Krupinski EA, Lopez AM, Erps KA, Bhattacharyya AK, Yagi Y, Gilbertson JR. *Human Pathol* 2009; **40**: 1057–69.
8. Treanor D. Virtual slide: An introduction. *Diagn Histopathol* 2009; **15**: 99–103.
9. Hamilton PW, Wang Y, McCullough SJ. Virtual microscopy and digital pathology in training and education. *APMIS* 2012; **120**: 305–15.
10. Al-Janabi S, Huisman A, Van Diest PJ. Digital pathology: Current status and future perspectives. *Histopathol* 2012; **61**: 1–9.
11. Potts SJ. Digital pathology in drug discovery and development: multisite integration. *Drug Discov Today* 2009; **14**: 935–41.
12. Lauro GR, Cable W, Lesniak A, Tseytlin E, McHugh J, Parwani A, Pantanowitz L. Digital pathology consultations- a new era in digital imaging, challenges, and practical applications. *J Digit Imaging* 2013; **26**: 668–77.
13. Park SL, Pantanowitz L, Sharma G, Parwani AV. Anatomic pathology laboratory information systems: A review. *Adv Anat Pathol* 2012; **19**: 81–96.
14. Pantanowitz L, Sinard JH, Henricks WH, Fatheree LA, Carter AB, Contis L, Beckwith BA, Evans AJ, Otis CN, Lal A, Parwani AV. Validating whole slide imaging for diagnostic purposes in pathology: Guidelines from the College of American Pathologists Pathology and Laboratory Quality Center. *Arch Pathol Lab Med* 2013; **137**: 1710–22.
15. Cheng CL, Azhar R, Sng SHA, Chua YQ, Hwang JSG, Chin JPF, Seah WK, Loke JCL, Ang RHL, Tan PH. (2016) Enabling digital pathology in the diagnostic setting: Navigating through the implementation journey in an academic medical centre. *J Clin Pathol* 1–9, doi:10.1136/jclinpath-2015-203600.
16. Vodovnik A. Diagnostic time in digital pathology: A comparative study on 400 cases. *J Pathol Inform* 2016; **7**: 4.
17. Pantanowitz L, Dickinson K, Evans AJ, Hassell LA, Henricks WH, Lennerz JK, Lowe A, Parwani AV, Riben M, Smith D, Tuthill JM, Weinstein RS, Wilbur DC, Krupinski EA, Bernard J. American telemedicine association clinical guidelines for telepathology. *J Pathol Inform* 2014; **5**: 39.
18. Parwani AV, Hassell L, Glassy E, Pantanowitz L. Regulatory barriers surrounding the use of whole slide imaging in the United States of America. *J Pathol Inform* 2014; **5**: 38.
19. Guidance for industry and Food and Drug Administration Staff. Technical performance assessment of digital pathology whole slide imaging devices. Division of Molecular and Genetics and Pathology. U.S. Food and Drug Administration April 20, 2016.
20. Treanor D, Jordan-Owers N, Hodrien J, Wood J, Quirke P, Ruddle RA. Virtual reality powerwall versus conventional microscope for viewing pathology slides: An experimental comparison. *Histopathol* 2009; **55**: 294–300.

21. Angelo M, Bendall SC, Finck R, Hale MB, Hiltzman C, Borowsky AD, Levenson RM, Lowe JB, Liu SD, Zhao S, Natkunam Y, Nolan GP. Multiplexed ion beam imaging of human breast tumors. *Nat Med* 2014; **20**: 436–42.
22. Mansfield JR. Multispectral imaging: A review of its technical aspects and applications in anatomic pathology. *Vet Pathol* 2014; **51**: 185–210.
23. Viray H, Li K, Long TA, Vasalos P, Bridge JA, Jennings LJ, Halling KC, Hameed M, Rimm DL. A prospective, multi-institutional diagnostic trial to determine pathologist accuracy in estimation of percentage of malignant cells. *Arch Pathol Lab Med* 2013; **137**: 1545–49.
24. Curry DG, Martinsen G, Hopper DG. Capability of the human visual system. *Proc SPIE* 2003; **5080**: 58–69.
25. Webster JD, Simpson ER, Michalowski AM, Hoover SB, Simpson RM. Quantifying histological features of cancer biospecimens for biobanking quality assurance using automated morphometric pattern recognition image analysis algorithms. *J Biomol Tech* 2011; **22**:108–18.
26. Gavrielides MA, Gallas BD, Lenz P, Badano A, Hewitt SM. Observer variability in the interpretation of HER2/neu immunohistochemical expression with unaided ad computer-aided digital microscopy. *Arch Pathol Lab Med* 2011; **135**: 233–42.
27. Shen DF, Wang SC. Measurements of JND property of HVS and its applications to image segmentation, coding and requantization. *Proc SPIE* 1996; **2953**: 113–21.
28. Camp RL, Chung GG, Rimm DL. Automated subcellular localization and quantification of protein expression in tissue microarrays. *Nat Med* 2002; **8**: 1323–27.
29. Jackson KR, Ramakrishnan L, Muriki K, Canon S, Cholia S, Shalf J, *et al.* Performance analysis of high performance computing applications on the amazon web services cloud. In IEEE Second Intl Conf. *Cloud Computing Technology and Science (CloudCom)*, 2010, pp. 159–68.
30. Wiesmann V, Franz D, Munzenmayer C, Palmisano R, Wittenberg T. Review of free software tools for image analysis of fluorescence cell micrographs. *J Microscop* 2015; **257**: 39–53.
31. Rogers R, Eastham-Anderson J, DeVoss J, Lesch J, Yan D, Xu M, Solon M, Hotzel K, Diehl L. Image analysis-based approaches for scoring mouse models of colitis. *Vet Pathol* 2016; **53**: 200.
32. Gerner MY, Kastenmuller W, Ifrim I, Kabat J, Germain RN. Histo-cytometry: A method for highly multiplex quantitative tissue imaging analysis applied to dendritic cell subset microanatomy in lymph node. *Immunity* 2012; **37**: 364–76.
33. Webster JD, Michalowski AM, Dwyer JE, Corps KN, Wei BR, Juopperi T, Hoover SB, Simpson RM. Investigation into diagnostic agreement using automated computer-assisted histopathology pattern recognition image analysis. *J Pathol Inform* 2012; **3**: 18.

34. McClelland RA, Finlay P, Walker KJ, Nicholson D, Robertson JFR, Blamey RW, Nicholson RI. Automated quantitation of immunocytochemically localized estrogen receptors in human breast cancer. *Cancer Res* 1990; **50**: 3545–50.
35. Lange H. Digital pathology: A regulatory overview. *Lab Med* 2011; **42**: 587–91.

6 The Future Role of Digital Imaging in the Pathology Workflow

Fumi Kawakami*, Naohiro Uraoka*,† and Yukako Yagi‡,§

*Department of Translational Molecular Pathology,
The University of Texas MD Anderson Cancer Center,
Houston, Texas, USA
† Department of Molecular Pathology,
Hiroshima University Institute of Biomedical and Health Sciences,
Hiroshima, Japan
‡ Department of Pathology,
Memorial Sloan Kettering Cancer Center, New York, USA
§ National Cancer Center Hospital East, Chiba, Japan

Contents

I. Introduction

During the last two decades, development and practical realization of whole-slide imaging (WSI) technology has expanded roles of digital imaging in pathology.[1] From 2000 to today, development of the modern robotic microscope, the automated WSI scanner, has been done by numerous vendors. WSI used to be in demand as a teaching material, for research use, and in telepathology, but clinical use of WSI will become more common in the near future including usage in primary diagnosis. To meet this need, there is demand for a wide range of devices for a wide range of purposes. Although usage of WSI is expanding from research lab and education institution to medium- to large-sized municipal hospitals, fundamental "grand challenges" intrinsic to WSI (for example, file size, interactivity, and image quality) as well as operational challenges that are developing as the use of digital slides in clinical labs (for example, validation and quality control) remain. In this chapter, we will review leveraging WSI to impact digital pathology in the pathology workflow. Finally, we will discuss challenges and solutions in the usage of WSI.

II. Leveraging Whole-slide Imaging

a. Education and training

One of the distinctive characteristics of WSI is its interactivity. Scanned images can be shared much more easily with multiple users anywhere at any time. Because of this characteristic, WSI is used as a teaching material commonly, from medical school to post-MD medical training program.

i. *Medical students, pathology residents, and fellows*

WSI is widely used as a standardized teaching material in medical schools.

WSI enables the sharing of slide images for the whole class, and this attribute promotes interactivity between students as well as between students and teachers, resulting in better educational effects. WSI is also a cost-effective technology since there is no need to support multiple physical slides and microscopes. Consequently, WSI has begun to replace conventional light microscope in many institutions.[2]

WSI is also used for training of pathology residents and fellows. In fact, one of the first uses of WSI was to create a series of interesting cases for residents or fellows in pathology and other specialties that require histological recognition of normal and pathological tissues.[3] Today, slides for teaching conferences and courses are often distributed with digitized images, and US national pathology boards have been held with fully digital images. Virtual tracking and tutoring, tele-education (teleconferencing), e-learning, and virtual workshops have become more common tools for learning.[4] The usage of digital slides in clinical conferences (tumor boards) and on specialty web sites is expanding.

ii. *Clinical Informatics Board*

Pathology informatics is subject to increasing demand with the expansion of WSI usage, especially for clinical application. In 2013, the American

Board of Pathology (ABP) administered the first board examination of Clinical Informatics. All candidates for certification in Clinical Informatics by the ABP must possess current certification by the ABP and must have completed 24 months professional training in an Accreditation Council for Graduate Medical Education (ACGME)-accredited Clinical Informatics fellowship program or the same level of education[5] including training in system design, software building, system installation, and system evaluation. Although, the number of applicant for both the board examination and the fellowship program is increasing, pathology informatics is still in its maturation period, and limited institutions provide Clinical Informatics fellowship programs. Expanded access to pathology informatics education and establishment of well-developed education system is desired.

b. Telepathology

Telepathology could be broadly defined as the spatial separation of slides from the location where the slides are interpreted. A key component of WSI-based telepathology is that, somewhere in the process, a pathologist sees the physical slide under the microscope. A variety of practices fit this general definition.

Remote intraoperative frozen section consultation is the current use of WSI. The Canadian Association of Pathologists formulated guidelines for establishing a telepathology service for anatomic pathology using WSI[6] providing basic information about how to implement and use this technology. Several large, published series[7,8] have shown that at least in neuropathology, WSI, when well validated and controlled, can be used successfully to support frozen sections with very acceptable diagnostic accuracy and turnaround times. Furthermore, it allows more pathologists to participate in difficult cases.

Telepathology can reduce turnaround times. With the combined use of teleconference and telepathology, pathologists can supervise gross examination and tissue sampling and make preliminary diagnoses using WSI from outside of the pathology laboratory. Sharing immunohistochemistry (IHC) WSI slides from reference laboratory to the pathologists is also a useful tool for a faster turnaround time for community pathologists. The reference laboratory, US Labs, has a "virtual laboratory" component. The pathologists send tissue blocks or unstained slides to the lab and the lab does the staining and sends the stained slide back to the pathologist both physically (through a courier service) and as a digital slide (through the internet from their server).[9]

Though not using WSI, a smartphone is also an appropriate device for telepathology. Smartphones are widely used around the world, including in developing countries, and provides easy to access mobile wireless network. Through sharing captured images, expert opinion can be made available even in low-resource areas.[10]

c. Primary diagnosis on digital slides

Several studies showed that pathological diagnosis made by digital systems generally shows good concordance with that made using glass slides.[11,12] In recent years, it has been demonstrated that primary histopathologic diagnoses can be rendered digitally using WSI.[13-15] One of the notable advantages of WSI over conventional light microscopy is the ability to make use of computer-aided diagnostic tools. The application of the tools, such as image analysis, on primary diagnosis will become more common in the near future.

To apply WSI for primary diagnosis, validation of WSI systems is recommended. For patient safety, pathologists can use a WSI system in clinical settings only when they can make an accurate diagnosis with WSI with the same or better level of ease as with a traditional microscope and without interfering artifacts or technological risks.[16] The College of American Pathologists Pathology and Laboratory Quality Center issued WSI validation guidelines. According to the guidelines, validation should address both technical and interpretative components, and a pathologist adequately trained to use the WSI system must be involved in the process.[17]

d. Research

WSI-based digital IHC analysis is commonly used in the research field. Although clinical use is still limited (e.g., HER2 scoring), it will be applied in clinical settings more broadly in the near future in histology labs. More details will be discussed in Section III.

WSI three-dimensional (3D) technique for the reconstruction of histological architecture is one of the challenging and intriguing projects using WSI. One of the authors is working on 3D reconstruction of WSI, and it has proven to be valuable in the visualization and accurate quantitative diagnosis of disease with accurate correlation between imaging modalities, such as MRI and CT, beyond the two-dimensional microscope slide.[18,19] Micro-CT imaging technology application can improve the accuracy of 3D analysis even further, since its reference data help to identify the correct site and amount of tissue damage during tissue processing[20]. 3D reconstruction of two-dimensional histological slides helps understand the anatomy and will provide new insights into pathological mechanisms of disease development.

III. Pathology Workflow in the Digital Imaging Era

In recent decades, pathologists have changed their diagnostic approach dramatically. In addition to histological evaluation using H&E stained slides, data integration from multiple sources, such as IHC, genetic data, and clinical information, becomes critical to make a clinically significant diagnosis which is directly linked to therapeutic approach. This trend poses several changes in the pathology workflow, especially dealing with the escalated data size and its complexity imposed by WSI and next-generation sequencing (NGS) is the big issue in every single institution. Thus, development of a well-designed database that can allow efficient navigation between multiple data types is necessary for both clinical and research purposes. Also, automation of process is necessary to handle complex operations of surgical specimens for concurrent histological and genetic examination to eliminate human error and manage workload. In this section, we review tools and methods which will be practically used in the near future in the pathology workflow.

a. Advanced database system

During the last decade, accessibility to patients' genetic information has dramatically improved, and in some tumors, proof of certain genetic alterations is crucial for tumor classification. Thus, interactive databases

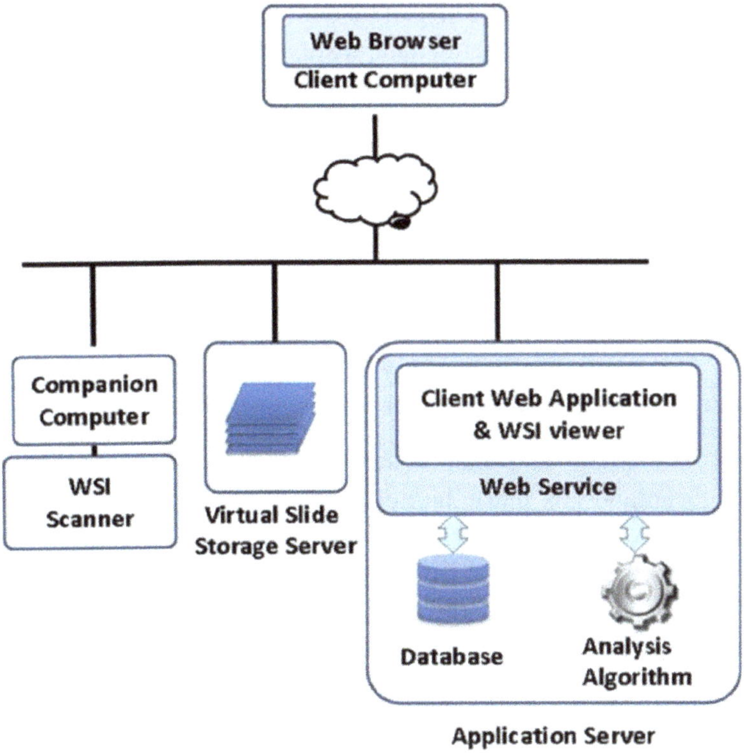

Figure 1. System architecture (from Ref. [21]).

that unify clinical, histologic, immunohistochemical, and genetic information is required, and such databass hold promise for multi-modality research and diagnosis. As an operative example, one of the authors makes test installations of the database system with the aims of (1) efficient management of textual information, (2) effective viewing of all varieties of stained WSI, and (3) assistance in evaluating WSI with computer-aided diagnosis[21] (Figure 1). The database supports modern pathologists' work, making diagnosis in an integrated fashion. And for research purpose, well-designed database system aid in correlation of multiple parameters, for example, different morphologic patterns with expression of various signaling proteins, DNA copy number changes, and DNA sequencing data. The system is unique because all the multiple systems for storing, viewing, and analyzing the data can be done in one system. The ability of this system to integrate imaging data, textual data, and the results of DNA sequencing information together is indeed helpful to researchers as well as diagnosticians.

For smooth operation of the system, close collaboration between pathologists (anatomic pathology, diagnostic molecular pathology, pathology informatics), engineers with pathology and informatics experience, and engineers with experience in optics, data management, image analysis, system management, and graphical user interfaces is needed.

b. Tissue handling techniques for digital pathology

Because slide quality impacts the quality of its digital image, tissue processing techniques to make homogeneous and high-quality slides are important.[22] Also, pathology labs should manage large-scale use of WSI.[23] Automated histology robotics could be a solution. Machines are available to automate virtually every step of the slide making process including processing, embedding, sectioning, staining, and cover-slipping. Although we still need to convey materials from one machine to another, all the process could be done automatically in the future histology lab.

i. *Tissue sectioning machine*

To minimize sectioning artifacts is a fundamental necessity to have high-quality WSI. An automated tissue sectioning machine is useful especially when you need serial sections for IHC or 3D analysis. Advantages of automated tissue sectioning are invariable thickness, uniform orientation, and fewer tissue sectioning artifacts.[24]

c. Properties of WSI in pathology workflow

WSI allows pathologists to use scanned slides as digital data. Scanned slide data is not only a fadeless image, but pathologists can also use it actively. For example, the color which is displayed can be virtually changed and quantitative pathology can be done in accurate using WSI.

i. *Color customization*

Digital systems can allow large, unexpected changes in the color display, and changes in color can result in better effect contrast and resolution. Thus, personalized color setting can be one of the unique properties of WSI. This peculiarity of digital imaging system benefits pathologists who suffer from are color weakness or blind.[25]

Digital staining is one of the unique examples of color customization which can be done in digital pathology. Digital staining is digital transformation of an H&E-stained specimen to an other special stain. One author reported digital transformation of an H&E-stained specimen to Masson's trichrome staining and showed its usefulness particularly in the visualization of specific tissue structure, fibrosis for instance, without performing the actual staining of tissue slides.[26]

ii. *Image analysis*

Introduction of quantitative pathology has been facilitated by digital technology.[27] Image analysis on pathology specimens results in improvement of reproducibility of pathological evaluation and numerical conversion of the result.

Automated IHC analysis is used in research and in a limited capacity in the clinical field. In breast cancer, HER2/neu immunohistochemical evaluation using Scanscope Aperio Technologies, Inc., (Aperio), has received clearance from the United States Food and Drug Administration (FDA).[28,29] In the near future, imaging analysis including automated IHC analysis and digital diagnosis will be more commonly used in the clinical field.

1. *Approach and applications*

The approach to image analysis is largely divided in two: pixel-level analysis and histocytometric analysis. In histocytometric analysis, cell segmentation and identification of cell compartments are done for analysis. In addition to open-source applications, a variety of applications is commercially available from some vendors, such as Genie from Aperio and TissueStudio from Definiens. However, tumor detection is still under development, and so annotation by pathologist is needed. The compatibility of the algorithm is also an issue. Currently, each algorithm works independently. Establishment of the integrated system will facilitate popular use of automated image analysis.

Beyond the automated scoring of IHC, image analysis is now applied to multiplex immunofluorescence (IF) and FISH analysis. Combined use of confocal WSI scanner will be of help in evaluation of the localization of the signal in cells.

2. *Immunoprofile*

Immunoprofiling using quantitative immunohistochemistry of tumor-infiltrating lymphocytes (TILs) and immune checkpoints has been useful for predicting response to immune therapies and prognosis in many types of cancer.[30,31] So far, chromogenic IHC has been used;[32] however, multiplex IHC has emerged as a far more potent tool, offering simultaneous detection of multiple markers, spatial distribution of each cell phenotype, and co-expression of several markers.[33]

The following is a description concerning an image analysis-based multiplex IF platform that one of the authors developed recently according to the instructions previously published[34] in order to assess the immunoprofiling of various solid tumors. It can be applied to formalin-fixed, paraffin-embedded tissues. Tyramide signal amplification (TSA) chemistry and multispectral microscopy (Vectra™, PerkinElmer, Waltham, MA) are used for simultaneous detection of up to seven markers.

First of all, to optimize the antibodies and generate spectral libraries, singleplex IF is performed prior to multiplex IF using the Opal Kit TSA detection system (PerkinElmer). In this system, signal amplification occurs as a result of covalent binding of fluorophore-labeled tyramide onto the tissue specimen in a highly specific manner. This method allows researchers to use antibodies from the same species for multiplex staining without cross-reactivity.[35] We constructed some panels composed of various markers including TILs and immune checkpoints. One of these panels, for instance, consists of programmed cell death protein 1 (PD-1), programmed cell death ligand 1 (PD-L1), CD3, CD8, CD68,

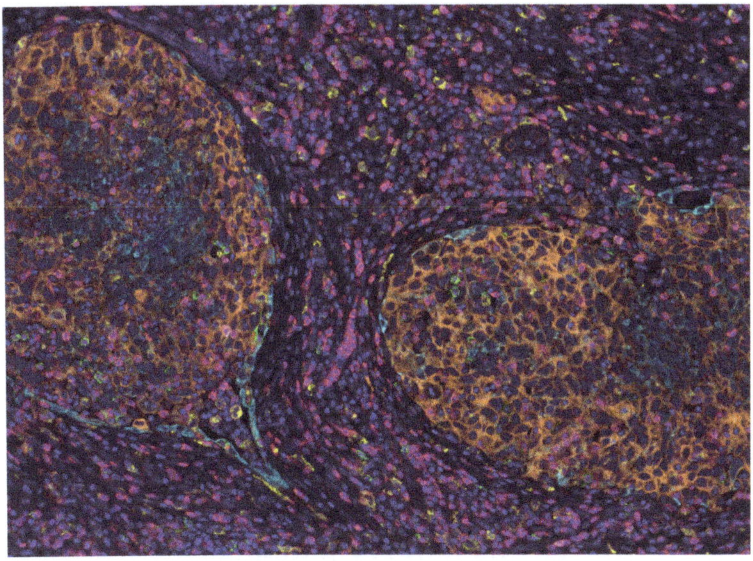

Figure 2. Image of multiplex immunohistochemistry in non-small cell lung cancer. Seven markers, including PD-L1 (orange), PD-1 (green), CD3 (red), CD8 (pink), CD68 (yellow) cytokeratin AE1/AE2 (cyan), and DAPI (blue), are simultaneously detected.

cytokeratin AE1/AE2, and DAPI (Figure 2). After singleplex and multiplex IF staining, each slide is imaged at 10 nm λ from 420 nm to 720 nm using the Vectra spectral imaging system (PerkinElmer). The slides are scanned at low magnification (×10, 1um.pixel) at first, and then five individual areas (0.3345 mm^2 each) are randomly selected for scanning at high magnification (×20) taking tumor heterogeneity into account.

Images captured by Vectra are then analyzed with InForm software (PerkinElmer). In this step, each unmixed staining is combined to efficiently quantitate each biomarker, as well as autofluorescence, based on their specific spectral properties obtained by singleplex IF. This software automatically calculates the amount of each marker and co-localization in the morphologies of interest (e.g., tumor or stroma) on a cell-by-cell basis, though users need train the software to recognize the tissue morphology and phenotypes.[36,37]

This method of multiplex IF analysis is a reproducible, reliable, and high-throughput tool. Furthermore, it is also helpful for efficient use of small tissue specimens, such as core needle biopsy. On the other hand, one problem of this method is that the process is time-consuming. Another problem is that, just like other image analysis tools, supervision by a pathologist is essential to obtain correct results.

Although there is still plenty of room for improvement, this methodology can offer much more information about the tumor microenvironment. Other vendors have also released imaging systems (e.g. Aperio from Leica Biosystems) and quantitative analysis tools (e.g. Indica Labs, Definiens, and Visiopharm) for multiplex IF. The immunoprofiling data obtained by multiplex IHC will be a great help in cancer patients on the road to accepting tailored immune therapies.

IV. Technical Challenges

In addition to existing technical challenges, technological evolution and prevalence of digital pathology bring emerging issues. With the expansion in the use of digital pathology in a variety of settings, the demand of technical specifications for standardized image quality, scanning speeds, color calibration, network system, and pathology informatics is increasing. Especially for clinical application, image quality should be strictly managed through appropriate color standardization and slide quality control. Guidelines which address technical minimum requirements for clinical usage of WSI are needed.

a. File size

The main difference between WSI file data and conventional digital image is the much larger file size due to multi-laminate images. WSI files can be extremely large and store pixel data "pyramids" at several different resolution levels.[38]

At the highest resolution level, it can store around 100000 × 100000 pixel image data, and each pixel may be represented using RGB values. In the result, a single WSI file can be around 30 GB when uncompressed. A typical WSI image file in compressed format may be as large as a few

GBs.[39] A large academic medical center such as the Massachusetts General Hospital generates ~2,500 slides per day. In case all the slides are scanned, this is at least a few TB/day. In the digital pathology era, pathology labs should have data storage facilities which can collect large-scale data for the long term and also possess appropriate data management system.

b. Interactivity

File exchangeability between devices is an important factor to promote usage of WSI in both clinical and research fields. In reality, individual, existing commercially available WSI scanners utilize unique digital data formats which rely upon their own meta-data and compression techniques. Therefore it becomes a very challenging task to develop generic applications which can work with data originating from one of these sources.[39] Although approval of Digital Imaging and Communications in Medicine (DICOM) supplement 145 in 2010 made it possible for histology laboratories to store digital pathology images in a form that is compatible with the DICOM archive, in reality, traditional medical imaging systems and Picture Archive and Communications Systems (PACS) operate on the principle of exchanging full image files between devices; however, the size of high-resolution WSIs makes this model impractical in digital pathology.[40] To address this issue, web-based high-resolution image management systems which accept multiple WSI file formats has started to develop (e.g., Digital Image Hub from Leica Microsystems).

In case WSI is used for diagnostic purpose, the longer time required for reviewing a case through the computer than when the physical slides are available in a microscope is a possible complaint. WSI systems have used a client–server system in which image files are structured as a multi-layered, from thumbnails to high-resolution images. Each layer is divided into equally sized tiles, and majority of the data stays in the server. According to a pathologist's requests (panning and zooming), tiles are sent across the network to the client. While this approach does allow the pathologist access to the digital slide without waiting for the entire data set to download and careful network design can improve performance,[41] it is still much slower than the response time when a pathologist moves a physical slide on a microscope stage.

c. Slide variances from the histology lab

The quality of the glass slides used impacts the quality of the digital image.[22] Because each phase of the tissue handling process, tissue sampling, processing, sectioning, and staining, is still manually run to varying degrees and each process causes variations in the slide, to have homogeneous high-quality slides, and flat and square sections placed flat and parallel to the slides without tissue folds or bubbles is challenging. Thus, a WSI scanner should be designed to incorporate complex focusing, tissue finding, and slide handling mechanisms to get around slide, tissue, and stain variations. No less than approximately 3–5% of the scanned slides need to be re-scanned because of defocus, partial defect of image, and so on.[42] Usage of automated histology robotics could be a solution to minimize slide variance. Virtually every step of the slide making process, including processing, embedding, staining, and cover-slipping, can be done by automated machines. These devices should help decrease slide variance and should therefore improve digital image quality.

These devices can be assembled to run in continuous flow modes, and it would not be unreasonable to consider placing a digital slide scanner at the end of the process. This would allow virtually all slides to be imaged, pre-diagnostically, with virtually no impact on the laboratory's throughput or the pathologist's schedule.

d. Color standardization

Color standardization in a digital imaging system has several important roles. Improvement of the image appearance by color regulation ameliorates the readability of the slides for both pathologists and automated image analysis. There are many factors which cause considerable color variation in digital pathology workflow such as the scanner, the compression board, the server, the workstation, the view software, the monitor, etc. In addition to that, there is a limitation in applying international standards in the digital communication of color[43] due to the fact that not all pathology systems support these standards. One of the authors propounded that a color-calibrated slide designed based on modified Macbeth color chart effectively works to normalize color variations in WSI.[44–46] Also, the staining condition correction method with a look-up table based on the dye amounts absorbed by the sample works for staining standardization for better automated image analysis application.[47]

e. Image quality

Image quality basically depends on the quality of the image captured. Especially, tissue finding, focus, and depth of field are important factors. Quick and accurate tissue finding is important particularly in the setting of pre-diagnostic WSI. Because the tissue area is usually much smaller than the total glass or cover-slip area, many WSI systems attempt to improve scanning times by identifying the tissue on the slide swiftly and precisely using a low-resolution snapshot to minimize the area scanned in high-resolution made. Especially for surgical cases, it is challenging because tissue fragments can be small, under- or over-stained, and can look remarkably like dirt, grease, or cover-slip edges at low resolution. Most importantly, failure to identify (and, by extension, failure to image) an area of tissue can be catastrophic if the area has unique, critical diagnostic information. Failure to present all the tissue to the pathologist will result in a lack of confidence in the system and would be a major barrier to large-scale clinical implementation. While systems have gotten much better at tissue finding in recent years, in our current experience, this still remains an important issue in many situations.

Focus and depth of field remain the most important issues in WSI. Like tissue finding, focus is something that is taken out of the pathologist's hands in an automated WSI environment, and so the systems must be a very capable substitute. The quality of focusing is limited by the range of focus movement, focus adjustment time, distance measurement methods, and movement resolution on the vertical *z*-axis.[38] Ideally, incorporating more focusing points manually or automatically and even using real-time focus scanning systems being able to digitize a slide at various focusing planes through the *z*-axis in a manner similar to the actual microscope helps to have fine focus control.

f. Infrastructure and costs

Since large-scale, pre-diagnostic WSI has just been started as a pilot program in confined institutions and is not generally available, building storage and network infrastructure is expensive and will have to compete for funding with other initiatives in pathology and across the hospital. But the cost (and the availability) of computational power, network connectivity, and storage space will continue to improve with general improvements in technology; cheap storage, fast networks and advanced robots needed to meet the challenge of large-scale, pre-diagnostic WSI will be available as part of the general hospital infrastructure.

V. Conclusion

Digital pathology has evolved over time from a purely education tool to having an increased presence in the clinical setting. Despite these advances, there are a number of challenges that remain for digital pathology to be fully integrated into the standard pathology workflow, including variations between pathology labs, color standardization, and costs. However, as the digital pathology field continues to grow and the cost of new technologies and digital storage decrease, these obstacles can be overcome to deliver high-quality care to patients.

References

1. Pantanowitz L, Valenstein PN, Evans AJ *et al.* Review of the current state of whole slide imaging in pathology. *J Pathol Inform* 2011; **2**: 36.
2. Saco A, Bombi JA, Garcia A, Ramirez J, Ordi J. Current Status of Whole-Slide Imaging in Education. *Pathobiol: J Immunopathol Molecular Cellular Biol* 2016; **83**(2–3): 79–88.
3. Hamilton PW, Wang Y, McCullough SJ. Virtual microscopy and digital pathology in training and education. *APMIS: Acta Pathologica, Microbiologica, et Immunologica Scandinavica* 2012; **120**(4): 305–15.
4. Pantanowitz L, Szymas J, Yagi Y, Wilbur D. Whole slide imaging for educational purposes. *J Pathol Inform* 2012; **3**: 46.
5. http://www.abpath.org/.
6. Canadian Association of Pathologists Telepathology Guidelines C, Bernard C, Chandrakanth SA *et al.* Guidelines from the Canadian Association of Pathologists for

establishing a telepathology service for anatomic pathology using whole-slide imaging. *J Pathol Inform* 2014; **5**(1): 15.

7. Evans AJ, Chetty R, Clarke BA *et al.* Primary frozen section diagnosis by robotic microscopy and virtual slide telepathology: The University Health Network experience. *Semin Diagnostic Pathol* 2009; **26**(4): 165–76.
8. Horbinski C, Wiley CA. Comparison of telepathology systems in neuropathological intraoperative consultations. *Neuropathol: Off J Jap Soc Neuropathol* 2009; **29**(6): 655–63.
9. http://www.uslabs.net/.
10. Fontelo P, Liu F, Yagi Y. Evaluation of a smartphone for telepathology: Lessons learned. *J Pathol Inform* 2015; **6**: 35.
11. Jara-Lazaro AR, Thamboo TP, Teh M, Tan PH. Digital pathology: Exploring its applications in diagnostic surgical pathology practice. *Pathol* 2010; **42**(6): 512–8.
12. Li X, Gong E, McNutt MA *et al.* Assessment of diagnostic accuracy and feasibility of dynamic telepathology in China. *Human Pathol* 2008; **39**(2): 236–42.
13. Fonyad L, Krenacs T, Nagy P *et al.* Validation of diagnostic accuracy using digital slides in routine histopathology. *Diagnostic Pathol* 2012; **7**: 35.
14. Al-Janabi S, Huisman A, Nap M, Clarijs R, van Diest PJ. Whole slide images as a platform for initial diagnostics in histopathology in a medium-sized routine laboratory. *J Clin Pathol* 2012; **65**(12): 1107–11.
15. Campbell WS, Lele SM, West WW, Lazenby AJ, Smith LM, Hinrichs SH. Concordance between whole-slide imaging and light microscopy for routine surgical pathology. *Human Pathol* 2012; **43**(10): 1739–44.
16. Evans AJ, Kiehl TR, Croul S. Frequently asked questions concerning the use of whole-slide imaging telepathology for neuropathology frozen sections. *Semin Diagnostic Pathol* 2010; **27**(3): 160–6.
17. Pantanowitz L, Sinard JH, Henricks WH, *et al.* Validating whole slide imaging for diagnostic purposes in pathology: Guideline from the College of American Pathologists Pathology and Laboratory Quality Center. *Archiv Pathol Lab Med* 2013; **137**(12): 1710–22.
18. Fonyad L, Shinoda K, Farkash EA *et al.* 3-dimensional digital reconstruction of the murine coronary system for the evaluation of chronic allograft vasculopathy. *Diagnostic Pathol* 2015; **10**: 16.
19. Hashimoto N, Bautista PA, Haneishi H, Snuderl M, Yagi Y. Development of a 2D image reconstruction and viewing system for histological images from multiple tissue blocks: Towards high-resolution whole-organ 3D histological images. *Pathobiol: J Immunopathol Molecular Cellular Biol* 2016; **83**(2–3): 127–39.
20. Senter-Zapata M, Patel K, Bautista PA, Griffin M, Michaelson J, Yagi Y. The role of micro-CT in 3D histology imaging. *Pathobiol: J Immunopathol Molecular Cellular Biol* 2016; **83**(2–3): 140–7.
21. Yagi Y, Riedlinger G, Xu X *et al.* Development of a database system and image viewer to assist in the correlation of histopathologic features and digital image analysis with clinical and molecular genetic information. *Pathol Int* 2016; **66**(2): 63–74.

22. Yagi Y, Gilbertson JR. A relationship between slide quality and image quality in whole slide imaging (WSI). *Diagnostic Pathol.* 2008; **3** (Suppl 1): S12.
23. Gilbertson J, Yagi Y. Histology, imaging and new diagnostic work-flows in pathology. *Diagnostic Pathol* 2008; **3** (Suppl 1): S14.
24. Onozato ML, Hammond S, Merren M, Yagi Y. Evaluation of a completely automated tissue-sectioning machine for paraffin blocks. *J Clin Pathol* 2013; **66**(2): 151–4.
25. Landini G, Perryer G. Digital enhancement of haematoxylin- and eosin-stained histological images for red-green colour-blind observers. *J Microscopy* 2009; **234**(3): 293–301.
26. Bautista PA, Abe T, Yamaguchi M, Yagi Y, Ohyama N. Digital staining for multispectral images of pathological tissue specimens based on combined classification of spectral transmittance. *Computerized Medical Imaging and Graphics: The Off J Computerized Medi Imaging Soci* 2005; **29**(8): 649–57.
27. Bacus JW, Grace LJ. Optical microscope system for standardized cell measurements and analyses. *Applied Optics* 1987; **26**(16): 3280–93.
28. Cantaloni C, Tonini RE, Eccher C *et al.* Diagnostic value of automated Her2 evaluation in breast cancer: a study on 272 equivocal (score 2+) Her2 immunoreactive cases using an FDA approved system. *Applied Immunohistochemistry & Molecular Morphology: AIMM/official publication of the Society for Applied Immunohistochemistry* 2011; **19**(4): 306–12.
29. Friedberg RC, Pantanowitz L. Practice evolution: decentralized computer-assisted immunohistochemical image analysis. *Archives of Pathology & Laboratory Medi.* 2009; **133**(4): 597–600.
30. Herbst RS, Soria JC, Kowanetz M, *et al.* Predictive correlates of response to the anti-PD-L1 antibody MPDL3280A in cancer patients. *Nature* 2014; **515**(7528): 563–7.
31. Fridman WH, Pages F, Sautes-Fridman C, Galon J. The immune contexture in human tumours: Impact on clinical outcome. *Nature Reviews Cancer* 2012; **12**(4): 298–306.
32. Parra ER, Behrens C, Rodriguez-Canales J *et al.* Image analysis-based assessment of PD-L1 and tumor-associated immune cells density supports distinct intratumoral microenvironment groups in non-small cell lung carcinoma patients. *Clinical Cancer Research: An Off J Am Associ Cancer Res* 2016.
33. Feng Z, Puri S, Moudgil T *et al.* Multispectral imaging of formalin-fixed tissue predicts ability to generate tumor-infiltrating lymphocytes from melanoma. *J Immunother Cancer* 2015; **3**: 47.
34. Stack EC, Wang C, Roman KA, Hoyt CC. Multiplexed immunohistochemistry, imaging, and quantitation: A review, with an assessment of Tyramide signal amplification, multispectral imaging and multiplex analysis. *Methods (San Diego, Calif)* 2014; **70**(1): 46–58.
35. Toth ZE, Mezey E. Simultaneous visualization of multiple antigens with tyramide signal amplification using antibodies from the same species. *The J Histochem Cytochem: Off J Histochem Soci* 2007; **55**(6): 545–54.
36. Mansfield JR. Multispectral imaging: A review of its technical aspects and applications in anatomic pathology. *Veterinary Pathol.* 2014; **51**(1): 185–210.

37. Huang W, Hennrick K, Drew S. A colorful future of quantitative pathology: validation of Vectra technology using chromogenic multiplexed immunohistochemistry and prostate tissue microarrays. *Human Pathol* 2013; **44**(1): 29–38.
38. Parwani AVF, M.; Balis, U.; Pantanowitz, L. Digital imaging. In *Pathology Informatics: Theory and Practice*. L.B. Pantanowitz, J.M. Tuthill, eds. Canada: ASCP Press, 2012, pp. 231–56.
39. Yildirim E, Foran D. Parallel vs. Distributed Data Access for Giga-pixel-resolution Histology Images: Challenges and Opportunities. *IEEE J Biomed Health Inform.* 2017 Jul; **21**(4): 1049–1057.
40. Wang F, Oh TW, Vergara-Niedermayr C, Kurc T, Saltz J. Managing and querying whole slide images. *Proc SPIE — The Int Soci Opti Eng* 2012; **8319**.
41. Lundin M, Szymas J, Linder E *et al.* A European network for virtual microscopy — Design, implementation and evaluation of performance. *Virchows Archiv: An International J Pathol* 2009; **454**(4): 421–9.
42. Gilbertson JR, Ho J, Anthony L, Jukic DM, Yagi Y, Parwani AV. Primary histologic diagnosis using automated whole slide imaging: A validation study. *BMC Clin Pathol* 2006; **6**: 4.
43. http://www.color.org.
44. Yagi Y. Color standardization and optimization in whole slide imaging. *Diagnostic Pathol* 2011; **6**(Suppl 1): S15.
45. Weinstein RS, Graham AR, Richter LC *et al.* Overview of telepathology, virtual microscopy, and whole slide imaging: Prospects for the future. *Human Pathol* 2009; **40**(8): 1057–69.
46. Bautista PA, Hashimoto N, Yagi Y. Color standardization in whole slide imaging using a color calibration slide. *J Pathol Inform* 2014; **5**(1): 4.
47. Bautista PA, Yagi Y. Staining correction in digital pathology by utilizing a dye amount table. *J Digit Imag* 2015; **28**(3): 283–94.

7 Machine Learning for the Integration of Pathology Images with Molecular Data

Martin C Stumpe*, Arran Schlosberg*, Laura Deming*, Lily Peng*, Graziella Solinas† and Jason Hipp*,†,‡

*Google Inc, Mountain View, CA, USA
†Verily Life Sciences, South San Francisco, CA, USA
‡Hipp@google.com

Contents

I. Introduction

When it comes to cancer diagnosis, therapy selection, and biomarkers of patient response to treatment, tissue represents an invaluable source of information. For cancer, several different patterns of the distribution of cellular and structural components of the tumor microenvironment (e.g., spatial localization of infiltrating immune cells, presence of extracellular matrix structures, number of suppressive stromal cells) may predict clinical outcome, and thus it is increasingly clear that comprehensive and multidimensional analyses of samples are necessary to provide an optimal understanding of tissue organization as visualized under the microscope.[1–4]

Currently, two methodologies are commonly used for the analysis of a patient's tumor tissue: microscopy and molecular assays. Microscopy is performed for all patients in the clinic. It consists of cutting very thin slices of tumor tissue (~4–6 microns thick), adhering them to glass slides, and staining to highlight tissue subtypes according to their affinity for particular dyes. A pathologist reviews the image under a microscope to assess the tissue morphology — sizes and shapes of cells and their spatial arrangement — in order to classify tumors by type, grade, and stage (providing unstructured data outputs that are often qualitative and semi-quantitative). Additionally, immunohistochemical (IHC) stains may be performed to assess biomarker expression that is used to assist in making the diagnosis and sometimes to recommend a therapeutic regimen.

Molecular assays are characterized by a chemical digestion of a portion of the tumor in order to isolate its DNA, RNA, and proteins for molecular identification and quantification by techniques such as polymerase chain reaction (PCR), next-generation sequencing (NGS), proteomics, and/or various other -omics. This molecular data can help a pathologist determine the clonality of a tumor or alternatively provide an oncologist with a list of mutations for which targeted therapies may exist. The data generated from molecular assays are often digital (structured such as a quantitative number, or a presence/absence indicator), which makes it easily interpretable by computer algorithms.

The data derived from these two workflows — images of tissue microscopy from pathology slides and molecular information — while derived from one common source, are analyzed relatively independently.

At most, molecular assays are often gated by the tissue-image data since the working diagnosis is first determined from the morphological features, which guide the molecular assay selection. In addition, the pathologist uses the tissue-image data to identify the tissue slice that has a certain percentage of tumor cells and the best viability for successful molecular analyses. This *modus operandi* is useful to identify the best available treatment for cancer patients, but it is far from providing adequate information for the selection of a truly personalized treatment.

Beyond the need for prior diagnosis and selection of viable regions of tissue, there are a number of reasons why tissue-image data is not integrated with its respective molecular results. First, tissue-image data is not routinely digitized. Second, current practices in analysis of morphological data do not necessarily account for all pixel-level data embedded in the image (often as high as 100 k × 100 k pixels per slide). In order to achieve the latter, human pathologists would need to account for complex pixel-level numeric measurements such as the size, shape, and number of every cell present in the image, as well as microenvironmental factors such as spatial relationships between tumor, immune cells, and extracellular matrix. Furthermore, it is possible that patterns, indiscernible by the human eye, may signal important molecular changes. Lastly, the correlation of all of these factors with high-dimensional molecular data is too complex for human analyses and computational tools are yet to be developed for this task. Consequently, diagnostic pipelines that analyze data independently do not fully utilize potential relations within the tissue image and molecular data, and thus may miss key insights into tumor biology.

To reach the ambitious aim of integrating tissue-image data with molecular data, two main challenges still need to be overcome: (1) increased quality and quantity of data generated from human samples (not in the scope of this chapter) and (2) creation of deep-learning capabilities to integrate multidimensional image and molecular data that go beyond handcrafted features (user/operator defined) for machine learning.

The current body of work for integrating pathology tissue and molecular data is nascent due to the recent advent of digital slide scanners for pathology and developments in machine learning. However, relevant advances in digital pathology do exist that begin to address the integration of tissue-image data sets (pathology images) with molecular data sets.

II. Digital Pathology

Digital pathology is enabled today by whole-slide imaging (WSI) scanners which convert glass slides into digital slides, thus converting pathology from an analog discipline to a digital one. With pathology in the digital space, computer algorithms can be utilized to analyze, quantify, and measure, in an image, morphologic and histologic features that cannot be performed with a regular microscope and glass slides alone. Image analysis algorithms are currently being utilized in clinical practice to quantify breast cancer immunohistochemical markers such as Her2, ER, and PR. They have also been used in translational studies to predict both response to checkpoint inhibitor therapy and recurrence of colon cancer by quantifying the number of T cells and their spatial location within the tissue.[5,6]

These algorithms are beginning to demonstrate new ways to think regarding cancer. For example, Beck *et al.* showed that breast cancer aggressiveness can be better predicted by using a computer algorithm to look at stromal features alone when compared to traditional methodology of assessing the tumor cells.[7] More recently, using data from a public pathology hackathon called the Camelyon16 Challenge, Liu *et al.* utilized a new approach using deep learning algorithms to identify metastatic breast cancer in lymph nodes.[8] Specifically, a convolutional neural net model based on Inception v3 by Szegedy *et al.*[9] was used and modified to classify abnormal tissue in image patches, without explicit feature extraction, from 269 digitized H&E slides (Hamamatsu NDPI, 3DHistech Pannoramic 250 Flash II) from unique lymph node specimens, of which 159 were normal (i.e., from breast cancer patients without lymph node metastasis) and 110 slides included lymph node metastasis of primary breast cancer. The authors then validated their findings on two datasets (A and B, as described in Table 1) that were independent from the training datasets.

WSI-level Receiver Operator Characteristics (ROC) analysis was performed, achieving an Area Under Curve (AUC) over 0.976 for both datasets. Additionally, as a measure of localization accuracy, Free-Response Receiver Operator Characteristics (FROC) analysis was performed on dataset A as summarized in Table 2.

Table 1. Dataset characteristics and WSI-level ROC performance metrics for initial validation testing.

Dataset characteristics				WSI-level ROC
Dataset	Total slides	# positive for LN metastasis	# negative for LN metastasis	WSI-level AUC: single network
A	128	48	80	0.977 (0.931, 100)
B	110	57	53	0.976 (0.936, 100)

Source: Reference 8

Table 2. Free-response ROC (FROC) analysis was performed for initial validation testing on Dataset A, following the methodology presented in Litjens *et al.* (https://www.ncbi.nlm.nih.gov/pubmed/27212078).

FROC score (excluding ITC)	Sensitivity @8FP	FROC score, biggest tumor per slide (micro, macro)	FROC score, macro
88.5 (84.3, 92.3)	92.4 (88.7, 95.7)	95.1 (89.1, 100)	100 (100, 100)

Notes: Sensitivity was calculated at predefined false-positives (FP) rates per tumor-negative image and combined for an overall FROC score. Additionally, the WSI were further categorized as containing micrometastasis (micro) or macrometastasis (macro), and FROC analysis was performed. Isolated tumor cells (ITCs) were excluded from the analysis (neither considered positive or negative).
Source: Reference 8

An example of a cervical lymph node metastasis (CLNM) heatmap from the initial validation testing is shown in the figure below. The tumor cells are a lighter purple than the surrounding cells in the left panel. A variety of artifacts are visible: the dark continuous region in the top left quadrant is an air bubble, and the white parallel streaks in the tumor and adjacent tissue are cutting artifacts. Furthermore, the tissue is hemorrhagic, necrotic, and poorly processed, leading to color alterations to the typical pink and purple of an H&E slide. The heatmap in the panel on the right accurately identifies the tumor cells while ignoring the various artifacts, including lymphocytes and the cutting artifacts running through the tumor tissue.[8]

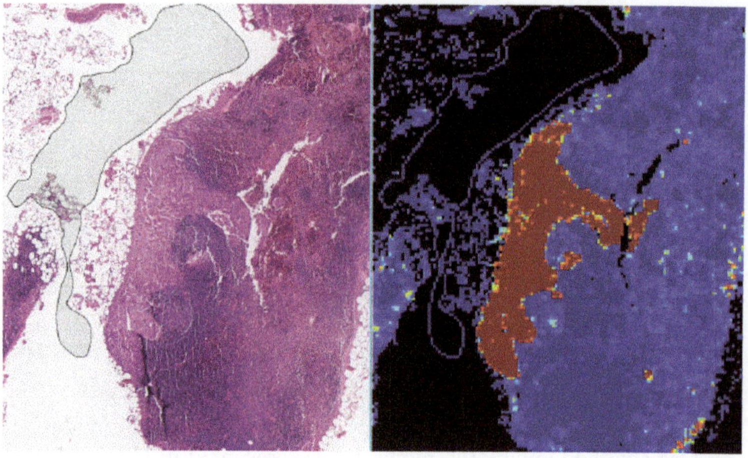

Furthermore, a recent study with H&E-stained WSIs of lung cancer used quantitative features (e.g., cell size, shape, texture, gradients) in the nuclei and cytoplasm along with classical machine learning techniques to predict patient survival better (AUC 0.85) than traditional grading and staging rubrics.[10] These studies, in addition to others described in the following section, are examples of the promise of digital pathology image analysis algorithms in providing greater insight into the molecular pathogenesis of cancer.

III. Integration of Tissue Images with Molecular Data

a. Transcriptional profiling of colon cancer and stromal contribution

Isella *et al.* utilized The Cancer Genome Atlas (TCGA) RNA-seq data to classify colon cancer subtypes and discovered the signature of the SSM (stem/serrated/mesenchymal) subtype derives from stromal cells (mainly cancer-associated fibroblasts) rather than tumor cells.[11] They confirmed the molecular findings by manually evaluating the corresponding digital pathology images for the amount of stromal tissue as reviewed by a pathologist and performed staining for the expression of SSM-related genes. This study represents a qualitative, non-computational way that 'integrates' findings from the tissue-image data with its corresponding RNA-seq data.

b. Morphometric analysis of TCGA glioblastoma multiforme data

Chang *et al.* developed an analytical pipeline for digital H&E tissue sections for glioblastoma multiforme (GBM) using publicly available digital pathology and molecular data from the TCGA website.[12] This pipeline converted images into morphometric indices (e.g., nuclear size, cellularity), which were then used for subtyping and genomic association. Using these indices, the authors identified a subtype that was shown to be a predictor of survival and associated with gene expression data that has been linked to (i) the immune response and AP-1 signaling pathways and (ii) IFNG, TGFB1, PKC, and MAPK14 hubs.[13]

Gutman *et al.* developed the Cancer Digital Slide Archive, a web platform to facilitate online slide image access and data integration from the TCGA.[14,15] They leveraged this approach and published several papers integrating pathology image analysis with TCGA genomics data. Their methodology starts with a morphometric analysis of tissue slide images, then they apply classical image-processing operations to detect and segment the nuclei, followed by computing nuclear features (e.g., shape, texture, gradients, and spatial arrangement) for the nuclei of each patient, and finally attempting to correlate the clustering of the mean and standard deviation of these properties with genomics or survival data. Using this approach, they found a correlation of the nuclear features with the membership in four distinct glioblastoma subtypes, defined via gene expression and genomics, with a classification accuracy of 80%. Similarly, they found that these nuclear features are correlated with patient survival.[16,17]

c. Predicting breast cancer molecular signatures of aggressiveness

Toi et *al.* utilized molecular data to predict aggressiveness and likelihood of recurrence as determined by a molecular assay.[18] The Bloom–

Richardson (BR) grade is a manual grading system performed by pathologists from the pathology slides (tissue image data) to determine the tumor's grade, and thus aggressiveness. BR takes into account mitotic index, tubule formation, and nuclear pleomorphism, and is known to be highly variable. The authors focused on tubule formation as a predictor of aggressiveness, and trained a deep neural network to determine the number of tubule to non-tubule nuclei in breast cancer images, and then to show a correlation between this score and the Oncotype DX risk category.[19]

d. Predicting mutational status via image analysis

Al-Ahmadie *et al.* showed that mutational status can result in specific morphological features in the context of the tissue-image data.[20] Specifically, the authors analyzed the prevalence of FGFR3 mutations in high-grade urothelial carcinoma. They found a set of histologic features unique to these mutants and used them to prospectively classify a small set of tumor samples, with no false negative results and a positive predictive value of 54%. Therefore, computational methods to predict mutation status from tissue image data are feasible to explore and may be useful in many domains.

Schaumberg *et al.* demonstrated an example of predicting molecular genotype from tissue image data alone.[21] The authors predicted whether a speckle-type POZ protein (SPOP) mutation was present in prostate cancer based on H&E-stained pathology slides (tissue image data) alone. The dataset was relatively small, with only 177 patient samples from the TCGA database, 20 of whom had the SPOP mutation. Using a classical feature-engineering approaches (i.e., the selection of salient regions), the authors used deep convolutional neural nets, a type of machine learning model successfully applied to label images in large databases such as ImageNet and CIFAR-10. The model achieved an AUC of 0.7128, showing that even a small amount of training data was sufficient to allow for detection of a signal. This paper highlights how an integration of tissue-image data with molecular data can shed new insights into tumor biology by identifying a correlation between morphology and molecular expression.

e. Laser microdissection workflow

Laser microdissection is a technique that enables individual cells to be cut from a slice of tissue, with the aid of lasers, for downstream molecular analysis.[22–24] The main bottleneck of this technology is the identification and selection of cells of interest, a time-consuming and laborious process that often requires the dedication of a domain expert such as a pathologist. Hipp *et al.* integrated a pattern-recognition algorithm — Spatially Invariant Vector Quantization (SIVQ) — into the workflow of the Arcturus XT laser microdissection machine to semi-automate this cellular identification and annotation step.[25,26] The benefit of having an image-analysis algorithm such as SIVQ guide the laser microdissection is that it improves the accuracy, speed, and reproducibility, thus enabling an increased throughput for molecular assays. Furthermore, it shifts the role of the pathologist from one of annotation of every cell of interest to a limited-contact supervisory role. Lastly, it provides a direct integration of image-analytics data from the cells of interest with its corresponding molecular data.[25]

IV. Conclusions

The integration of tissue image with molecular data has the potential to provide new insights into tumor biology and support the development of revolutionary methodologies to predict patient response to anti-cancer therapies. At this time, the integration of key pathology image and molecular data is often performed cognitively by a pathologist. As highlighted above, a limited number of computational tools/methods are presently available to assist in this task. However, other disciplines have been developing novel computational tools and methods involving machine learning to solve similar tasks. Recent examples for object detection and recognition in real life applications include applications from Google Street View, where deep neural networks have shown the ability to detect ambiguous objects such as business storefronts in Street View images, as well as successfully distinguishing between fine grained classes such as "restaurant" vs. "cafe" or "nail salon" vs. "hair salon", or also emerging self-driving car technology which algorithms for identifying pedestrians and other vehicles and drive to a specific destination.[27,28] Interestingly, this is very similar to the tasks required above which also require a form of computational image analysis and data integration to make decisions.

As the recent studies described above have shown, computer algorithms can learn from these integrated data sets and even predict molecular phenotypes from tissue images alone, potentially bypassing, or at least triaging, the need for expensive and tissue-dependent molecular tests. A very promising direction will be to leverage deep learning even further to improve beyond the current models which are mostly relying on human-selected features to compute or to select image regions to analyze. In that scenario, the algorithm is trained end-to-end, including the selection of relevant features. This will eventually facilitate advancement of our understanding of clinically relevant histological features that might be a predictor of a patient's genotype or mortality risk.

References

1. http://www.nature.com/nature/journal/v515/n7528/abs/nature13954.html
2. https://www.ncbi.nlm.nih.gov/pubmed/23034130
3. http://www.nature.com/nrc/journal/v6/n5/full/nrc1877.html
4. http://www.nature.com/nmat/journal/v13/n10/pdf/nmat4009.pdf
5. http://www.nature.com/nature/journal/v515/n7528/abs/nature13954.html
6. https://www.ncbi.nlm.nih.gov/pubmed/23034130
7. http://stm.sciencemag.org/content/3/108/108ra113
8. https://arxiv.org/abs/1703.02442
9. http://arxiv.org/abs/1512.00567
10. http://www.nature.com/articles/ncomms12474#s1
11. http://www.nature.com/ng/journal/v47/n4/abs/ng.3224.html
12. http://bmcbioinformatics.biomedcentral.com/articles/10.1186/1471-2105-12-484
13. http://tcga.lbl.gov/
14. http://cancer.digitalslidearchive.net/
15. https://www.ncbi.nlm.nih.gov/pubmed/23893318
16. https://www.ncbi.nlm.nih.gov/pubmed/20656651
17. https://www.ncbi.nlm.nih.gov/pubmed/25098236
18. https://www.ncbi.nlm.nih.gov/pubmed/20564629
19. http://www.nature.com/articles/srep32706
20. https://www.ncbi.nlm.nih.gov/pmc/articles/PMC3235805/
21. http://biorxiv.org/content/biorxiv/early/2016/07/21/064279.full.pdf
22. http://science.sciencemag.org/content/274/5289/998.long
23. http://www.nature.com/nprot/journal/v1/n2/full/nprot.2006.85.html

24. https://www.ncbi.nlm.nih.gov/pubmed/15578881?dopt=Abstract
25. Hipp J, Cheng J, Hanson JC, Yan W, Taylor P, Hu N, Rodriguez-Canales J, Hipp J, Tangrea MA, Emmert-Buck MR, Balis U. SIVQ-aided laser capture microdissection: A tool for high-throughput expression profiling. *J Pathol Inform.* 2011; **31**: 2–19, doi: 10.4103/2153-3539.78500.
26. https://www.ncbi.nlm.nih.gov/pubmed/25078867
27. https://arxiv.org/pdf/1512.05430.pdf
28. http://www.cvfoundation.org/openaccess/content_cvpr_2015/papers/Movshovitz-Attias_Ontological_Supervision_for_2015_CVPR_paper.pdf

8 Artificial Intelligence in Pathology and Genomic Medicine

Philip A. Pappalardo* and Christopher B. Townsend†

*Clinical Pathologist
†IBM Corporation, Armonk NY, USA

Contents

I. Introduction

Artificial intelligence (AI) is quickly becoming pervasive in our society and offering greater functionality for internet, entertainment, and smartphone-assisted search; speech recognition; autonomous driving cars; and computer vision (i.e., computer high-level understanding of digital images or video).[1] Advances in AI are crossing over to many fields in science and medicine including genomics, pharmacology, radiology, pathology, dermatology, and subspecialties.[2–6] The remarkable potential of AI has inspired highly talented startup companies and academic groups to find applications in pathology. Companies such as PAIGE.AI of New York, NY, PathAI from Cambridge, MA and Proscia of Baltimore, MD are spearheading AI in digital imaging and pathology.[7,8] AI has helped to uncover functions within the human genome provided by the human genome project and the availability of -omics data sets to decipher complex regulatory elements in non-coding regions of the genome far from the encoding genes.[9] Several new startup companies in genomics are offering services, including Deep Genomics in Toronto, Canada, among the pioneers of neural networks (nets), now working at the cutting edge of genomic science.[10] The latest techniques in machine-learning and neural net digital imaging (or AI imaging) are being applied to state-of-the-art and award-winning results in diagnostic histopathology,[11–15] offering further degrees of automation and improved accuracy in prognostic pathology.[16–23]

Looking ahead in pathology, consider the newly addressable domain in histopathology as H&E and immunohistochemically stained tissue slides go digital through conversion to image data sets via whole slide imaging (WSI). The application of AI to WSI has been made possible as computer processing speeds and storage become exponentially faster and cheaper (every 18 months according to Moore's Law), in conjunction with the availability of high-performance graphical processing units (GPUs) that process large-scale images at high power magnification. These large-scale digital WSI data sets are now fully accessible by automated machine-learning, computational algorithms employing statistical methods for the quantitative analysis of automatically extracted features (into moderately high-dimensional sets) from selected digital images and, most recently, by AI, computational algorithms within the most successful branch of AI to date — artificial neural networks (ANNs). The ability of neural nets, often stacked in interconnected layers referred to as deep neural nets (deep learning), to process pathology images with automated extraction of features (into very high-dimensional sets) to learn a hierarchy to differentiate histomorphologic patterns has ushered in an era of precision histology.[24]

II. Screening and Diagnosis

Computational tools in pathology promise to serve as a valuable assistant to the pathologist, for example, screening lymph nodes or normal tissue with high sensitivity or identifying salient microscopic features such as mitotic figures.[24] In a recent imaging challenge, [Camelyon16 (https://camelyon16.grand-challenge.org/); Grand Challenges in Biomedical Imaging, https://grand-challenge.org/], deep learning achieved state-of-the-art results and improved the ability to screen for breast cancer metastasis in cases with tumor foci as small as 100 × 100 pixels in gigapixel images while achieving a sensitivity of 92.4% for tumor detected, exceeding a 73.2% sensitivity achieved by a human pathologist.[11] Despite a higher false positive rate, such AI tools could significantly reduce false negatives in detecting cancer metastasis, critical to clinical staging and treatment planning. An award winning paper at the time of the Camelyon16 challenge demonstrated that combining the deep learning system with the pathologist led to a major reduction in error rate from 3% to less than 1% (Area Under the Receiver Operating Characteristic (ROC) Curve (AUC) from 0.966 (pathologist alone) to 0.995 (combined)) for correctly detecting the slides which contain breast cancer within lymph nodes, suggesting that integrating deep learning into the pathology work-flow may lead to improvements in both accuracy and reproducibility.[12] The Camelyon17 challenge advances WSI deep learning imaging for lymph node metastasis with patient-level analysis of multiple lymph node slides (five slides/patient) as well as clinical outcome prediction with the pathologic (p) regional lymph node (N) (pN) staging system (https://camelyon17.grand-challenge.org/results/). The Camelyon17 dataset pools images from five different medical centers (seven total when combining the Camelyon16 dataset for training) with limited ground truth labeling (annotation) of breast cancer metastasis slides (50/1,000 slides). Highest performance algorithms (94% slide level and staging accuracy, Lunit Inc., South Korea) overcome technical challenges with differing slide staining procedures and WSI scanning variations aggregating slide results to classify sizes of detected lymph node metastasis (i.e., macro-, micro-, and isolated tumor cells (ITCs) clusters). Thus, Camelyon17 goes beyond highly sensitive metastasis localization (detection) and slide-level identification (classification) in Camelyon16 with robust, high performance initial results for the clinical application of pN staging in breast cancer metastasis, which is a workload-intensive, time-consuming task for pathologists. Evaluation in further studies and in a clinical setting with participation of additional pathologist's assessments will be required for validation of the challenge competition's promising results.

Computational models have emerged which can distinguish benign from malignant intraductal proliferations of the breast, sometimes a challenge for even experienced pathologists.[25] Improvements in the computational methods by this research group have led to higher levels of accuracy and automation, which may lead to assistance in diagnosis or a validation tool for pathologists in the future.[26] Diagnostic capabilities are at their inception, but quickly improving by acquiring techniques from deep learning applied to computer vision (e.g., ImageNet, a large visual database, and ImageNet Challenge, an annual competition for research teams to develop algorithms for visual recognition tasks such as classification and segmentation of objects).[27] Some recent technical advances in AI pathology digital imaging include automated (unsupervised) feature extraction, color normalization (improving classification performance and extensibility), expansion to multiclass classification from two class (binary) classification allowing deep learning algorithms to classify subtypes of cancer, transfer learning (machine learning of a new task applied from the transfer of knowledge of a similar task), and visualization techniques (e.g. in the last hidden layer of neural nets) to identify specific features with biological insight.[3,13,18,28,29]

III. Prognosis

The ability of computational digital imaging to improve prognostic accuracy in pathology has been demonstrated in several cancers types.[16–23]

A recent machine-learning imaging study in WSI showed compelling results incorporating survival statistics, successfully predicting survival in non-small cell lung carcinoma (NSCLC).[16] This study introduced an automated system which can extract thousands of image features for analysis with machine-learning, making full access of the vast amounts of information in WSI.[16] Whereas the diagnostic capabilities of these tools are only approaching the ability level of pathologists, the prognostic capabilities have exceeded human pathologists in stage 1 lung carcinoma, such as this case in NSCLC. The authors point out this capability in lung cancer is superior to current practice whereby pathologists assess images in terms of grade and stage, as these patients may have diverse survival outcomes even with the same grade or stage.[16] Importantly, not only predictive of survival rates in lung carcinoma, the system is applicable to prognostic studies in other organs and is scalable.[16] Moreover, supplementing the prognostic study with additional molecular data may provide supporting biological evidence (i.e., upregulation of cellular adhesion processing and extracellular matrix remodeling associated with a poor prognosis).[17]

Histology images and genomic data (biomarkers) have recently been incorporated into a single unified deep learning framework within survival models creating genomic survival convolutional neural networks (GSCNNs) that have been shown to improve prognostic accuracy.[47] This deep learning model also addresses tumor heterogeneity by sampling high-power images for training the GSCNNs. The GSCNNs are analyzable with visualization of the histologic patterns (features) through the predicted heat maps. By visualizing the predictive features, pathologists can confirm pathologic grading (such as cell density, nuclear pleomorphism, microvascular proliferation) as well as potentially identify novel prognostic features associated with survival or patient outcomes.[47] In the future, GSCNN models may be able to incorporate any number of data (e.g. genomic or proteomic biomarkers) in combination with histology image analysis to even further improve prediction of patient outcomes.[47]

Another recent prognostic study successfully predicts prostate cancer recurrence using a novel two-stage, automated deep learning approach in mid-grade, intermediate-stage prostate cancer (PCa), the most commonly encountered in patients with PCa.[18] This study predicted a higher probability of recurrence than commonly used clinical staging criteria (i.e.,

Table 1.

Possible invention	Histology image screener	Metastatic cancer lymph node screen	Diagnostic validator	Histological bio-marker	Interactive phenotyping	Skin cancer diagnostic hand held device
	Histological bio-marker (MC)					(i.e. Malignant melanoma)
Date	2017	2017	2017	2016	2017	2017
Field	Anatomic pathology	Anatomic pathology	Anatomic pathology	Anatomic pathology	Anatomic pathology	Dermatology
Study	Brain and colon CA	Metastatic breast CA	Proliferative breast lesions	Lung Adeno- and squamous cell CA	Gliomas	Cutaneous skin cancer
Method	DL CNN	DL CNN	ML	ML	ML	DL CNN
Automatic	Yes	Yes	Yes	Yes	Yes	Yes
Medium	WSI	WSI	WSI	WSI	WSI	Digital Image
Application	Diagnostic	Lymph node screening	Diagnostic	Prognostic	Prognostic	Diagnostic
Ancillary Study	No	No	No	No	Yes	No
Reference	#13	#11	#26	#16	#35	#3
Author	Xu *et al.*	Liu *et al.*	Radiya-Dixit E, Zhu D and Beck AH	Kun-Hsing Yu *et al.*	Nalisnik M *et al.*	Esteva A *et al.*

Note: Abbreviations: DL = Deep Learning, CNN = Convolutional Neural Network, ML = Machine Learning, CA = Carcinoma, MC = Multiclass Classification, WSI = Whole Slide Images.

Gleason scoring, PSA level, age (AUC 0.81 (deep learning) vs. 0.59 (common clinical criteria)).[18] The study first applied a neural net to annotated nuclear morphology (PCa, adjacent stroma, and epithelial nuclei), while a second neural net was applied to annotated PCa tumor images (centered on the nuclei from stage 1) to learn a hierarchy of image features. This method was used to train both images from patients who had biochemical recurrence and no recurrence of PCa following radical prostatectomy. Test cases were then used to predict the probability of recurrence/non-recurrence. The authors note this methodology uses automatic feature extraction compared to the previous commonly used machine-learning methods applying more labor-intensive, hand-crafted feature extraction. The authors also emphasize their strategy of focusing on nuclear morphology that may often provide key predictive features of tumor grade and stage. This is important for their two-step deep learning approach, as the second net focuses on a centered nucleus in PCa tumor image patches in conjunction with trained clinical outcome data. The authors conclude that when validated, this method may be particularly useful given the difficulty in predicting clinical response to therapy in these mid-grade PCa patients, offering more effective treatment planning.[18]

Prognostic stromal features in cancer represent an emerging and exciting area of computational pathology, as historically pathology has relied primarily on epithelial morphology for grading cancer.[19] In PCa, tumor nuclear morphology as well as the adjacent architecture of the normal-appearing benign fields (i.e., the most predictive features of recurrence within the cancerous and benign fields or tumor associated with benign signature, TABS) has also been shown recently to improve prognostic modeling accuracy.[23] There could also be biological insight within these prognostic histomorphologic features and structures warranting further histological and correlative studies including molecular and -omics analysis. This discovery potential is explored below. The technological advancements in hardware and software along with advances in computational pathology and AI algorithms (machine-learning and deep learning) occurring within just the past few years have increased the likelihood of new diagnostic tools to aid the pathologist, along with new prognostic histopathological biomarkers that promise to help guide and improve precision medicine (Table 1).

IV. Big Data

Also of interest is the prospect of how the rapidly evolving functionality of AI will be applied to pathology in creative ways as big data continues to grow exponentially. Consider that more data is created each year than in all prior years of human history. The Internet of Things (IOT) will create even more data to leverage. IOT devices can be beacons of real-time status of body functions to fuel AI engines evaluating after-action care. Further, with the rise of personal wearable health technology, IOT data recorded prior to diagnosis and treatment can provide information that AI can statistically use in recommending additional therapy. Similarly, many exciting possibilities exist for data integration within an AI-driven world for pathology, including advances in three-dimensional scanning in WSI and data fusion algorithms between medical records, radiology, pathology, and –omics.[28] As advances in digitization have furthered the transformation of pathology, and while more data is being made machine readable (i.e., digital), the recording, enriching, and sharing of this data is a critical component of current advances.

Some might argue that this is in fact more important than the impact of AI. Combining data from seemingly unrelated sources and fields was historically rarely done because of the technical challenges and uncertainty of result. Now data are pouring into common platforms. AI statistical models can leverage platforms of disparate data to improve diagnostic results.[5] For example, IBM's medical imaging and analytics division, with Eyes of Watson, is leveraging clinical information in medical records to improve diagnosis in radiology through deep learning algorithms (i.e., mammography).[5]

V. Literature

Natural language processing (NLP) represents a field in computer science utilizing algorithms for human language processing encompassing conversion to a digital voice signal trained on massive datasets to understand language as it is spoken for speech recognition.[1] Natural language understanding (NLU), a subset of NLP for machine reading comprehension, is another boost to the medical data revolution, unlocking unstructured data (medical and biological data extracted from the literature including full-text articles), making it part of an AI data platform. The medical literature is being digitized, not just the words, which have been available for decades, but the meaning. In a 2016 study, The University of Ottawa reported that ~2.5 million new scientific papers are published each year, adding to the sum total of 50 million written since 1665.[30] The impossibility of reading and retaining all this information is obvious; yet this is what AI can do by sorting through large volumes of data in the literature to provide evidence that assists in decision-making and problem solving. For example, through Quest Diagnostics and Illumina, mutational analysis with next-generation sequencing (NGS), whole-genome DNA sequencing and RNA sequencing (RNA-Seq) are being offered using formalin-fixed paraffin-embedded tumor samples form patients, with IBM's Watson genomic analytics (WGA) further analyzing and annotating the results by reviewing large volumes of relevant information in the medical literature and clinical trials database resources.[4] With NGS data, WGA can prioritize somatic variants, copy number alterations, and differential gene expression (from RNA-Seq), identifying clinically actionable mutations in cancer-associated genes, and suggest available treatment options including on-going clinical trials.[4,31]

VI. Artificial Intelligence and Laser Capture Microdissection

It has been envisioned that AI pathology imaging may allow for a preliminary diagnosis in the future with an opportunity for automated tissue sampling soon after slides are cut, H&E stained, and scanned for WSI.[24] If achieved, triage of -omics, immunohistochemical staining, or ancillary testing could significantly improve costs and turnaround time in pathology.[24] A prototype for the application of semi-automated LCM in the field of computational pathology and WSI was demonstrated as SIVQ-aided LCM[32] with a protocol available on JoVE.[33] This system envisioned the potential of the vast biochemical surface area accessible in WSI for morphogenomics and morphoproteomics, demonstrating gene expression profiling down to the individual cell and nucleus level.[32] Combining machine-learned and AI-derived histomorphologic features and histological structures with Laser Capture Microdissection (LCM) may have potential to offer new insights and efficiencies in pathology.[36] Recovery and molecular analysis of precise histological targets could offer new molecular biological insights to disease, including cancer. For example, AI analysis and LCM sampling of histomorphologic targets for -omics studies using WSI of H&E-stained tissue sections in the clinical laboratory can provide increased sensitivity for identifying histomorphologic targets at high resolution, e.g., metastatic carcinoma cells during lymph node screening. Alternatively, discovery studies could focus on the histopathological progression of precancerous lesions and cancer staging (e.g., the extended Vogelgram of colon cancer or in prostate carcinoma). The discriminatory power of some of the newest AI diagnostic methods (using neural net techniques adapted from ImageNet) was demonstrated in colon cancer, not only thoroughly differentiating areas of tumor and normal (in binary classification) but also specific histomorphologic differentiation within each of several different colon cancer subtypes (in multiclass classification).[13] Recent state-of-the-art deep learning imaging methods offer discrimination of subtle (fine-grained) differences in complex histological patterns allowing for potential biological investigation across multiple cancer subtypes, demonstrated in colon, prostate, and breast carcinoma.[13,14,18,29]

Biological targets for LCM may identify different types of clinically relevant breast cancer that could not be appreciated prior to AI imaging diagnostics.[14,29] A recent investigation from the finalists for NVIDIA's 2017 Global Impact Award revealed two subtypes of breast carcinoma using AI imaging and molecular characterization, HER2-positive breast carcinoma and triple-negative breast carcinoma, in the same patient specimen.[14,29] This was not discernible by light microscopy examination, which detected only one dominant subtype. Heterogeneous patches containing mixtures of both subtypes were also identified on the AI imaging detectors (predicted heat maps) and confirmed by -omics analysis.[29] These common breast cancer subtypes require different therapies; thus, their identification would be critical to proper treatment of the patient. These studies highlight the potential for the improved sensitivity of the discriminatory capacity of AI pathology imaging combined with a larger addressable tumor volume in WSI for clinical and research applications.

As the cost of DNA (and RNA) sequencing decreases one or more orders of magnitude from the current trend of exponential decline, adoption of LCM into the clinical work-flow could further drive large scale genomics with AI-derived high-quality ground truth (benchmark) diagnostic and prognostic datasets in pathology. A study illustrating the potential of deep learning in genomics predicted survival risk factors in cancer with large scale integrated genomic profiling datasets from The Cancer Genome Atlas (TCGA).[48] This study identified clinical and molecular risk factors or features (e.g. mutations, RNA transcripts, proteins) matching many features from the WHO classification for diffuse gliomas, and new survival risk features with gene set and molecular pathways data analysis (e.g. TGFB signaling and epithelial mesencymal transition (EMT)-related transcription factors associated with poor survival).[48] The authors point out the need for yet even larger datasets for deep learning prognostic studies in genomics, as seen with improved performance in image classification (ImageNet) and language processing tasks (NLP).[1,48] To meet this challenge, large collaborators (e.g. the NIH, the Broad Institute, and Google) are teaming up for data science and cloud computing increasing the quantity of data and availability of software and computational power for AI in large and diverse cohort studies in health and disease.

Looking ahead, it will be interesting to investigate and define the extent of molecular subspecificity (molecular signatures) that can be discerned through -omics characterization of cancer progression using AI pathology imaging (diagnostic and prognostic). Of particular interest may be molecular analysis of the critically important early stages of disease such as carcinoma *in situ* (CIS), microinvasion, and early metastatic disease (lymphovascular invasion, tumor-associated stromal features, TABS).[34] One approach may include utilizing LCM for comparative molecular analysis in prognostic risk stratification with Kaplan–Meier survival analysis, already a component of prognostic imaging studies (addressable features in low-intermediate, and high-risk groups), to identify biological differences. For example, investigation with LCM may assess the molecular basis of highly predictive prognostic stromal features in breast cancer to identify biomarkers (candidate metastasis driver genes or proteins) within the tumor microenvironment during invasion and metastasis (the crossing of basement membranes during the transition from in situ to invasive carcinoma involving adhesion to the basement membrane, proteolytic degradation of the extracellular matrix, and invasion (de-adhesion, reversal of proteolysis and migratory, invasive or mesenchymal stem cell-like properties)).[19,46] As the molecular mechanisms of these critical steps are now being defined for anti-invasion therapies (in possible combination with anti-proliferative therapies) AI-derived phenotypes and LCM could potentially be used to further correlate molecular mechanisms in different cancer types and subtypes.[46]

An open-source interactive machine-learning model, HistomicsML, classifies histologic phenotypes (AI-derived histomorphologic features) through an active learning classification model between machine-learned features and a domain expert (i.e., pathologist).[35] HistomicsML allows for direct user feedback to improve classification performance of prognostic phenotypes through manageable navigation of large-scale digital images.[35] Classified image features or phenotypes can vary by abundance, morphology/morphometrics (cell, nuclear shape, color (intensity), texture, or histologic structure), and higher-level hierarchical features including spatial (proximity of cells, nuclei, stroma *e.g., relative border between spindled and round stromal nuclei), relational/contextual (between cell types, nuclei, histologic structures, *thin cords of epithelial tumor cells through stroma), or diffuse/global (immune cellular, inflammatory cells, *infiltrates in stroma).[19,35] Clinical and digital pathology imaging datasets from the Cancer Genome Atlas were integrated with HistomicsML (machine-learning) to accurately predict survival from endothelial phenotypes (subtle changes in microvasculature representing angiogenesis) and progression in patients with gliomas independent of grade and molecular subtype.[35] Thus, in addition to grade and molecular subtype, these angiogenic prognostic phenotypes may be important for treatment planning as well as genotype–phenotype associations.[35] The authors point out that these investigations are timely in the era of precision medicine whereby the genomic characterization (molecular basis) of novel prognostic phenotypes has not been defined in emergent genomic classification studies.[19,35] Furthermore, histology contains critical spatial and morphological information of biological and prognostic value (e.g., tumor/stroma microenvironments, immune response, angiogenesis) that may be difficult or impossible to ascertain through genomic assays alone.[35] A feature transparent machine-learning method for prognostic phenotyping (similar to C-Path described by Beck for prognostic stromal features associated with survival in breast cancer) capable of defining phenotypic boundaries (segmentation boundaries) with high accuracy (through active learning) may be critical to investigating phenotype–genotype associations in AI prognostic pathology studies (e.g., survival and response to therapy).[35]

HistomicsML and similar AI-imaging models may also link downstream -omics applications with LCM, originated for highly precise tissue- and morphology-based molecular pathology investigation.[19,35,36] For example, as novel prognostic phenotypes are identified, validated (by annotation), and categorized (i.e., stored and searched in a database), LCM could be used to microdissect prognostic histologic phenotypes for the increasing utilization of molecular analysis in clinical trials (i.e., RNA expression analysis, NGS, or -omics) and/or additional prognostic tumor profiling efforts for discovery.

(* = Highly statistically significant stromal features predictive of survival in breast cancer identified by C-Path machine-learning imaging system).[19]

VII. Liquid Biopsies: Correlation and Biomarker Discovery

Recent developments in liquid biopsy such as direct detection of circulating cell-free tumor DNA (ctDNA) of early-stage cancer for early diagnosis[37] or circulating tumor cells (CTCs) could be correlated with molecular analysis from precision histopathology, if biopsy tissue is available. Precision histopathology and molecular analysis may also offer potential for early intervention to guide therapy, as shown recently with the liquid biopsy.[38] Prognostic AI-derived histomorphologic features represent a type of surrogate biomarker for a clinical outcome in a similar manner as somatic mutation biomarkers are used in liquid biopsy and molecular pathology. If prognostic histomorphologic markers in AI pathology can accurately predict how tumors behave clinically, they may yield biomarkers (e.g., somatic mutation biomarkers directly linked to pathogenesis). Thus, along with the collection of clinical follow-up data for urgently needed AI prognostic studies for continued improvements in treatment planning, the technical advances in AI pathology and WSI discussed may have potential for biomarker discovery, for example, in the pursuit of critical early-stage cancer or occult metastatic disease, which is perhaps more vulnerable to modern therapy such as targeted therapies and/or immunotherapies (fewer drug-resistant clones prior to therapy).[34]

Using Illumina's TruSight (TM) Tumor NGS in large sequencing panels of hundreds of cancer-associated genes (170 genes presently, 500 genes in development), in conjunction with Watson Genomic Analytics (WGA), represents a potentially rapid, high-throughput approach for the detection of clinically actionable mutations within cancer-associated genes to suggest available treatment options in clinical trials.[4,31,39] This compares to use of targeted sequencing panels with high redundancy of known clinically actionable mutations available in molecular diagnostic laboratories.[31,40] Genomic analysis requires manual and semi-automated methods of compilation and prioritization of somatic variant data (e.g., searching structured databases).

Determining between clinically significant mutations (in cancer driver genes) or inconsequential variants (passenger mutations) requires statistical evaluation in large patient cohorts. Recent developments highlight the consideration of computational variables of the underlying cancer mutation rates (affecting specificity, false-positive calls) in large-scale sequencing studies.[41] A recent study comparing machine and human genomics analysis showed significant time saving using WGA (10 minutes) compared to a bioinformatics team (160-person hours).[31] Both groups provided a comparable high-quality analysis for available clinically actionable therapies in a patient with brain cancer. The authors note that machine analytics could alleviate a bottleneck in cancer genomics for the practical application of NGS in clinical practice.[31] Cancer researchers are also performing similar studies now on conventional light microscopy samples of tumors, validating Illumina NGS panels and WGA to determine new genes involved in cancer pathogenesis.[4,39] AI pathology-derived histomorphologic samples, both diagnostic and prognostic, may also be informative in these studies. Eventually, molecular analysis of histomorphologic markers could then help connect the dots in parallel to testing the liquid biopsy, providing an improved road map for staging, in particular, as clinical outcome data can be integrated with more prognostic AI pathology imaging studies. These efforts may complement efforts by innovative companies such as Personal Genome Diagnostics (PGDX), and new entrants such as Grail, which join the field for earlier diagnosis, treatment, and prevention.[37,42,43] These concepts highlight the potential of the molecular characterization of machine-learned and AI-derived histomorphologic biomarkers (including quantitative features, semi-quantitative, and qualitative histological structures).

VIII. Challenges Ahead

Recent developments in AI have enormous potential benefits to healthcare, but there are also immense challenges ahead. For example, AI pathology imaging methods require validation of qualitative and quantitative differences within imaging datasets across institutions (independent data set validation) and within daily practice.[18,23,28,44] Competitive annual image challenges are offering innovative methods with greater yet diverse levels of performance (classification and segmentation, color normalization) and functionality (two-class, multiclass classification).[13,18,28] These methods require standardization and regulatory approval before promising applications such as high-sensitivity lymph node screening or validation tools for challenging diagnostic applications (proliferative lesions of the breast) can be implemented in pathology.[11,12,25,26] Robust methods with the ability to scale WSI data sets are needed, as well as strategies for high-quality and abundant ground truth annotation. ImageNet, with millions of images and identified objects, offers a promising model for AI pathology,[27] already providing highly successful translatable techniques, despite the challenges of limited image data sets with rare disease types and highly complex, large-scale images in pathology.[13]

Further confirmatory results in prognostic studies are needed to define the range of accuracy seen in early studies for continued improvements in treatment planning.[16,18,19,23] Automated sampling for molecular analysis in AI pathology requires developing software in WSI for feasibility studies, while advances in machine analytics for genomics require completion of clinical validation and further studies of candidate genes associated with cancer progression. Clinical evidence for defining molecular sub specificity by improving clinical outcome and survival within clinical trials is needed, similar to existing targeted therapies, as new drugs are developed. This strategy represents a major focus of cancer research following The Cancer Genome Atlas (TCGA), a U.S. program to genetically profile 10,000 tumors.[45] Clinical translation of tumor profiling may require incorporating functional models (the role of tumor-initiating cancer stem cells and therapeutic strategies such as combining conventional, targeted, and immunotherapies in responsive cancer types and patient populations).[45]

Despite these challenges, the rewards are likely to be great, as deep learning methods gain traction and participants such as PathAI show confidence in bringing imaging applications forward in pathology, in particular, to areas with limited expertise.[7] AI represents a nascent, rapidly growing field with broad reach.[1] Attempting to connect several types of AI (imaging, bioinformatics, genomics, analytics, drug discovery) along with developments in traditionally guarded clinical data management will also benefit precision medicine. Most important to implementing AI systems with pathology imaging and genomics (including cancer genomics) will be the individuals, teams, and institutions that integrate computational, medical, and scientific disciplines together.

References

1. LeCun Y, Bengio Y, Hinton G. Deep learning. *Nature* 2015; **521**(7553):436–44. doi:10.1038/nature14539.
2. Ma J, Sheridan RP, Liaw A, Dahl GE, Svetnik V. Deep neural nets as a method for quantitative structure-activity relationships. *J Chem Inf Model.* 2015; **55**(2): 263–74, doi: 10.1021/ci500747n. Epub 2015 Feb17.
3. Esteva A, Kuprel B, Novoa RA, Ko J, Swetter SM, Blau HM, Thrun S. Dermatologist-level classification of skin cancer with deep neural networks. *Nature* (London, United Kingdom) 2017; **542**(7639): 115–18. doi. 10.1038/nature21056. Epub 2017 Jan 25.
4. Watson Genomics, https://www.ibm.com/watson/health/oncology-and-genomics/
5. Health Informatics: IBM Research, Eyes of Watson, https://www.research.ibm.com/haifa/dept/imt/mia.shtml
6. Modern Medicine, https://www.modmed.com/company/
7. PathAI, https://pathai.com/
8. Proscia, https://proscia.com/

9. Roadmap Epigenomics Consortium, Kundaje A., Meuleman W, et al. Integrative analysis of 111 reference human epigenomes. Nature. 2015;518(7539):317-30. doi: 10.1038/nature14248.
10. Deep Genomics, https://www.deepgenomics.com/
11. Liu Y, Gadepalli K, Norouzi M, Dahl GE, Kohlberger T, Boyko A, Venugopalan S, Timofeev A, Nelson PQ, Corrado GS, Hipp JD, Peng L, Stumpe MC. Detecting Cancer metastases on gigapixel pathology images, replace, with . arXiv:1703.02442.
12. Wang D, Khosla A, Gargeya R, Irshad H, Beck A. Deep learning for identifying metastatic breast cancer. arXiv:1606.05718.
13. Xu Y, Jia Z, Wang L-B, Ai Y, Zhang F, Lai M, I-Chao Chang E. Large scale tissue histopathology image classification, segmentation, and visualization via deep convolutional activation features. *BMC Bioinformatics*. 2017; **18**: 281, doi: 10.1186/s12859-017-1685-x.
14. Nvidia 2017 Competition,. https://blogs.nvidia.com/blog/2017/04/03/ai-pathologist-helps-zero-in-on-correct-cancer-diagnosis/
15. The MICCAI 2014 Brain Tumor Digital Pathology Challenge https://wiki.cancerimagingarchive.net/display/Public/MICCAI+2014+Grand+Challenges
16. Yu K-H, Zhang C, Berry GJ, Altman RB, Ré C, Rubin DL, Snyder M. Predicting non-small cell lung cancer prognosis by fully automated microscopic pathology image features. *Nature Commun*. 2016; **7**: 12474, doi:10.1038/ncomms12474.
17. Casanova R, Xia D, Rulle U, Nanni P, Grossmann J, Vrugt B, Wettstein R, Ballester R, Astolfo A, Weder W, Moch H, Stampanoni M, Beck AH, Soltermann A. Morphoproteomic Characterization of lung squamous cell carcinoma fragmentation, a histological marker of increased tumor invasiveness. *Cancer Res*. 2017; **77**(10): 2585–93, doi: 10.1158/0008-5472.
18. Kumar N, Verma R, Arora A, Kumar A, Gupta S, Sethi A, Gann PH. Convolutional neural networks for prostate cancer recurrence prediction In Medical Imaging 2017: Digital Pathology; (Vol. 10140). [101400H] SPIE. doi: 10.1117/12.2255774.
19. Beck AH *et al*. Systematic analysis of breast cancer morphology uncovers stromal features associated with survival. *Sci. Transl. Med*. 2011; **3**: 108ra113, DOI: 10.1126/scitranslmed.3002564.
20. Sertel O *et al*. Computer-aided prognosis of neuroblastoma on whole-slide images: Classification of stromal development. *Pattern Recognit*. 2009; **42**: 1093–103, https://doi.org/10.1016/j.patcog.2008.08.027.
21. Sertel O, Kong J, Catalyurek UV *et al*. *J Sign Process Syst Sign Image Video Technol* 2009; **55**: 169, https://doi.org/10.1007/s11265-008-0201-y.
22. Sabo E *et al*. Computerized morphometry as an aid in determining the grade of dysplasia and progression to adenocarcinoma in Barrett's esophagus. *Lab. Invest*. 2006; **86**: 1261–71, DOI: 10.1038/labinvest.3700481.
23. Lee G, Veltri RW, Zhu G, Ali S, Epstein JI, Madabhushi A nuclear shape and architecture in benign fields predict biochemical recurrence in prostate cancer patients

following radical prostatectomy: Preliminary findings. *Eur Urol Focus*. 2016; pii:S2405–4569(16)30060–8, doi: 10.1016/j.euf.2016.05.009.

24. Djuric U, Zadeh G, Aldape K, Diamandis P. Precision histology: How deep learning is poised to revitalize histomorphology for personalized cancer care. *NPJ Precision Oncology*, 2017; **22**, doi: 10.1038/s41698-017-0022-1.
25. Dong F, Irshad H, Oh E-Y, Lerwill MF, Brachtel EF, Jones NC, Knoblauch NW, Montaser-Kouhsari L, Johnson NB, Rao LKF, Faulkner-Jones B, Wilbur DC, Schnitt S, Beck AH. Computational pathology to discriminate benign from malignant intraductal proliferations of the breast. *PLoS One*. 2014; **9**(12): doi: 10.1371/journal.pone.0114885.
26. Radiya-Dixit E, Zhu D, Beck AH. Automated classification of benign and malignant proliferative breast lesions. *Sci Rep*. 2017; **7**(1): 9900. doi: 10.1038/s41598-017-10324-y.
27. ImageNet, http://www.image-net.org/.
28. Madabhushi A, Lee G. Image analysis and machine learning in digital pathology: Challenges and opportunities. *Med Image Anal*. 2016; **33**: 170–5, doi: 10.1016/j.media.2016.06.037.
29. Verma R, Kumar N, Sethi A, Gann PH. Detecting multiple sub-types of breast cancer in a single patient. 2016 *IEEE International Conference*. Image Processing (ICIP), 2016 — Proceedings (Vol. 2016-August, pp. 2648-2652). [7532839] IEEE Computer Society. doi: 10.1109/ICIP.2016.7532839.
30. Jinha A. Article 50 million: An estimate of the number of scholarly articles in existence. Learned Publishing. 2010; **23**: 258–63, doi: 10.1087/20100308.
31. Wrzeszczynski KO, Frank MO, Koyama T, *et al*. Comparing sequencing assays and human-machine analyses in actionable genomics for glioblastoma. *Neurology: Genetics*. 2017; **3**(4): e164. doi:10.1212/NXG.0000000000000164. eCollection 2017 Aug.
32. Hipp J, Cheng J, Hanson JC *et al*. SIVQ-aided laser capture microdissection: A tool for high-throughput expression profiling. *J. Pathol. Inform*. 2011; **2**: 19, doi:10.4103/2153-3539.78500.
33. Hipp JD, Cheng J, Hanson JC, Rosenberg AZ, Emmert-Buck MR, Tangrea MA, Balis UJ. SIVQ-LCM Protocol for the ArcturusXT Instrument. *J. Visual. Exp*. 2014; **89**: 51662, doi:10.3791/51662.
34. Bert V, Lecture Video, *The Earlier Detection of Cancer, Summer Symposium*, The Koch Institute, MIT, 2016, https://www.youtube.com/watch?v=43N75VjQ708.
35. Nalisnik M, Amgad M, Lee S, Halani SH, Velazquez Vega JE, Brat DJ, Gutman DA, Cooper LAD Interactive phenotyping of large-scale histology imaging data with HistomicsML. *Sci Rep*. 2017; **7**(1): 14588, doi: 10.1038/s41598-017-15092-3.
36. Emmert-Buck MR, Bonner RF, Smith PD, Chuaqui RF, Zhuang Z, Goldstein SR, Weiss RA, Liotta LA. Laser capture microdissection. *Science* 1996; **274**(5289): 998–1001.
37. Phallen J. *et al*. "Direct detection of early-stage cancers using circulating tumor DNA," *Science Translational Medicine* 2017; **9**(403): pii:eaan2415, doi 10.1126/scitranslmed.aan2415.

38. Goodall J. *et al.* TOPARP-A investigators. Circulating cell-free DNA to guide prostate cancer treatment with PARP inhibition. *Cancer Discov.* 2017; **7**(9): 1006–17. doi: 10.1158/2159-8290.CD-17-0261. Epub 2017 Apr27.
39. Validation of Watson for Genomics in Cancer Research, https://www.sciencedaily.com/releases/2017/08/170807110400.htm
40. Begg CB, Ostrovnaya I, Geyer FC, Papanastasiou AD, Ng CKY, Sakr RA, Bernstein JL, Burke KA, King TA, Piscuoglio S, Mauguen A, Orlow I, Weigelt B, Seshan VE, Morrow M, Reis-Filho JS. Contralateral breast cancers: Independent cancers or metastases?. *Int. J. Cancer.* 2017, doi: 10.1002/ijc.31051.
41. Tokheim CJ, Papadopoulos N, Kinzler KW, Vogelstein B, Karchin R. Evaluating the evaluation of cancer driver genes. *Proc. Nat. Acad. Sci. USA*. 2016; **113**(50): 14330–335, doi:10.1073/pnas.1616440113.
42. Personal Genome Diagnostics (PGDX), http://www.personalgenome.com/
43. GRAIL, Detecting Cancer Early, When It Can Be Cured https://grail.com/about/
44. Cruz-Roa A, Gilmore H, Basavanhally A *et al.* Accurate and reproducible invasive breast cancer detection in whole-slide images: A Deep Learning approach for quantifying tumor extent. *Scientific Reports*. 2017; **7**: 46450. doi:10.1038/srep46450.
45. Heidi L. End of cancer-genome project prompts rethink. Geneticists debate whether focus should shift from sequencing genomes to analysing function. *Nature* 2015; **517**: 128–9, doi:10.1038/517128a.
46. Liotta LA. Adhere, Degrade, and move: The Three-step model of invasion. *Cancer Res*. 2016; **76**(11): 3115–7, doi:10.1158/0008-5472.CAN-16-1297.
47. Mobadersany P, Yousefi S, Amgad M, *et al.* Predicting cancer outcomes from histology and genomics using convolutional networks. Proceedings of the National Academy of Sciences Mar 2018, 201717139; doi: 10.1073/pnas.1717139115
48. Yousefi S, Amrollahi F, Amgad M, *et al.* Predicting clinical outcomes from large scale cancer genomic profiles with deep survival models. Sci Rep. 2017; **7**(1):11707, doi: 10.1038/s41598-017-11817-6

Molecular Analysis — Issues, Challenges, and Successes

9 Applying High-Throughput Molecular Techniques to FFPE Clinical Tissues — DNA and mRNA

Chih-Jian Lih,* Hala Makhlouf,† P. Mickey Williams,* Scott M. Lawrence,‡ and Rodrigo Chuaqui†

*Molecular Characterization Laboratory, Frederick National Laboratory for Cancer Research
Frederick, MD, USA
†Cancer Diagnosis Program (CDP)
Division of Cancer Treatment and Diagnosis (DCTD)
National Cancer Institute
Rockville, MD, USA
‡Pharmacodynamics Assay Section (PADIS)
Laboratory of Human Toxicology and Pharmacology
Applied/Developmental Research Directorate
Leidos Biomedical Research, Inc.
Frederick National Laboratory for Cancer Research
Frederick, MD, USA

Contents

I. Introduction

Precision medicine encompasses the use of genomic, epigenomic, and other data to define individual molecular patterns of disease, potentially leading to better individualized treatment.[1] In addition, correlative science studies evaluate the relationship between these molecular biology data (i.e., biomarkers such as genes and proteins) and clinical outcomes (i.e., disease progression or response to therapy). Thus, bridging the gap between basic researchers and clinical investigators to advance research on clinical correlations is key to improving therapeutic approaches. Precision medicine is a major goal at the National Cancer Institute (NCI). Two exciting new initiatives include the NCI Molecular Analysis for Therapy Choice Program (MATCH) & Pediatric MATCH trial and the Exceptional Responders Initiative (ERI) (http://deainfo.nci.nih.gov/advisory/ctac/1113/

Precision%20Cancer%20Medicine.pdf). The NCI MATCH Trial (http://www.cancer.gov/about-cancer/treatment/clinical-trials/nci-supported/nci-match) includes small, phase II trials enrolling adults with advanced solid tumors and lymphomas whose tumors are no longer responding to standard therapy. Biopsies from tumors from as many as 3,000 patients will undergo next-generation DNA sequencing (NGS) to identify individuals whose tumors have genetic abnormalities that may respond to selected targeted drugs. Exceptional responders (http://www.cancer.gov/news-events/press-releases/2014/ExceptionalRespondersQandA) are patients who have a unique response to treatments that are not effective for most other patients. The goal is to analyze their samples to understand the molecular bases of exceptional responses to treatment, primarily via chemotherapy, in cancer patients.

Correlative studies in clinical trials can either be "integral" or "integrated".[2] Integral assays are those that must be performed before the clinical trial can proceed. The results of these assays may inform about eligibility, stratification, or treatment assignment. Integrated assays for research are used for studying "proof of mechanism", such as to assess whether the drug has affected its target. The results of such assays will not influence the treatment of patients in the trial. In addition, retrospective studies can be conducted after the trial goals have been met, and they benefit from the clinical data collected at the trial, such as the treatment regimens, outcomes, toxicity, etc. The molecular analysis of the patients' samples is essential in all these types of studies. Thus, analyzing clinical samples becomes a central aspect in the development and validation process of clinical markers and assays.

However, applying molecular techniques to clinical samples presents several challenges.[3] Typical clinical samples, such as biopsies, provide limited amounts of material; they present a degree of biomolecule degradation due to clinical handling and processing, and human tissues are heterogeneous (including tumor, normal, inflammatory cell populations, and necrotic material, among others). All these issues may lead to difficulties in designing appropriate controls and reproducing results, especially when compared to other experimental models such as cell lines or animal models. Additional confounding factors are the variability within the types of tissues and samples, as well as the range of conditions under which they are procured in the clinical setting, all variables, which from the experimental perspective are difficult to control.

One of the most difficult challenges when applying molecular techniques to clinical samples is defining quality and quantity thresholds the samples need to meet to obtain consistent technical results. Some of these difficulties derive from the following facts: the quality needed varies depending on the biomolecule analyzed (DNA, RNA, proteins) and the protocol implemented or the sensitivity/specificity of the kit utilized. Each platform (i.e., mRNA microarrays vs. NanoString) might have different thresholds for sample quality and input. Even within the same platform, the sensitivity/specificity might vary according to the individual signature that is being studied. A major hurdle is that quality assessment (QA) and control (QC) tools are not standardized between laboratories. Betsou *et al.*[4] reviewed over 600 publications using QA/QC tools for their samples. Included were studies in a clinical setting, studies in research settings, as well as studies in environmental biospecimen science. Elements extracted from each publication included, among others: type of sample (liquid/solid), a set of pre-analytical variables and their ranges, QC tools (markers) and their ranges, whether the method was quantitative or semi-quantitative, simple or multiplex, as well as the reference controls used. The most readily applicable tools were those where a threshold for the pre-analytical variation and a reference interval of the QC analyte are known. Few meaningful markers that met these criteria were identified. Usually the sample quality (and quantity) thresholds are set up for each particular project individually, and this process is mostly empirical, strongly relying on previous experience of each researcher group.

Therefore, the quality assessment and control of clinical samples is a crucial step before the downstream molecular techniques are applied. Quality assessment of the histologic material used in downstream molecular analysis presents various challenges, as will be discussed in the present chapter. A key to success is the close work and communication between the pathologist and the molecular team to understand each other's language and goals. Also included will be an overview of the molecular techniques that are currently being applied routinely on formalin-fixed, paraffin-embedded tissues (FFPE).

II. Histology Assessment of Clinical Tissue Samples for Downstream Analysis

a. Overview

Histological assessments play an important role in the QA and QC of clinical tissue samples.[5] Successful molecular oncology testing depends on the accurate assessment of malignant cell percentages in the analyzed tissue samples from fresh, frozen, or FFPE tissues. Some studies showed that only 10% of the samples are unsuitable for molecular analysis because of sampling problems with the tissue (e.g., no tumor, insufficient tumor cells, and insufficient viable tumor cells).[6–8] Contamination with non-tumor cells can significantly affect the ability to identify molecular tumor signatures in clinical tissue samples.[9] Therefore, an effective method to assess the tumor cell content is essential prior to the analysis of clinical tissue samples.[10]

b. Specimen quality/pre-analytics for molecular testing

QA/QC reviews can have a significant impact in molecular research. Since the advent of precision medicine-biopsy driven clinical trials, NCI performs a high volume of biomarker testing. Most are genomic testing. Such trials depend on adequate tissue collection, availability of quality-controlled, high-throughput molecular assays, and the ability to screen large numbers of tumors to find those with the desired molecular alterations.[11] Accurate pre-analytical assessment of precious clinical samples is key to guarantee a wise use of the sample. Various specific challenges are faced when analyzing clinical samples such as low quantity or quality of the material, cellular heterogeneity, control and reproducibility issues, etc.

Typically, QA/QC evaluation involves assessing three variables: quality, quantity, and purity of the material. Limited information on the molecular quality of a sample is obtained from microscope analysis solely (e.g., some processing artifacts or lysis of the cells), and mostly solicited from the history of sample processing. Quantity and purity include specifying the percent tumor surface area or the percent of tumor cells in the section, along with the percent necrosis/fibrosis or other elements such as mucin formation. More important is the determination of the percent of tumor nuclei or the percentage of neoplastic cells within the tumor/ selected area, or the so-called "purity". This is necessary because a tumor may be infiltrated by a large proportion of inflammatory cells, which may have impact on some assays. It is useful to mark the tumor area within the tissue section in order to enrich the specimen by macrodissection (Figure 1). When most of the sample is involved by the tumor, this is usually

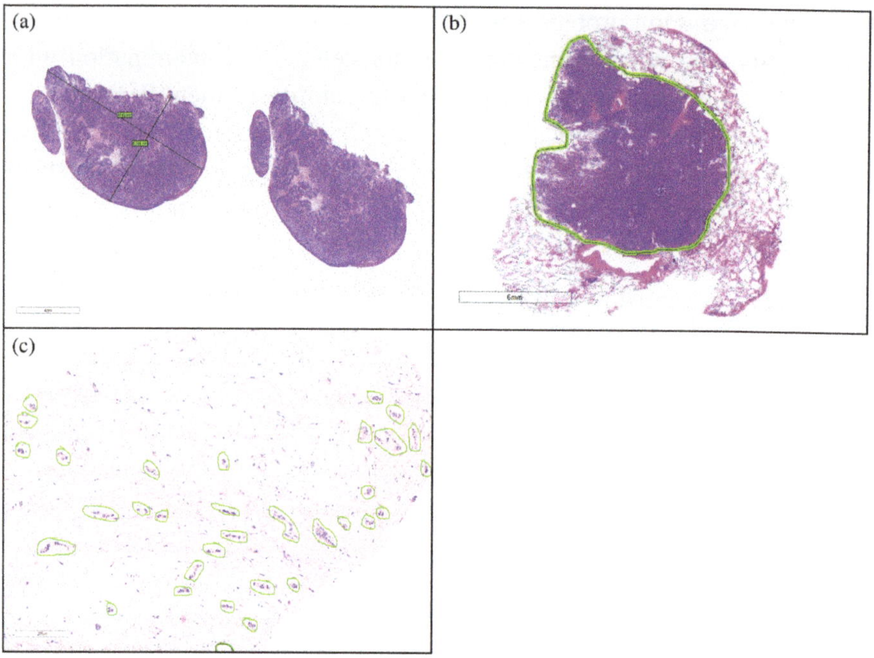

Figure 1. Three histologic tumor patterns often encountered when assessing tumor content in a tissue section. Each scenario may imply different approaches for using the tissue for molecular downstream analysis. (a) Most of the section is occupied by tumor, and the tumor shows low percentage of necrosis or little inflammatory infiltrate. Even encircling the tumor itself might not be necessary for determining what area to use for biomolecule extraction. (b) Well-defined tumor nodule within a larger tissue section. Encircling the tumor nodule for macrodissection (trimming off surrounding non-tumor tissue) might be necessary to reach purity thresholds required for downstream analysis. (c) As is frequently the case, the tumor infiltrates in small foci throughout the section, admixed with stroma, normal cells, or inflammatory infiltrates. Microdissection might be needed for downstream analysis.

not necessary. Occasionally, however, Laser Capture Microdissection (LCM) is needed when the tumor cells are fewer and/or intermingled with surrounding normal or inflammatory cells. The technique, however, is costly, laborious, and has low yield. Specimens with poor tumor cellularity may lead to false negative results, particularly if assays with lower analytic sensitivity are implemented. Accurate evaluation of tumor cellularity not only ensures sufficient material for required analytic sensitivity but also provides an independent QA measure of the molecular assays.[12]

A successful strategy is to perform histological quality assessment by more than one pathologist who reviews the respective H&E sections of the tumor tissue samples. Tumor cell content is evaluated by the cellular percentage. When a consensus is reached during the evaluation, the average percentage of tumor cells for each tumor tissue sample is reported by the pathologists. Discrepancies may arise on how pathologists determine tumor cellularity. When assessing tumor percent, a good communication between pathologists and the molecular team is essential. Pathologists must clarify whether the percent of tumor cells refers to the total specimen, or more frequently, to an encircled area of the tissue containing the tumor focus. It is not surprising that pathologists may overestimate tumor cellularity; a sheet of large cancer cells, for instance, is visually more daunting than a small lymphocyte, yet the total DNA in both tumor cell and lymphocyte may be about the same. Assays with higher analytic sensitivity will help identify mutations in specimens with lower tumor cellularity; NGS platforms may be ideally suited for this purpose.[12,13]

Reasons for histopathology QA/QC failures include samples with too little viable tumor, high degree of necrosis, small sample size, or non-cancer tissue.[14] Most tissues containing tumor also contain a mixture of cell types such as non-neoplastic epithelial cells, mesenchymal tissue, inflammatory cells and acellular material such as mucin, all of which have an influence on subsequent processes including nucleic acid isolation, PCR amplification, and NGS.[15] Therefore, in most studies it is important to determine the tumor nuclei content by visually/semi-quantitatively evaluating the percentage of tumor cells among all cells. If that falls below a certain threshold, it may be helpful to enrich the tumor cell content of the sample by macrodissection or microdissection (i.e., LCM) in order to (a) make the sample suitable depending on the sensitivity of the test and

(b) make the molecular analysis as broad as possible to be able to identify "clonal disease".[16]

Macrodissection is a technique that can be used to enrich for the desired tumor cells. It aims to increase the proportion of neoplastic cell nuclei generally by removing non-tumor-containing regions, not only facilitating the detection of a relevant mutation if it exists but also helping assure that a mutant signature originates from the isolated malignant cells. Microdissection, however, enriches samples by directly procuring the cellular component of interest and excludes nondesirable elements, such as necrotic areas, blood, or contaminating normal cells.

Although sensitivity of molecular tests is highly variable, accurate determination of neoplastic cellularity is critical for achieving reliable results. Specimens with poor tumor cellularity may lead to false-negative results, particularly if assays with lower analytic sensitivity, such as Sanger sequencing, are used to detect the mutation.[15,17] Some assays, like Sanger or NGS methods may require that malignant cells represent between 25% and 40% of the total cellular content.[18] In one study, it was determined that tissue was of sufficient quality and quantity to conduct NGS when it met the following criteria: greater than or equal to 50% tumor nuclei and 50% or less necrosis was selected.[19]

The NCI-MATCH clinical trial requires a tumor biopsy before enrollment for targeted NGS supplemented by immunohistochemical or fluorescence *in situ* hybridization (FISH) assays to assign treatment (http://www.cancer.gov/about-cancer/treatment/clinical-trials/nci-supported/nci-match). The MPACT (Molecular Profiling-based Assignment of Cancer Therapy trials, (https://clinicaltrials.gov/ct2/show/NCT01827384) aims to guide specific treatment options for specific tumor gene variations. In both trials, tumor content requirements was set at >50%. In cases where the tumor content is less than 50%, pathologists mark the area for tumor enrichment via macrodissection. In some projects, only a few cases can meet these tumor percent requirements. Reducing the tumor requirements is sometimes necessary to recruit the desired number of cases in a project. For example, The Cancer Genome Atlas (TCGA) project initially set a quality requirement of 80% tumor nuclei of samples included in the cohort.[9] However, this threshold was later reduced to 60%. An overestimation of the percentage of malignant cells could lead to false-negative results if the true percentage of malignant cells in the specimen fell below the particular assay's analytic sensitivity and could subsequently lead to incorrect selection of therapy. Estimation of percent neoplastic cellularity can also affect care in other diagnostic settings, including assessment of copy number variation, chromosomal translocation, determination of residual cancer burden after neoadjuvant therapy, and other genomic aberrations that can be affected by contaminating normal DNA.

Absolute quantitation, however, is not typically required for pharmacodynamic (PD) immunoassays, as the most important data reported from these assays are comparisons of the pre-dose and post-dose clinical samples. NCI's Division of Cancer Treatment and Diagnosis (DCTD) develops and validates PD assays to obtain accurate information about drug effect on intended molecular targets in first-in-human clinical trials and inform clinical development decisions.[20] The most fundamental issue we encountered when working to address the effects of biopsy variability and heterogeneity on PD measurements was the high rate of biopsies with insufficient material for analysis. Initially, sufficiency was defined by the tumor biopsy pair (before and after treatment) and both tissues were required to have sufficient tumor content to provide for a minimum of three 20× image fields of approximately 80% viable tissue. The analysis initially required a minimum of three slides to be analyzed, separated by a minimum of 5 μm to avoid evaluating the same cells more than once, so that quantitative biomarker data was reported from a minimum of nine quality image fields of both the pre-dose and post-dose tissue pairs. The initial pathology review criteria implemented for sufficiency vs. insufficiency of the tumor biopsy pairs led to a failure rate of over 60% for the first clinical trial in which the PD assay was used. After a review of the tissue failures and an improved understanding of the high frequency of heterogeneous specimens, fundamental changes were made to the tissue QC requirements. The minimum image field requirement per tissue section was decreased to one, with successful analyzability and reporting of PD data depending on analysis of at least one image in each of eight slides separated by a minimum of 5 μm. These changes significantly improved the success rate of PD biomarker reporting. Again, a key to this improvement was the diligent communication among the pathologists, assay operators, and scientific team to reach a common understanding of each other's objectives.

As can be seen in the few examples shown above, setting quality thresholds for samples in order to perform a given assay or identify a marker in clinical samples is a difficult process. The thresholds are usually set empirically by each molecular laboratory according to their experience, and it will depend on a variety of factors, including: is it a DNA, RNA, or a protein assay; is the assay quantitative or on/off; what reagents are being used; what tissue type is being analyzed; is it a uniplex or multiplex assay; etc.

c. Automatic histology assessment of tissues

Few methods exist for determining the percentage of malignant nuclei. Counting the total nuclei in a tissue section is extremely time consuming and not practical for clinical use. Estimation of the percentage of malignant nuclei by the pathologists may not be always precise or accurate. This method has been shown to be highly subjective and results in a low level of inter-observer reproducibility. Previous studies have evaluated pathologist agreement when estimating percentage of malignant cells (concordance) and have illustrated the difficulty in using visual assessment to quantify absolute numbers of cells. These studies were relatively limited and included small numbers of cases, and in particular small numbers of participants.[12,21] In contrast, automated digital image analysis promises objective and reproducible determination of tumor cellularity for molecular testing.[21–23]

Image analysis software provides a tool to analyze a multitude of different components within tissues, from large, macro-structures including stroma, tumor, and necrosis through minute structures within cells including nuclei and cellular membranes. When the software is combined with whole-slide digital imaging, analysis workflows are significantly streamlined to quickly generate answers for straightforward questions (amount of necrosis) and providing insights into more complex relationships (spatial distance between distinct cell populations). At the core of the analysis is the segmentation of phenotypic structures by morphological features and/or by the use of phenotyping biomarkers while scoring different metrics and relationships between/across classifications and maintaining objective criteria.

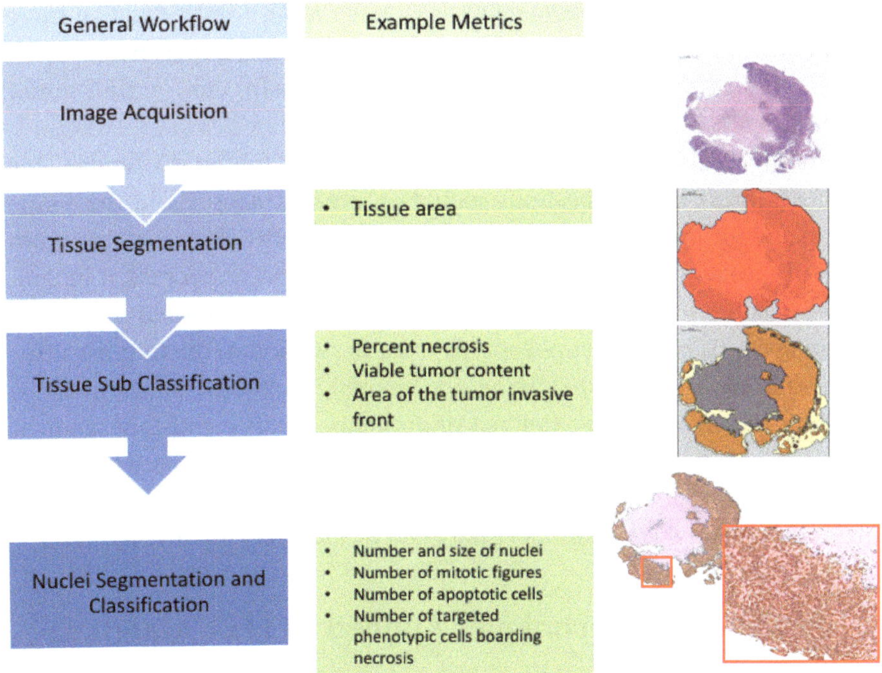

Figure 2. General workflow of an image analysis software.

Examples of image analysis software commonly used include Definiens, Matlab, PathXL Tissuemark, Leica/Aperio Analysis, Visiopharm, and MediaCybernetics Image-Pro, and many microscope/scanner vendors also have their own homegrown packages as well. The general workflow of an image analysis software is shown in Figure 2. Typically, the first round of segmentation involves separating the tissue from the surrounding glass in the image. Once the tissue(s) are classified as such, further segmentation within the tissue class can result in sub-classification into areas of viable tumor, necrosis, stroma, and normal tissue. Much of the sub-classification can be done exclusively on morphological features but can be made more robust with the addition of specific phenotyping markers for the regions of interest. For instance, separating viable tumor from stroma can be done by selecting for the density of hematoxylin-stained cells in an H&E slide scan, but with the addition of tumor-specific markers, segmentation is made more robust and can more accurately segment the tumor population with less supervision. After tissue sub-classification, the cells (nuclei, cytoplasm, etc.) within tumor subclasses can be segmented and further classified.

Hierarchical segmentation and classification within Definiens can quickly address broad questions including percent necrosis in the tissue sample and the amount of viable tumor to stroma ratio. With further exploration of the sample, more complicated questions can be addressed including what is the number of mitotic figures in tumor vs. normal tissue and how many cells of a target phenotype exist in relation to the tumor's invasive front. With validated staining and imaging methods, image analysis software can be utilized to conduct preliminary screening of thousands of samples or used to discover novel phenotypic characterization when combined with Big Data modeling algorithms. The granularity and depth of the analysis is readily scaled to meet the need of the users and the research.

III. Molecular Techniques Applied to FFPE Materials

Major challenges of using FFPE samples for molecular profiling include low quantity and poor quality of the biomolecules, as discussed above. Nucleic acids are vulnerable to treatments during tissue processing, endogenous nuclease digestion, physical conditions (heat), chemical modifications (deamination), and cross-linking to other molecules, all of which dramatically affect their integrity and create challenges for subsequent molecular analysis. FFPE nucleic acid samples are typically fragmented into small pieces (DNA 150–250 bp, RNA tends to be smaller), and this is an essential factor to be considered in the design of molecular assays. The quality of FFPE nucleic acid samples varies significantly when using clinical samples, depending on the tissue type used, preparation factors such as warm ischemia time (clamping time during surgery), cold ischemia time (from the excision to fixation), fixative reagents used, extraction methods, etc. It is believed that FFPE tissue samples are stable over long periods; however, the quality of nucleic acids can be affected by the storage time and conditions. Also, as mentioned above, the heterogeneity in both histology and genetics usually adds another layer of complexity, which needs to be addressed in the assay system.

Many molecular assays have been developed and used for various clinical needs, both for frozen as well as fixed tissues. Recently, an excellent review comparing the accuracy of FFPE and matched frozen tissues across multiple assay platforms has been published.[24]

In this section, we will focus on a number of major molecular technology platforms to study nucleic acids in fixed material and how to overcome the challenges that these samples present for molecular studies, emphasizing the advantages and limitations of the use of archival material for downstream molecular analysis.

a. Sanger sequencing

Sequencing PCR amplicons from FFPE DNA by dye termination followed by capillary electrophoresis technology, so-called Sanger sequencing, was the gold standard for mutation detection for many years. While it is robust and accurate for germline mutations, there are limitations in detecting somatic mutations in FFPE samples. The major limitation is that the sensitivity of detection is approximately 20–25% allele frequency in mutation alleles. In addition, due to the nature of the technology, it is not suitable for multiplex analysis of numerous mutations in a single test. The preparation of multiple amplicons is time consuming and labor intensive, and high input of DNA is required. In a clinical setting, FFPE DNA quantity is usually limited.

b. PCR-based molecular applications

These approaches were soon applied to FFPE nucleic acid samples after the introduction of real-time quantitative PCR (qPCR).[25] qPCR works well with small amplicons (even less than 100bp) which is suitable for fragmented nucleic acid yielded by FFPE tissue.

i. *DNA*

Numerous PCR-based technologies have been developed recently for detecting DNA mutations in FFPE samples. Co-amplification at lower denaturation temperature PCR (COLD-PCR) uses a lower melting temperature of the mismatch from mutated base to selectively amplify the minor alleles. It was shown to increase the detection sensitivity up to 100 fold.[26] High-resolution melting (HRM) assay analyzes the melting curve generated from specifically designed primers and enabled to detect mutations to 3.1–12.5% percentage.[27] Amplification refractory mutation system (ARMS-PCR) selectively amplifies the mutation allele by introducing a stopper at the 3′ end of primer to block the extension of the wild type allele at 3′ end of primers.[28] Competitive Allele-Specific TaqMan (CAST-PCR) was able to detect as low as 0.1% allele frequency (one copy of mutant allele out of 1,000 copies wild type allele) by blocking primer extension from the wild-type allele-specific primer.[29] Droplet digital PCR (ddPCR) has been shown to provide high sensitivity and specificity using FFPE DNA.[30]

PCR-based technologies are easy to apply, sensitive, accurate, and require low DNA input amount for mutation detection. However, the major bottleneck is the need to multiplex mutations in a single test. Most PCR-based assays study a single allele (analyte), and even for the same position, different probes are necessary for each base substitution. For example, in order to cover all possible alterations at KRAS G12 codon, it requires up to 16 different probes to match two frequently mutated bases. This limitation is difficult to overcome by the chemistry and probe design. Though some panels in array or droplet format are available on the market, it is just a paralleled process of multiple single analyte assays and is not measuring the multiple analytes in the same pool of DNA molecules. Nevertheless, PCR technologies are not only the foundation for NGS assays, but also the core technology for determining the quality (fraction of amplifiable molecule) of FFPE DNA samples (see details in NGS technologies).

ii. *RNA*

Quantitative measurement of the expression level of gene transcripts was the first application of real-time quantitative reverse transcription PCR (qRT-PCR). Advantages included high sensitivity, accuracy, reproducibility, and wide dynamic range. A few modifications were made to accommodate the FFPE RNA samples for qRT-PCR assay, such as random priming for cDNA synthesis and designing shorter amplicons that target

3′ regions of the transcript. Unlike DNA, RNA molecules are highly vulnerable to damage during FFPE tissue preparation, making inter-sample comparison more difficult if the FFPE specimens are not prepared under the same conditions. Thus, normalization using housekeeping genes is critical for expression-level comparison across multiple samples. Many vendors (ThermoFisher Scientific, Qiagen, BioRad) offer multi-gene qRT-PCR assays in pathway specific panels. Oncotype DX® (Genomic Health) is a qRT-PCR-based multiple gene test that was developed to monitor gene expression of 21-genes (five of which are housekeeping genes for normalization) for breast cancer patients that are ER positive and HER2 negative and have negative lymph nodes.[31] The gene expression data is then calculated by a proprietary algorithm to obtain an index value, a recurrence score to predict likelihood of recurrence of tumor in next 5–10 years. It was demonstrated by many clinical studies that only patients with high recurrence score received benefits from chemotherapy and hormonal combination therapy and that it was unnecessary to treat patients with low scores. The TAILORx study, a multi-center prospective trial of more than 10,000 women with early-stage breast cancer, was published recently and demonstrated that patients with low recurrence score from Oncotype DX® test have no difference (less than 1%) in survival and distal recurrence (5 years) between groups who received hormonal therapy alone and chemotherapy and hormonal combination therapy.[32] This pivotal study showed that a solid clinical test can change the medical decision and save unnecessary medical expenses. It was truly a useful case to demonstrate the value of companion diagnostics.

Profiling gene expression level by qRT-PCR is suitable for smaller panels of targeted genes with known clinical or biological utility. For larger panels (greater than several hundreds), other platforms (i.e., Microarray, NanoString, and NGS) are more cost-effective. Furthermore, qRT-PCR is useful to determine FFPE RNA quality prior to performing high-throughput assays.

c. Multiplexed methods

To increase the possibility of detecting multiple mutations, MassARRAY™ (Sequenom®) technology combined multiplexed PCR, primer extension, and MALDI-TOF (matrix assisted laser desorption/ionization time-of-flight) mass spectrometry.[33] A panel (OncoCarta®) consisting of 238 mutations in 19 genes were used to screen FFPE specimens of 1,000 patients[34] and showed novel discovery in mutation landscapes in various types of tumors. On the other hand, the SNapShot assay took a similar approach with multiplex PCR, primer extension, but used capillary electrophoresis-based Sanger sequencing[35] for detection. After multiplex PCR, a single-base extension with fluorescent label ddNTP was performed for a mixture of pre-designed mutation-specific probes. These probes were designed to be of different length and were subsequently separated by capillary electrophoresis by automated DNA sequencers. The mutation is identified by the position and its corresponding fluorescent color in the locus. A panel of SNapShot assay that enables to detect up to 120 mutations in 13 genes has been used to profile hundreds of FFPE specimens from various types of cancer and was shown to be a highly sensitive and accurate method.[35,36]

d. Array-based technologies

Assay-based technologies have been successfully applied for single-nucleotide polymorphism (SNP) and copy number variation detection using FFPE DNA. Affymetrix® SNP arrays enabled the interrogation of more than 900K SNPs and copy number changes,[34] while Agilent's inkjet printing oligo (60mer) CGH arrays offered copy number change profiling in the whole genome.[37]

Relative to FFPE DNA, FFPE RNA samples have shown even poorer quality due to the higher vulnerability of RNA to chemical and heat treatments during tissue processing. One of the challenges is the RNA labeling method. While the traditional Eberwine oligo (dT)-based method[38] was successfully and widely used in labeling for good quality RNA, the input amount (at microgram level) and the 3′ end orientated becomes a hurdle for FFPE RNA.[39] An isothermal linear amplification named Ribo-SPIA® (NuGen Technology) that uses proprietary random primers to capture any transcript fragments requires much lower input (~50ng). The linear amplification property maintains the relative abundance of expression among genes consistent after amplification. Using FFPE RNA samples, this method shows similar classification results when compared to the gold standard, which is fresh RNA labeling by Eberwine method, for subtyping ABC (activating B cell) vs. GCB (germinal center B cell) in DLBCL (diffuse large B cell lymphoma) patients.[40,41] Recently, Affymetrix launched the latest version of the FFPE RNA amplification and labeling method: SensationPlus™ FFPE Amplification. Based on linear amplification, this technology generates high-quality amplified RNA from degraded FFPE RNA primed by both oligo dT and random primers. We have been able to profile poor-quality FFPE RNA that was previously not possible by Ribo-SPIA®, by using SensationPlus™ FFPE Amplification (personal communication). A pivotal study published recently, demonstrated that treatment of a bruton tyrosine kinase inhibitor, Ibrutinib, selectively benefited ABC subtype patients that were stratified by SensationPlus™ FFPE Amplification-mediated microarray profiling.[42]

Microarray is an ideal platform for exploratory studies (i.e., identifying new expression signatures), but it was limited by the relatively long and complicated experimental processes, relative high expense in reagents, and potentially inconsistent assay performance when compared with other emerging technologies (i.e., NanoString).

e. NanoString technology

This technology counts the number of transcripts in FFPE specimens using multiplexed gene-specific fluorescently color-coded probe pairs. Since there are no enzymatic reactions used in the procedure, the quantity of captured probes correlates linearly with the expression level of targeted transcripts. The probes were designed to hybridize the 3′ regions of transcripts to accommodate the nature of FFPE RNA samples. The proprietary reagents allow for the hybridization of probes with cell lysate from FFPE specimens without extracting RNA molecules.[43] NanoString instruments consist of an automatic liquid handing robot and a confocal microscopic scanner, which can easily process up 48–96 samples a day with minimal operation time. These features, including the high sensitivity, accuracy, ease of handling, and fast turnaround time, make this platform ideal for clinical diagnostics applications. An analytical validation study has shown outstanding performance with high reproducibility and sensitivity.[44]

NanoString (nCounter®) platforms have been used to develop diagnostic assays for a couple of well-defined gene signatures with clinical utility.

PAM50 is a 50-gene expression signature that can classify a breast cancer patient into one of four intrinsic subtypes: Luminal A, Luminal B, HER2, and Basal Like. The subtype information has been shown to be of prognostic significance in patients with or without adjuvant hormonal therapy. Analytical validation of NanoString-based PAM50 assay using FFPE specimens showed high inter-lab concordance (97%) in subtype results.[45] Lymph2Cx assay, another NanoString-based diagnostic assay, was validated to determine cell origin in DLBCL patients using FFPE specimens. This assay measured expression level of 20 genes to assign ABC or GCB subtype for a DLBCL patient. Since numerous new drugs (i.e., ibrutinib) showed selective effect in ABC subtype, the Lymph2Cx assay is expected to be a companion diagnostic assay for the several ABC-targeting agents in the pipeline. Lymph2Cx assay showed high inter-laboratory concordance in subtype classification. The outcome of patients stratified by Lymph2Cx was comparable with the results stratified by gold standard (array-based) assays.[46]

Though NanoString technology has many advantages, in particular for multiplex gene expression signature-based diagnostics, it requires well-characterized and robust gene signatures. Currently, 800 genes per panel is the maximum number that this platform can measure in one hybridization.

f. Next-generation sequencing (NGS)

The emergence of NGS technologies has dramatically transformed biomedicine research, especially cancer research, by aiding in the rapid discovery of genetic aberrations in a large number of patient tumors.[47–50] NGS technologies continue to evolve and are becoming comprehensive platforms for diagnostic assay development as well as for screening for somatic mutations in FFPE tumor specimens in the research and clinical settings. Several NGS technologies have been applied in FFPE specimen-based diagnostics assays.

i. *DNA target amplicon sequencing*

While large numbers of genetic alterations have been identified in various types of cancer, only few mutations have actionable value to guide the medical decision, such as serving as drug targets (i.e., BRAF p.V600E for BRAF inhibitor vemurafenib) or conferring resistance to drug activity (KRAS G12 mutation for anti-EGFR cetuximab). NGS methods that target the so-called "hotspot" regions were developed as diagnostic assays for screening cancer patients and treatment assignment. A couple of targeted panel NGS assays are commercially available for DNA extracted from FFPE specimens. The Ion Torrent Hotspot Cancer Panel™ (Thermo Fisher) employs a proprietary primer chemistry for multiplex PCR named Ampliseq technology, containing 799 hotspots in 46 frequently mutated cancer genes. Illumina TruSeq Custom Amplicon (TSCA) developed a similar panel using their primer extension ligation method (GoldenGate, Illumina), but the TSCA technology requires high DNA input quantity (250 ng) which might not be feasible in many clinical specimens. Ion Torrent Ampliseq only requires 20 ng (Illumina recently launched a low input 10 ng method). The length of amplicons in a targeted panel is usually designed to be shorter than 200 base pairs to accommodate the fragmented FFPE DNA.

NCI developed an NGS-targeted sequencing assay for NCI-MPACT trial (Molecular Profiling-Based Assignment of Cancer Therapy) (https://clinicaltrials.gov/ct2/show/NCT01827384) using a customized panel by Ampliseq technology and Personal Genome Sequencer.[51] Both FFPE and frozen DNA samples were used in the analytical validation for the NCI-MPACT assay. The results showed that the average read depth (how many times a base has been sequenced) in FFPE DNA samples is slightly higher than that in frozen DNA samples, indicating that the fragmentation in FFPE genomic DNA is not an issue for targeted amplicon sequencing. The analytical performance and preliminary screening results of NCI-MPACT assay validation study demonstrated that sequencing FFPE DNA is a reliable method, especially if the assay is performed with good quality control, including pre-analytical steps.

For DNA extracted from tissue specimens without pre-analytical control of variables (time to fixation, time of fixation, type of fixative, and storage time), DNA quality assessment prior to NGS assay became crucial for consistent results. The FFPE DNA Kapa QC (Kapa Biosystems) assay is a commercially available qPCR-based method to quantify the fraction of amplifiable DNA. Three amplicons (41 bp, 129 bp, and 351 bp, respectively) with the same 3′ end anchor primer were used to perform qPCR. Since the amplification of longer fragments is less efficient due to the DNA degradation, the Ct value for longer amplicons is higher. The ratio of the quantity among three amplicons (interpolated from the standard curve), Q129/41 and Q315/41, named Q scores, were used to represent the quality of a FFPE DNA. Pre-NGS assay DNA quality control is important to identify poor-quality samples and elevate the successful rate of the downstream analysis.

Target amplicon sequencing technologies using FFPE DNA have been widely adapted in the diagnostic field. Advantages such as high read depth coverage, quick turnaround time, streamlining of data analysis pipeline, accuracy, and low cost made it ideal as a screen tool to assign patients to matched treatment. NCI-MATCH (NCI-Molecular Analysis for Therapy Choice) (http://www.cancer.gov/about-cancer/treatment/clinical-trials/nci-supported/nci-match) is one of largest precision oncology clinical trials that uses target amplicon NGS sequencing. Up to 3000 patients are planned to be screened to assign 30 to 40 targeted treatments within three years.[52] The NCI-MATCH assay can measure up to 4066 mutations of interest in 143 genes, and they include SNVs, Indels, CNVs, and gene fusions. Though the targeted amplicon sequencing-based assays are prevalent, they are limited to the detection of known mutations and are not suitable for large exploratory research purposes (i.e. discovery for new biomarkers).

ii. *DNA-exome sequencing*

Exome sequencing technology enriches the target regions by oligo-probe hybridization before sequencing reaction. The oligo probes can cover the entire coding regions of all genes (whole exome sequencing) or selected genes (targeted exome sequencing) (Agilent, Nimblegen,[53]). In order to facilitate the hybridization reaction, high molecular weight genomic DNA is required to be randomly sheared to 200–250 bp size range fragments before library preparation. In contrast to fixed start and end positions in each amplicon in the target amplicon sequencing, the captured fragments displayed high complexity with unique start and end positions. By aligning the sequencing reads against the reference genome, in addition to SNV and Indels, the exome sequencing technologies were able to detect CNVs, structural variants, and mutations in exon–intron splicing junctions.

Comparing amplicon sequencing, exome capture sequencing generates more consistent results in challenge regions such as GC-rich, homopolymeric, repetitive motifs. Poor quality and chemical modifications in FFPE DNA used to be impediments for exome sequencing. Yet, it has been shown feasible to sequence FFPE DNA by exome sequencing with reasonable performance.[54] A commercialized laboratory development NGS assay that is capable of detecting somatic mutations in 287 cancer-related genes using FFPE DNA exome capture sequencing (Foundation Medicine,[55]) has been used for sequencing tens of thousands of cancer patients.

In order to understand the effect of FFPE specimen storage time on the NGS sequencing quality, a study was performed by the Surveillance Epidemiology and End Results (SEER) Program at NCI. Fifty-nine FFPE DNA samples extracted from 59 high-grade serous adenocarcinomas stored from 3 to 32 years were studied using whole exome sequencing.[18] Specimens stored for longer periods of time had lower coverage of the target region (6% lower per 10 years) and lower average read depth (40× lower per 10 years). However, 90% of the specimens provided usable NGS data regardless of storage time, showing that FFPE DNA samples are a promising resource for retrospective population-based studies. It was also found that pre-analytical FFPE DNA Kapa QC results show good correlations with specimen age, successful sequencing rate, total reads, percentage of duplicated read (complexity), and transition versus transversion (Ti/TV) ratio. These data supported that quality scores for pre-NGS assay as a quality metric is important to conduct successful exome sequencing for FFPE DNA samples.

While exome capture technologies are applied in many applications, whole exome sequencing is not sensitive in detecting variants with low allele frequency (below 10%) because ~250× or above average read depth is required (due to the trade-off between read depth and coverage region). If the sequencer throughput and reagent cost are limited, the targeted exome sequencing is a better option for high sensitivity assay.

IV. Conclusions

The studies mentioned above have demonstrated that it is feasible to generate consistent genetic information from FFPE DNA and RNA. Applying molecular techniques to clinical samples offered the possibility of new promising biomarker discovery that may identify patients who would benefit from specific therapeutic strategies. Even though there are various challenges in molecular analysis of FFPE tissues, the recent advancements in technology have made the dream possibly come true — treating patients using their own genetic/genomic information profiled from FFPE samples.

A few key points need to be stressed when developing molecular assays using nucleic acids from FFPE samples.

(1) Taking the molecular nature of FFPE nucleic acid samples into account. Due to the fragmentation, the amplicon size should not be too long (<200 bp) and the use of random primers to capture any fragmented RNA transcript is recommended, instead of oligo dT for the 3'end. Lack of these basic understandings in the assay design may result in severe failure or poor performance in molecular assays.
(2) Performing quality control evaluation prior, the molecular assays and establishing QC sample metrics for the viability of the assay. Spectrophotometry (Nanodrop, Qubit) readouts usually are not a good indication of the quality of FFPE nucleic acid samples, i.e., samples with same readout may have large quality differences. qPCR-based quality assays that measure the fraction of amplifiable nucleic acid molecules reflect quality more accurately. Most molecular assays are expensive and time consuming, and therefore identifying and eliminating the poor-quality samples can improve the successful rate, saving reagent costs and avoiding wasting valuable clinical samples.
(3) Including pre-analytical tissue processing information as part of assay system. The variations in pre-analytic stage could significantly affect the results of molecular assay. Changes in pre-analytic conditions such as time to fixation or time of fixation should be validated and approved first.
(4) Choosing robust and appropriate platforms for your application. Each technology platform has its pros and cons, not one fits all. For example, while PCR-based methods are simple, fast, and sensitive for

DNA mutations, they require multiple assays (probes) to cover multiple alleles (different substitutions) at the same base. Such problems do not exist in NGS-based platforms; however, NGS platforms tend to be more expensive, with a longer turnaround time. Therefore, PCR-based platforms are good choice for single or fewer analyte assays, but NGS platforms are suitable for a panel of target gene with multiplex analyte assays.

Technological advancements have allowed for the performance of high-throughput techniques with just few hundred nanograms of nucleic acids. Biopsies are becoming smaller and more data can be obtained from less material.[56] Many opportunities exist for conducting correlative laboratory studies that may be immediately relevant to cancer treatment. In addition, opportunities exist for the molecular characterization of pre-cancer and early-stage cancer through analysis of specimens from treatment and prevention trials. Clinical specimens are collected at different time points and serve as the guide for treatment. Multidimensional high-throughput techniques including PCR, NGS, NanoString, and array technologies are used to analyze these "next-generation biospecimens".[57] Protein extraction protocols from archival material have been developed.[58] and promising results are being published on the use of FFPE for proteomic analysis,[59] including quantitative proteomics, phospho- and glyco-proteomics, as well as mass spectrometry. Companion diagnostic tests show which patients could be helped by a drug and which patients would not benefit. Understanding the quality and quantity thresholds that are involved in using those samples for downstream molecular analysis is key. The higher histological variation seen in these small samples adds a further challenge to their annotation for molecular analysis. Communication between pathologists conducting the histologic assessment of clinical tissue samples and the molecular team performing the assays is essential.

References

1. Dubbinka JZ, Deans ZC, Tops BBJ, van Kemenade FJ, Koljenovic S, van Krieken HJM, Blokx WAM, Dinjens WNM, Groenen PJTA. Next generation diagnostic molecular pathology: Critical appraisal of quality assurance in Europe. *Mol Oncol* 2014; **8**: 830–839.

2 SchilskyRL, Doroshow JH, LeBlanc M, Conley BA. Development and use of integral assays in clinical trials. *Clin Cancer Res* 2012; **18**: 1540–6.
3. Tangrea MA, Gillespie JW, Falig MJ, Gannot G, Knezevic V, Emmert-Buck MR, Chuaqui RF. Methods of Microdissection and New technologies. In *Dissecting the Molecular Anatomy of Tissue — Principles and Practices*, MR Emmert-Buck, JW Gillespie, RF Chuaqui (eds). New York, Berlin, Heidelberg, Tokyo, Springer Verlag, 2004.
4. Betsou F, Gunter E, Clements J, DeSouza Y, Goddard KAB, Guadagni F, Yan W, Skubitz A, Somiari S, Yeadon T, Chuaqui R. J. Identification of evidence-based biospecimen quality control tools. *Mol Diagn* (*J Mol Diag*, online Nov 2012).
5. Le Page C, Kobel M, de Ladurantaye M, *et al.* Specimen quality evaluation in Canadian biobanks participating in the COEUR repository. *Biopreserv Biobank* 2013; **11**: 83–93.
6. Bellon E, Ligtenberg MJ, Tejpar S, *et al.* External quality assessment for KRAS testing is needed: Setup of a European program and report of the first joined regional quality assessment rounds. *Oncologist* 2011; **16**: 467–478.
7 Lamy A, Blanchard F, Le Pessot F, *et al.* Metastatic colorectal cancer KRAS genotyping in routine practice: Results and pitfalls. *Mod Pathol* 2011; **24**: 1090–1100.
8. Qualman SJ, France M, Grizzle WE, *et al.* Establishing a tumour bank: Banking, informatics and ethics. *Br J Cancer* (2004); **90**: 1115–1119.
9. Aran D, Sirota M, Butte AJ. Systematic pan-cancer analysis of tumor purity. *Nature Communications* 2015; **6**: 8971.
10. Haixin Li, Yan Guo, Baocun Sun, and Kexin Chen. Histological assessment of tumor tissue samples via the mirror image method. *Biopreserv. Biobank.* 2015; **13**: 25–30.
11. Abrams J, Conley B, Mooney M, *et al.* National Cancer Institute's Precision Medicine Initiatives for the new National Clinical Trials Network. *Am Soc Clin Oncol Educ Book.* 2014: 71–6.
12. Dudley JC, Gurda GT, Tseng LH, *et al.* Tumor cellularity as a quality assurance measure for accurate clinical detection of BRAF mutations in melanoma. *Mol Diagn Ther.* 2014; **18**(4): 409–418.
13. Viray H, Li K, Long TA, *et al.* A prospective, multi-institutional diagnostic trial to determine pathologist accuracy in estimation of percentage of malignant cells. *Arch Pathol Lab Med.* 2013; **137**(11): 1545–1549.
14. Grizzle WE, Sexton KC, and Bell WC. Quality assurance in tissue resources supporting biomedical research. *Cell Preserv Technol.* 2008; **6**(2): 113–118.
15. Angulo B, Conde E, Suárez-Gauthier A, Plaza C, Martínez R, Redondo P, *et al.* A comparison of EGFR mutation testing methods in lung carcinoma: Direct sequencing, real-time PCR and immunohistochemistry. *PLoS One.* 2012; 7.
16. Hamilton PW, Wang Y, Boyd C, *et al.* Automated tumor analysis for molecular profiling in lung cancer. *Oncotarget.* 2015; **6**(29): 27938–27952.
17. Weichert W, Schewe C, Lehmann A, *et al.* KRAS genotyping of paraffin-embedded colorectal cancer tissue in routine diagnostics: Comparison of methods and impact of histology. *J Mol Diagn.* 2010; **12**(1): 35–42.

18. Carrick D, Altekruse S, Mehaffey M, Sachs MC, Makhlouf H, *et al.* Robustness of next generation sequencing on older formalin-fixed paraffin-embedded tissue. *PLoS One*. 2015; **10**(7): e0127353.
19. Makhlouf H, Lubensky I, Altekruse S, *et al.* Effect of storage time of FFPE ovarian cancer tissue on DNA quality for next generation sequencing. *Mod Pathol* 2015; **28**(Suppl 2), 519A.
20. Ferry-Galow K, Lawrence S, Navas T, Makhlouf H, Butcher B, Gouker B,Yutzy W, Ji J, Kinders R, Parchment RE, Kummar S, Tomaszewski JE, Doroshow JH. Establishing robust pharmacodynamic (PD) immunofluorescence assays on clinical biopsies at the national cancer Institute: Optimized quality control procedures for the evaluation of DNA damage response and epithelial mesenchymal transition (EMT) biomarkers in FFPE tissues. *Cancer Res*, 2015; **75**: 5279.
21. Viray H, Coulter M, Li K, Lane K, *et al.* Automated objective determination of percentage of malignant nuclei for mutation testing. *Appl Immunohistochem Mol Morphol*, 2014; **22**(5): 363–371.
22. Konsti J, Lundin M, Joensuu H, *et al.* Development and evaluation of a virtual microscopy application for automated assessment of Ki-67 expression in breast cancer. *BMC Clin Pathol.* 2011; **11**: 3.
23. van Krieken JH, Jung A, Kirchner T, *et al.* KRAS mutation testing for predicting response to anti-EGFR therapy for colorectal carcinoma: Proposal for an European quality assurance program. *Virchows Arch.* 2008; **453**(5): 417–431.
24. Greytak SR, Engel KB, Bass BP and Moore HM. Accuracy of molecular data generated with FFPE biospecimens: Lessons from the Literature. *Cancer Res* 2015; **75**: 1541–1547.
25. Higuchi R, Fockler C, Dollinger G and Watson R. Kinetic PCR analysis: real-time monitoring of DNA amplification reactions. *Biotechnology (NY)* 1993; **11**: 1026–1030.
26. Li J, Wang L, Mamon H, Kulke MH, Berbeco R and Makrigiorgos GM. Replacing PCR with COLD-PCR enriches variant DNA sequences and redefines the sensitivity of genetic testing. *Nat Med* 2008; **14**: 579–584.
27. Ney JT, Froehner S, Roesler A, Buettner R and Merkelbach-Bruse S. High-resolution melting analysis as a sensitive prescreening diagnostic tool to detect KRAS, BRAF, PIK3CA, and AKT1 mutations in formalin-fixed, paraffin-embedded tissues. *Arch Pathol Lab Med* 2012; **136**: 983–992.
28. Thelwell N, Millington S, Solinas A, Booth J and Brown T. Mode of action and application of Scorpion primers to mutation detection. *Nucleic Acids Res* 2000; **28**: 3752–3761.
29. Didelot A, Le Corre D, Luscan A, Cazes A, Pallier K, Emile JF, Laurent-Puig P, Blons H. Competitive allele specific TaqMan PCR for KRAS, BRAF and EGFR mutation detection in clinical formalin fixed paraffin embedded samples. *Exp Mol Pathol* (2012); **92**: 275–280.
30. Garcia-Murillas I, Lambros M, Turner NC. Determination of HER2 amplification status on tumour DNA by digital PCR. *PLoS One* 2013; **8**: e83409.

31. Paik S, Shak S, Tang G, Kim C, Baker J, Cronin M, Baehner FL, Walker MG, Watson D, Park T, *et al.* A multigene assay to predict recurrence of tamoxifen-treated, node-negative breast cancer. *N Engl J Med* 2004; **351**: 2817–2826.
32. Sparano JA, Gray RJ, Makower DF, Pritchard KI, Albain KS, Hayes DF, Geyer CE, Jr., Dees EC, Perez EA, Olson JA, Jr., *et al.* Prospective Validation of a 21-Gene Expression Assay in Breast Cancer. *N Engl J Med* 2015; **373**: 2005–2014.
33. Fumagalli D, Gavin PG, Taniyama Y, Kim SI, Choi HJ, Paik S, Pogue-Geile KL. A rapid, sensitive, reproducible and cost-effective method for mutation profiling of colon cancer and metastatic lymph nodes. *BMC Cancer* 2010; **10**: 101.
34. Thomas RK, Baker AC, Debiasi RM, Winckler W, Laframboise T, Lin WM, Wang M, Feng W, Zander T, MacConaill L, *et al.* High-throughput oncogene mutation profiling in human cancer. *Nat Genet* 2007; **39**: 347–351.
35. Dias-Santagata D, Akhavanfard S, David SS, Vernovsky K, Kuhlmann G, Boisvert SL, Stubbs H, McDermott U, Settleman J, Kwak EL, *et al.* Rapid targeted mutational analysis of human tumours: a clinical platform to guide personalized cancer medicine. *EMBO Mol Med* 2010; **2**: 146–158.
36. Su Z, Dias-Santagata D, Duke M, Hutchinson K, Lin YL, Borger DR, Chung CH, Massion PP, Vnencak-Jones CL, Iafrate AJ, *et al.* A platform for rapid detection of multiple oncogenic mutations with relevance to targeted therapy in non-small-cell lung cancer. *J Mol Diagn* 2011; **13**: 74–84.
37. de Jong D, Verbeke S, Meijer D, Hogendoorn PC, Bovee JV, Szuhai K. Opening the archives for state of the art tumour genetic research: Sample processing for array-CGH using decalcified, formalin-fixed, paraffin-embedded tissue-derived DNA samples. *BMC Res Notes* 2011; **4**: 1.
38. Van Gelder RN, von Zastrow ME, Yool A, Dement WC, Barchas JD, Eberwine JH. Amplified RNA synthesized from limited quantities of heterogeneous cDNA. *Proc Natl Acad Sci USA* 1990; **87**: 1663–1667.
39. Farragher SM, Tanney A, Kennedy RD, Paul Harkin D. RNA expression analysis from formalin fixed paraffin embedded tissues. *Histochem Cell Biol* 2008; **130**: 435–445.
40. Williams PM, Li R, Johnson NA, Wright G, Heath JD, Gascoyne RD. A novel method of amplification of FFPET-derived RNA enables accurate disease classification with microarrays. *J Mol Diagn* 2010; **12**: 680–686.
41. Linton K, Howarth C, Wappett M, Newton G, Lachel C, Iqbal J, Pepper S, Byers R, Chan WJ, Radford J. Microarray gene expression analysis of fixed archival tissue permits molecular classification and identification of potential therapeutic targets in diffuse large B-cell lymphoma. *J Mol Diagn* 2012; **14**: 223–232.
42. Wilson WH, Young RM, Schmitz R, Yang Y, Pittaluga S, Wright G, Lih CJ, Williams PM, Shaffer AL, Gerecitano J, *et al.* Targeting B cell receptor signaling with ibrutinib in diffuse large B cell lymphoma. *Nat Med* 2015; **21**: 922–926.
43. Geiss GK, Bumgarner RE, Birditt B, Dahl T, Dowidar N, Dunaway DL, Fell HP, Ferree S, George RD, Grogan T, *et al.* Direct multiplexed measurement of gene expression with color-coded probe pairs. *Nat Biotechnol* 2008; **26**: 317–325.

44. Reis PP, Waldron L, Goswami RS, Xu W, Xuan Y, Perez-Ordonez B, Gullane P, Irish J, Jurisica I, Kamel-Reid S. mRNA transcript quantification in archival samples using multiplexed, color-coded probes. *BMC Biotechnol* 2011; **11**: 46.
45. Nielsen T, Wallden B, Schaper C, Ferree S, Liu S, Gao D, Barry G, Dowidar N, Maysuria M, Storhoff J. Analytical validation of the PAM50-based Prosigna Breast Cancer Prognostic Gene Signature Assay and nCounter Analysis System using formalin-fixed paraffin-embedded breast tumor specimens. *BMC Cancer* 2014; **14**: 177.
46. Scott DW, Wright GW, Williams PM, Lih CJ, Walsh W, Jaffe ES, Rosenwald A, Campo E, Chan WC, Connors JM, *et al.* Determining cell-of-origin subtypes of diffuse large B-cell lymphoma using gene expression in formalin-fixed paraffin-embedded tissue. *Blood* 2014; **123**: 1214–1217.
47. Cancer Genome Atlas Research N. Comprehensive genomic characterization defines human glioblastoma genes and core pathways. *Nature* 2008; **455**: 1061–1068.
48. Cancer Genome Atlas Research N. Integrated genomic analyses of ovarian carcinoma. *Nature* 2011; **474**: 609–615.
49. Cancer Genome Atlas Research N. Comprehensive genomic characterization of squamous cell lung cancers. *Nature* 2012; **489**: 519–525.
50. Cancer Genome Atlas N. Comprehensive molecular portraits of human breast tumours. *Nature* 2012; **490**: 61–70.
51. Lih CJ, Sims DJ, Harrington RD, Polley EC, Zhao Y, Mehaffey MG, Forbes TD, Das B, Walsh WD, Datta V, *et al.* Analytical validation and application of a targeted next-generation sequencing mutation-detection assay for use in treatment assignment in the NCI-MPACT trial. *J Mol Diagn*. 2015.
52. Brower V. NCI-MATCH pairs tumor mutations with matching drugs. Nat *Biotechnol* 2015; **33**: 790-791.
53. Hedges DJ, Guettouche T, Yang S, Bademci G, Diaz A, Andersen A, Hulme WF, Linker S, Mehta A, Edwards YJ, *et al.* Comparison of three targeted enrichment strategies on the SOLiD sequencing platform. *PLoS One* 2011; **6**: e18595.
54. Wagle N, Berger MF, Davis MJ, Blumenstiel B, Defelice M, Pochanard P, Ducar M, Van Hummelen P, Macconaill LE, Hahn WC, *et al.* High-throughput detection of actionable genomic alterations in clinical tumor samples by targeted, massively parallel sequencing. *Cancer Discov* 2012; **2**: 82–93.
55. Frampton GM, Fichtenholtz A, Otto GA, Wang K, Downing SR, He J, Schnall-Levin M, White J, Sanford EM, An P, *et al.* Development and validation of a clinical cancer genomic profiling test based on massively parallel DNA sequencing. *Nat Biotechnol* 2013; **31**: 1023–1031.
56. Fairley JA, Gilmour K, Walsh K. Making the most of pathological specimens: Molecular diagnosis in formalin-fixed, paraffin embedded tissue. *Curr Drug Targets* 2012; **13**: 1475–1487.
57. Basik M, Aguilar-Mahecha A, Rousseau C, Diaz Z, Tejpar S, Spatz A, Greenwood CM, Batist G. Biopsies: Next-generation biospecimens for tailoring therapy. *Nat Rev Clin Oncol*. 2013; **10**(8): 437–50.

58. Shi SR, Liu C, Balgley BM, Lee CL, Taylor CR. Protein extraction from formalin-fixed, paraffin-embedded tissue sections: Quality evaluation by mass spectrometry. *J Histochem Cytochen* 2006; **54**: 739–743.
59. Gustafsson OJR, Arentz G, Hoffmann P. Proteomic developments in the analysis of formalin-fixed tissue. *Biochim Biophys Acta* 2014; **1854**: 559–580.

10 Proteomic Analysis of Archival Samples

Adele R. Blackler

Adalheidis Consulting, LLC, Denver, Co, USA
adele@adalheidisconsulting.com

Contents

I. Introduction

Formalin fixation causes unpredictable modifications and cross-linking of proteins,[1] and thus for many years it was assumed that protein analysis of formalin-fixed, paraffin-embedded (FFPE) tissue was unrealistic. However, a number of advancements in antigen retrieval, sample

Table 1. Proteomic studies on FFPE tissue listed in this chapter

Study description	Microdissection method	Solubilization/ digestion method	Number of recovered proteins	Quantitation method	Year published	Citation
Comparison of prostate cancer tissue to benign hyperplastic prostate tissue	LMD	Liquid Tissue	684 (from FFPE liver)	Trypsin-mediated 18O isotope labeling	2005	[2]
Comparison of normal epithelium and HNSCC	LCM	Liquid Tissue	391–866	None	2008	[3]
Comparison of oral squamous cell carcinoma and adjacent normal tissue	LMD	Liquid Tissue	NA-focused only peaks that were significantly different between the two groups	Label-free quantitation	2009	[4]
Analysis of FASP method for proteomics	LCM	FASP	5000 (FFPE mouse liver)	SILAC	2010	[5]
Comparison of whole protein extraction from heart tissue	None	Several different detergent based buffers and heating	~100	None	2010	[6]
Analysis of early-stage lung cancer	LMD	Liquid Tissue	600	Label-free quantitation	2010	[7]
Quantitation of differentially expressed proteins in early-stage lung cancer	LMD	Liquid Tissue	NA	SRM	2010	[8]

Imaging mass spectrometry of gastric carcinoma	None	On-tissue trypsin digestion	NA	Label-free quantitation	2010	[9]
Whole protein extraction from FFPE tissues	None	Qproteome	examined total protein amounts	None	2011	[10]
Analysis of endometrial cancer tissue	LCM on membrane slides	20% MeCN	~800	None	2011	[11]
MRM on pre-eclampsia tissue	LMD	Liquid Tissue	not reported	SRM	2011	[12]
Analysis of normal and diabetic glomeruli	LMD	Liquid Tissue	170	iTraq	2012	[13]
MRM on EGFR in lung cancer	LMD	Liquid Tissue	not reported	SRM	2012	[14]
Enrichment of nuclei from FFPE rat brain using expression microdissection	xMD	TFE	~750	SRM	2013	[15]
Whole protein extraction from FFPE tissues and analysis by RPPA	None	Qproteome	differentiation of NSCLC subtypes by RPPA	Fluorescence (RPPA)	2013	[16]
Analysis of human aortas	None	SDS with heat and/or pressure	100–600	SRM	2013	[17]
Quantitation of several biomarkers in archival FFPE tissue	LMC	Liquid Tissue	NA	SRM	2013	[18]

(*Continued*)

Table 1. (*Continued*)

Study description	Microdissection method	Solubilization/ digestion method	Number of recovered proteins	Quantitation method	Year published	Citation
Analysis of pancreatic cancerous and non-cancerous tissue	LMD	Liquid Tissue	1229	SRM	2013	[19]
Comparison of different sample preparation method	None	FASP, direct trypsin, in-solution digestion	1693(FASP), 1510 (DT), 1358 (ISD)	Label-free quantitation	2014	[20]
Analysis of glomeruli in hypertensive models	LCM on membrane slides	SDS and 1D SDS page	1121	Label-free quantitation	2014	[21]
Analysis of histone PTMs in breast cancer	None	SDS/Heat-based antigen retrieval	NA-focused only on histones	SILAC	2015	[22]
Comparison of normal and cancerous colorectal samples	LCM on membrane slides	FASP	10900	Label-free quantitation	2015	[23]
Analysis of HER2 in archival tissue	None	MS-compatible detergent	NA	SRM	2015	[24]
Analysis of HER2 in breast cancer	LMD	Liquid Tissue	NA	SRM	2016	[25]
Analysis of ALK in lung cancer	LMD	Liquid Tissue	NA	SRM	2016	[26]

preparation techniques, and new platforms such as nanoflow-liquid chromatography coupled with tandem mass spectrometry (LC/MS/MS) and Imaging Mass Spectrometry (IMS) have demonstrated the possibility of high-throughput protein analysis from FFPE tissue. This chapter reviews techniques and platforms that can be used to characterize and quantify proteins from FFPE tissue, and it demonstrates the utility of proteomic analysis of FFPE samples.

High-throughput proteomic analysis using LC/MS/MS involves taking a complex protein mixture and enzymatically digesting the proteins into peptides before separation using liquid chromatography and automated peptide analysis with tandem mass spectrometry. This method is compatible with formalin-fixed tissue as the enzymatic digestion, usually trypsin, alleviates the need for total reversal of cross-linking. As long as some protein domains are accessible to trypsin, peptides from those domains can be identified and positive identification of the protein can be made. Many proteomic studies of FFPE tissue have been published, demonstrating the utility of this technique for understanding protein alterations in disease states, and some of these are listed in Table 1.

The use of microdissection techniques, such as laser microdissection (LMD), laser capture microdissection (LCM), and expression microdissection (xMD), allows for homogenous cell populations to be isolated from FFPE samples for proteomic analysis. The combination of microdissection with high-throughput proteomic techniques facilitates identification and quantitation of thousands of proteins in a single analysis, and can aid in the elucidation of protein alterations involved in disease etiology.

II. Epitope Degradation and Antigen Masking

Formalin is known to modify several amino acids, including lysines, tryptophans, histidines, cysteines, and arginines.[1,27] Several types of modifications can occur, including side-chain formylation, hydroxymethyl adducts, and stable methylene bridges. The major result of these covalent modifications is protein cross-linking, where proteins within close proximity of each other are covalently bonded together upon introduction of formalin. This cross-linking causes difficulty with downstream protein analysis by masking epitopes, causing reduced efficiency of antibodies, and by interfering with enzymatic digestion, making downstream proteomic analysis with mass spectrometry challenging.

The invention to use antigen recovery techniques, such as boiling the tissue in an acidic buffer, has greatly increased the utility of FFPE for protein analysis. Antigen recovery not only allows for epitope recovery and improved analysis with antibodies, but also reverses much (but not all) of the crosslinking that hinders proteases and subsequent analysis of peptides by mass spectrometry.

III. Analysis of Post-translational Modifications

Post-translational modifications (PTMs) of proteins, such as phosphorylation, glycosylation, and methylation, can control the activity and localization of proteins. Thus, to truly understand protein behavior it is necessary to understand modifications and identify any changes in PTMs due to disease states. Proteomic analysis of PTMs from FFPE tissue is limited; however, some reports show that PTMs are preserved by FFPE and can be quantitatively analyzed. Ostasiewicz *et al.*[5] reported quantitative recovery of phosphorylated and glycosylated sites from FFPE tissue, when compared with frozen tissue. This suggests that formalin fixation alone does not affect labile sites such as phosphorylated sites and does not affect glycosylation. It is important to note however that this study was done in mouse livers that were extracted and formalin fixed in-house for this study; thus, the authors were able to minimize any potential variability in PTM preservation caused by hypoxic conditions or over-fixation. It remains to be seen if this study is truly repeatable in archival human samples.

One study examined histone modifications, such as acetylation and methylation, and the preservation of modifications in FFPE tissue.[22] Similar to studies examining phosphorylation and glycosylation, the authors noted that histone modifications were similar between frozen and FFPE tissue. While some formylations and methylations were detected only in FFPE tissue samples, not in frozen ones, suggesting that these modifications were artifacts due to formalin fixation, the rate of these artifactual modifications was quite low and did not interfere with the overall analysis of modifications from FFPE tissue. Using this technique, the authors were able to identify differences in modification state of histones between different subtypes of breast cancers; notably several differences were identified between the ER/PR-positive Luminal-A subtype and triple-negative subtype.

While more research is warranted into the examination of PTMs in FFPE archival tissue, initial reports show that this field is very promising.

IV. Considerations for Microdissection for Proteomic Analysis

One challenge for proteomic analysis of FFPE tissue is the relative lack of ability for upfront purification of organelles and proteins. Standard biochemical techniques used for fresh frozen tissues, such as centrifugation or detergent extraction for organellar enrichment and immunoprecipitation for isolation of specific proteins and their binding partners, are not available for FFPE samples. Due to the complexity of proteins in a cell (each cell contains roughly 30,000 proteins) and the dynamic range of protein expression, identification of all proteins in a population of cells is currently not feasible, and generally very low abundance proteins are missed in these analyses.

Various microdissection techniques allow for very high-resolution dissection of purified cell populations. Several published reports are available that use three main techniques: (1) Laser capture microdissection (LCM), where cells are captured onto a polymer cap,[28] (2) laser microdissection (LMD), where cells are isolated off of special DIRECTOR slides,[2] and (3) expression microdissection (xMD),[29] where cells are first stained by IHC and then these stained cells are isolated onto a polymer film.

Many published proteomic analyses use LCM to first isolate purified cell populations. LCM has been used to dissect out normal epithelial cells and compare them to cancerous epithelial cells from head and neck squamous cell carcinoma (HNSCC). By being able to isolate specific cell populations, the complexity was reduced enough that the authors were able to identify proteins that were unique to HNSCC and not identified in normal epithelial cells.[3]

LMD differs slightly from LCM in that the isolated cells are not collected onto a solid substrate. This has the advantage of not having to resolubilize cells and remove them from the substrate. One group used LMD to dissect out tumor tissue from squamous cell cancer of the tongue and compared the proteomic profile to adjacent normal tissue dissected from the same slide.[4] Comparing normal and tumor tissue from the same sample has the advantage of removing any inherent variability in protein expression that exists in a diverse population, but is not related to disease state.

While both LMD and LCM require an expert to mark up areas of interest (usually a pathologist identifying diseased or normal tissue), xMD removes this time consuming step by marking the regions of interest using antibodies and chromogens to stain tissue expressing a specific protein marker. One concern for proteomic analysis of xMD tissue is how much antibody interference would be present; however, reports to date show no interference from antibodies and normal recovery of peptides after IHC staining.[30] One distinct advantage that xMD currently has over LCM or LMD is much finer resolution that is dependent on the localization of the stain and not the resolution of the laser. This allows for dissection of subcellular organelles, such as nuclei, and results in enrichment of proteins localized to that organelle.[15]

As alluded to previously, one challenge with xMD and LCM is the capture of cells and proteins onto a solid substrate. For efficient protein digestion and analysis, it is necessary to remove the proteins from the solid substrate for analysis. This can be done by dissolving the substrate, either through heat[5] or solvents,[15] or by using harsh lysis buffers to remove proteins from the substrate.[23,31] One challenge with separating sample from the substrate is that the buffers and solvents must be compatible with downstream analysis, or extensive sample cleanup may be necessary.

V. Considerations for Sample Preparation

An important consideration for sample preparation and protein solubilization is the downstream application. Many detergents are incompatible with LC/MS/MS systems and their use in sample solubilization may require sample cleanup, which can cause sample loss and increase variability in recovery. Samples that will be run on SDS-PAGE gels for protein analysis can handle large amounts of detergent, which can aid in crosslinking reversal. However, whole protein separation and analysis from FFPE tissue is not always successful. One comparison of several detergent-based solubilization methods found that proteins are not readily resolved by SDS-PAGE, and while proteins could be identified and western blotting was successful, many proteins did not migrate correctly and were found in several different molecular weight ranges on the gel.[6] This suggests that enough protein crosslinking remained even after harsh solubilization conditions that protein separation by SDS-PAGE is a technique with limited success; thus separation at the peptide level is still considered to be the standard technique.[32]

More recent reports of the use of a commercial kit, the QProteome FFPE kit (Qiagen, Hilden, Germany), had much improved whole protein extraction and reported protein yields from a variety of FFPE tissues that were comparable to fresh frozen samples. Several Western blots show that this kit is superior to other whole protein extraction method for FFPE tissue,[10] and is amenable to analysis by Reverse Phase Protein Array (RPPA).[16] If gel-based analysis and whole protein separation is desirable for analysis, the QProteome kit may be a viable option that is readily and commercially available.

Regardless of the buffer system used, all sample preparation methods use high temperatures for crosslinking reversal. Heat-induced antigen retrieval (AR) was first introduced as a means of unmasking epitopes to improve antibody-based protein detection with IHC, and has since been translated over to sample preparation for either peptide or whole protein solubilization and analysis. Generally, AR techniques for proteomic analysis are similar to AR techniques for IHC.[27] Samples are placed in a buffered solution and heated to at least 95°C for at least 20 minutes, sometimes this heating step repeated multiple times. As many different buffer systems are used, either acidic or basic, with detergent or without, the exact makeup of the buffer seems to matter less than the heat for downstream proteomic analysis.

For samples that will be analyzed by LC/MS/MS, efficient proteolytic digestion is the most important consideration for sample preparation. Trypsin, the most commonly used enzyme, is generally very robust and stable, and thus works in many different buffer systems, the exception being that large amounts of detergents (especially SDS) will dramatically hinder trypsin's activity.

Most reports of LC/MS/MS analysis of microdissected FFPE tissue use Liquid Tissue buffer (Expression Pathologies, Inc, Rockville Maryland).[3,12,13,26] Liquid tissue is a non-detergent buffer system compatible with trypsin and downstream mass spectrometry analysis that does not require any sample cleanup prior to analysis.

Other reports use organic solvents, such as trifluoroethanol (TFE) or acetonitrile (MeCN) in combination with ammonium bicarbonate (AmBic) buffer.[11,15] Organic solvents reduce the polarity of the buffer solution and can aid in keeping non-polar proteins, such as membrane proteins, in solution during digestion. Most organic solvents are compatible with downstream LC/MS/MS, so cleanup is also minimal with these buffer systems.

Some groups have developed protocols that use SDS and heat for protein solubilization and then remove the SDS prior to digestion with trypsin.[17,23] The extra step of SDS removal has not been shown to have an impact on protein recovery, in fact one report shows improved protein identification with this method; however, the SDS removal step may decrease recovery of high molecular weight protein.[20]

VI. Proteomic Applications for Microdissected FFPE Tissue

Reports of proteomic analysis of FFPE tissue focus primarily on the discovery of new biomarkers, with quantitative methods being used to verify differential expression of biomarkers (Table 1).

Many discovery efforts are now able to reproducibly identify thousands of proteins from limited amounts of FFPE tissue, facilitating the identification of potential biomarkers for disease progression or prognosis. A recent report focusing on changes between normal and cancerous colorectal tissue identified over 10,000 proteins and found large-scale expression changes representative of large changes in cell structure and biology.[23] Large increases in proteins localized to nucleoli suggested an increase in the number of these organelles in cancer cells, and increases in DNA replication proteins point to cancer cells increased speed of division and replication.

While the majority of reports are focused on the field of cancer, these techniques can be applied to archival tissue from any disease state. Using a model of hypertension,[21] a report of discovery proteomics found 23 proteins that were differentially expressed in glomeruli of normal and hypertensive rats.

The most common method for discovery proteomics efforts is LC/MS/MS; however, advances in technology have allowed for some reports of successful proteomic imaging mass spectrometry (IMS) experiments on FFPE tissue. IMS starts with an intact tissue section, with on-tissue trypsin digestion that is controlled to not allow special spread of the tryptic peptides, but instead keeps them localized during digestion. After digestion, the intact tissue is placed into the IMS and is "scanned" by the mass spectrometer. The mass spectrometer records the masses identified at each part of the tissue, and then software is used to generate heat maps of the tissue and masses identified at different parts.[33] Due to the specialized equipment and technical expertise required, few labs in the world are using this technique; however, it can be very useful for high-resolution maps of protein distribution in tissue. Few reports have applied this to protein and FFPE tissue; however, one analysis using tissue microarrays of gastric cancer samples successfully found histone 4 to be increased in poorly differentiated cancer tissue.[9]

The use of selective reaction monitoring (SRM) is the standard proteomic method for determining quantitative changes in a selected set of proteins. In SRM experiments, a set of peptides to quantitate is identified, and isotopically labeled synthetic peptides are used as internal standards.[26] The synthetic peptides are spiked into samples for analysis, and the mass spectrometer is programmed to look for the predetermined set of peptides, both the native and synthetic peptides, and the amount of native peptide in the sample is determined from the spiked-in synthetic peptide. SRM is highly sensitive and selective, and can provide very precise and accurate quantitation of selected peptides. Several reports are published of known biomarkers, such as HER2 in breast cancer, that are easily targetable and quantifiable using SRM.[18,24]

There are many reports of labs engaging in discovery efforts and then using SRM or antibody-based techniques to validate potential biomarkers identified during the discovery phase experiments. A publication examining protein expression in early-stage lung cancer identified more than 500 proteins, with 81 of them being correlated with a specific cancer stage.[7]

The lab then took these 81 candidates and built an SRM assay to examine a subset of these candidates and determine if the differential expression could be replicated. Final analysis confirmed two proteins as useful for exact stage diagnosis and found that samples positive for Napsin A had a better outcome than those that were negative.[8]

Another discovery effort looking at pancreatic adenocarcinoma identified over 1200 proteins, and found 170 of them to be candidate biomarkers. SRM experiments verified that 35 out of 42 candidate proteins were overexpressed in cancer and that 14 proteins were differentially expressed between patients with good and poor outcome.[19] Four proteins were replicated using both SRM and IHC, suggesting that these four protein candidates are robust markers of prognosis in pancreatic cancer.

VII. Conclusions and Looking to the Future

Proteomic analysis of FFPE tissue has the ability to quantify large numbers of proteins and identify potential candidates involved in disease progression and potential therapeutic targets. As understanding of the proteomic changes in disease states increases, multiplexed assays that can examine multiple protein biomarkers simultaneously will become necessary for further personalized treatment and care. Quantitative mass spectrometric assays have already demonstrated their equivalence to the current standard of care,[14,18,26] IHC, and in some cases have demonstrated superior patient stratification capabilities.[25] As the body of evidence grows for the utility of mass spectrometry-based proteomics, standard techniques, such as SRM, and emerging techniques, such as IMS, will become integrated in the pathology workflow and start to displace older techniques that do not have the throughput necessary for characterization of multiple protein biomarkers from a single sample. In the future, proteomic workflows will be the standard of care for proper diagnosis and care management.

References

1. Giusti L, Lucacchini A. Proteomic studies of formalin-fixed paraffin-embedded tissues. *Expert Rev Proteomics* 2013; **10**(2): 165–77.
2. Hood BL, Darfler MM, Guiel TG, Furusato B, Lucas DA, Ringeisen BR, Sesterhenn IA. Conrads TP, Veenstra TD, Krizman DB. Proteomic analysis of formalin-fixed prostate cancer tissue. *Mol Cell Proteomics* 2005; **4**(11): 1741–1753.

3. Patel V, Hood BL, Molinolo AA, Lee NH, Conrads TP, Braisted JC, Krizman DB, Veenstra TD, Gutkind JS. (2008). Proteomic analysis of laser-captured paraffin-embedded tissues: a molecular portrait of head and neck cancer progression. *Clin Cancer Res* **14**(4): 1002–1014.
4. Negishi A, Masuda M, Ono M, Honda K, Shitashige M, Satow R, Sakuma T, Kuwabara H, Nakanishi Y, Kanai Y, Omura K, Hirohashi S, Yamada T. Quantitative proteomics using formalin-fixed paraffin-embedded tissues of oral squamous cell carcinoma. *Cancer Sci* 2009; **100**(9): 1605–1611.
5. Ostasiewicz P, Zielinska DF, Mann M, Wisniewski JR. Proteome, phosphoproteome, and N-glycoproteome are quantitatively preserved in formalin-fixed paraffin-embedded tissue and analyzable by high-resolution mass spectrometry. *J Proteome Res* 2010; **9**(7): 3688–3700.
6. Azimzadeh O, Barjaktarovic Z, Aubele M, Calzada-Wack J, Sarioglu H, Atkinson MJ, Tapio S. Formalin-fixed paraffin-embedded (FFPE) proteome analysis using gel-free and gel-based proteomics. *J Proteome Res* 2010; **9**(9): 4710–4720.
7. Kawamura T, Nomura M, Tojo H, Fujii K, Hamasaki H, Mikami S, Bando Y, Kato H, Nishimura T. Proteomic analysis of laser-microdissected paraffin-embedded tissues: (1) Stage-related protein candidates upon non-metastatic lung adenocarcinoma. *J Proteomics* 2010; **73**(6): 1089–1099.
8. Nishimura T, Nomura M, Tojo H, Hamasaki H, Fukuda T, Fujii K, Mikami S, Bando Y, Kato H. Proteomic analysis of laser-microdissected paraffin-embedded tissues: (2) MRM assay for stage-related proteins upon non-metastatic lung adenocarcinoma. *J Proteomics* 2010; **73**(6): 1100–1110.
9. Morita Y, Ikegami K, Goto-Inoue N, Hayasaka T, Zaima N, Tanaka H, Uehara T, Setoguchi T, Sakaguchi T, Igarashi H, Sugimura H, Setou M, Konno H. Imaging mass spectrometry of gastric carcinoma in formalin-fixed paraffin-embedded tissue micro-array. *Cancer Sci* 2010; **101**(1): 267–273.
10. Wolff C, Schott C, Porschewski P, Reischauer B, Becker KF. Successful protein extraction from over-fixed and long-term stored formalin-fixed tissues. *PLoS One* 2011; **6**(1): e16353.
11. Alkhas A, Hood BL, Oliver K, Teng PN, Oliver J, Mitchell D, Hamilton CA, Maxwell GL, Conrads TP. Standardization of a sample preparation and analytical workflow for proteomics of archival endometrial cancer tissue. *J Proteome Res* 2011; **10**(11): 5264–5271.
12. Guzel C, Ursem NT, Dekker LJ, Derkx P, Joore J, van Dijk E, Ligtvoet G, Steegers EA, Luider TM. Multiple reaction monitoring assay for pre-eclampsia related calcyclin peptides in formalin fixed paraffin embedded placenta. *J Proteome Res* 2011; **10**(7): 3274–3282.
13. Nakatani S, Wei M, Ishimura E, Kakehashi A, Mori K, Nishizawa Y, Inaba M, Wanibuchi H. Proteome analysis of laser microdissected glomeruli from formalin-fixed paraffin-embedded kidneys of autopsies of diabetic patients: nephronectin is associated with the development of diabetic glomerulosclerosis. *Nephrol Dial Transplant* 2012; **27**(5): 1889–1897.

14. Hembrough T, Liao WL, Hartley CP, Ma PC, Velcheti V, Lanigan C, Thyparambil S, An E, Monga M, Krizman D, Burrows J, Tafe LJ. Quantification of Anaplastic Lymphoma Kinase Protein Expression in Non-Small Cell Lung Cancer Tissues from Patients Treated with Crizotinib. *Clin Chem* 2016; **62**(1): 252–261.
15. Blackler AR, Morgan NY, Gao B, Olano LR, Armani MD, Romantseva E, Kakareka JW, Bonner RF, Mukherjee S, Xiao B, Tran K, Pohida TJ, Emmert-Buck MR, Tangrea MA. Markey SP. Proteomic analysis of nuclei dissected from fixed rat brain tissue using expression microdissection. *Anal Chem* 2013; **85**(15): 7139–7145.
16. Assadi M, Lamerz J, Jarutat T, Farfsing A, Paul H, Gierke B, Breitinger E, Templin MF, Essioux L, Arbogast S, Venturi M, Pawlak M, Langen H, Schindler T. Multiple protein analysis of formalin-fixed and paraffin-embedded tissue samples with reverse phase protein arrays. *Mol Cell Proteomics* 2013; **12**(9): 2615–2622.
17. Fu Z, Yan K, Rosenberg A, Jin Z, Crain B, Athas G, Heide RS, Howard T, Everett AD, Herrington D, Van Eyk JE. Improved protein extraction and protein identification from archival formalin-fixed paraffin-embedded human aortas. *Proteomics Clin Appl* 2013; **7**(3–4): 217–224.
18. Hembrough T, Thyparambil S, Liao WL, Darfler MM, Abdo J, Bengali KM, Hewitt SM, Bender RA, Krizman DB, Burrows J. Application of selected reaction monitoring for multiplex quantification of clinically validated biomarkers in formalin-fixed, paraffin-embedded tumor tissue. *J Mol Diagn* 2013; **15**(4): 454–465.
19. Takadate T, Onogawa T, Fukuda T, Motoi F, Suzuki T, Fujii K, Kihara M, Mikami S, Bando Y, Maeda S, Ishida K, Minowa T, Hanagata N, Ohtsuka H, Katayose Y, Egawa S, Nishimura T, Unno M. Novel prognostic protein markers of resectable pancreatic cancer identified by coupled shotgun and targeted proteomics using formalin-fixed paraffin-embedded tissues. *Int J Cancer* 2013; **132**(6): 1368–1382.
20. Tanca A, Abbondio M, Pisanu S, Pagnozzi D, Uzzau S, Addis F. Critical comparison of sample preparation strategies for shotgun proteomic analysis of formalin-fixed, paraffin-embedded samples: insights from liver tissue. *Clin Proteomics* 2014; **11**(1): 28.
21. Finne K, Vethe H, Skogstrand T, Leh S, Dahl TD, Tenstad O, Berven FS, Reed RK, Vikse BE. Proteomic analysis of formalin-fixed paraffin-embedded glomeruli suggests depletion of glomerular filtration barrier proteins in two-kidney, one-clip hypertensive rats. *Nephrol Dial Transplant* 2014; **29**(12): 2217–2227.
22. Noberini R, Uggetti A, Pruneri G, Minucci S, Bonaldi T. Pathology tissue-quantitative mass spectrometry analysis to profile histone post-translational modification patterns in patient samples. *Mol Cell Proteomics* 2016; **15**(3): 866–877.
23. Wisniewski JR, Dus-Szachniewicz K, Ostasiewicz P, Ziolkowski P, Rakus D, Mann M. Absolute proteome analysis of colorectal mucosa, adenoma, and cancer reveals drastic changes in fatty acid metabolism and plasma membrane transporters. *J Proteome Res* 2015; **14**(9): 4005–4018.
24. Steiner C, Tille JC, Lamerz J, Kux van Geijtenbeek S, McKee TA, Venturi M, Rubbia-Brandt L, Hochstrasser D, Cutler P, Lescuyer P, Ducret A. Quantification of HER2 by

targeted mass spectrometry in formalin-fixed paraffin-embedded (FFPE) breast cancer tissues. *Mol Cell Proteomics* 2015; **14**(10): 2786–2799.

25. Nuciforo P, Thyparambil S, Aura C, Garrido-Castro A, Vilaro M, Peg V, Jimenez J, Vicario R, Cecchi F, Hoos W, Burrows J, Hembrough T, Ferreres JC, Perez-Garcia J, Arribas J, Cortes J, Scaltriti M. High HER2 protein levels correlate with increased survival in breast cancer patients treated with anti-HER2 therapy. *Mol Oncol* 2016; **10**(1): 138–147.
26. Hembrough T, Thyparambil S, Liao WL, Darfler MM, Abdo J, Bengali KM, Taylor P, Tong J, Lara-Guerra H, Waddell TK, Moran MF, Tsao MS, Krizman DB, Burrows J. Selected reaction monitoring (SRM) analysis of epidermal growth factor receptor (EGFR) in formalin fixed tumor tissue. *Clin Proteomics* 2012; **9**(1): 5.
27. Shi SR, Taylor CR, Fowler CB, Mason JT. Complete solubilization of formalin-fixed, paraffin-embedded tissue may improve proteomic studies. *Proteomics Clin Appl* 2013; **7**(3–4): 264–272.
28. Emmert-Buck MR, Bonner RF, Smith PD, Chuaqui RF, Zhuang Z, Goldstein SR, Weiss RA, Liotta LA. Laser capture microdissection. *Science* 1996; **274**(5289): 998–1001.
29. Tangrea MA, Chuaqui RF, Gillespie JW, Ahram M, Gannot G, Wallis BS, Best CJ, Linehan WM, Liotta LA, Pohida TJ, Bonner RF, Emmert-Buck MR. Expression microdissection: operator-independent retrieval of cells for molecular profiling. *Diagn Mol Pathol* 2004; **13**(4): 207–212.
30. Tangrea MA, Mukherjee S, Gao B, Markey SP, Du Q. Armani M. Kreitman MS. Rosenberg AM. Wallis BS. Eberle FC. Duncan FC. Hanson JC. Chuaqui RF. Rodriguez-Canales J. Emmert-Buck MR. Effect of immunohistochemistry on molecular analysis of tissue samples: implications for microdissection technologies. *J Histochem Cytochem* 2011; **59**(6): 591–600.
31. Du Q, Yan W, Burton VH, Hewitt SM, Wang L, Hu N, Taylor PR, Armani MD, Mukherjee S, Emmert-Buck MR. Tangrea MA. Validation of esophageal squamous cell carcinoma candidate genes from high-throughput transcriptomic studies. *Am J Cancer Res* 2013; **3**(4): 402–410.
32. Guo T, Wang W, Rudnick PA, Song T, Li J, Zhuang Z, Weil RJ, DeVoe DL, Lee CS, Balgley BM. Proteome analysis of microdissected formalin-fixed and paraffin-embedded tissue specimens. *J Histochem Cytochem* 2007; **55**(7): 763–772.
33. Stauber J, MacAleese L, Franck J, Claude E, Snel M, Kaletas BK, Wiel IM, Wisztorski M, Fournier I, Heeren RM. On-tissue protein identification and imaging by MALDI-ion mobility mass spectrometry. *J Am Soc Mass Spectrom* 2010; **21**(3): 338–347.

11 microRNAs in Tissue-Based Diagnosis and Prognosis: Possibilities and Limitations

Marc K. Halushka

Department of Pathology, Johns Hopkins University SOM, Baltimore, MD, USA

Contents

I. Defining microRNAs

microRNAs (miRNAs) are small (17–22 base pair [bp]) RNAs that serve as critical regulators of protein translation from mRNAs.[1] miRNAs are altered by cell stressors and malignant states, and changes in their expression have been strongly associated with human disease.[2] These changes suggest that miRNAs could be useful biomarkers of disease states if we can understand and harness their discerning capabilities. To get there, one needs a reasonable understanding of miRNAs, which begins with a brief explanation of their morphogenesis.

The formation of miRNAs is a complex action that occurs throughout the cell. A miRNA begins its lifecycle when it is transcribed from the host DNA in the nucleus. This transcriptional segment is known as a pri-miRNA and can be 5–100+ kilobases long.[3] The pri-miRNA can contain only the miRNA (or a collection of miRNAs) or it may be part of coding genes or long noncoding RNAs.[4] Many miRNAs are found within gene introns (miRtrons) while others are in bicistronic or polycistronic clusters. The smaller miRNA hairpins, known as pre-miRNAs (generally ~70 bp in length) are cleaved from the larger pri-miRNA through the actions of Drosha and DGCR8. The pre-miRNA then leaves the nucleus via Exportin 5 and is acted upon by Dicer. Dicer cleaves the hairpin structure to yield two mostly complementary mature miRNAs 17–22 bp in length from the 5p and 3p ends of the hairpin. These are known as the leading and passenger strand miRNAs. The leading strand enters into an RNA-induced silencing complex (RISC), and this larger protein structure then finds a mRNA binding partner. The passenger strand miRNA is not loaded into RISC and is generally quickly degraded.[3] Only rare miRNAs have reasonably abundant levels of both 5p and 3p miRNAs in which the less thermodynamically stable strand avoids immediate degradation.[5] The miRNA in the RISC complex binds a specific complementary sequence (generally) on the 3'UTR region of a mRNA. The most important part of this binding is a 6-8bp "seed sequence" starting at the +1 nucleotide position of the miRNA. Additional complementary binding across the length of the miRNA helps to stabilize the interaction.[6] Depending on the extent of Watson–Crick base pairing, this interaction may either cause translational repression by blocking translational machinery or it may result in degradation of the mRNA. These are the generally worked out mechanisms of miRNA maturation and activity; however, there are exceptions to just about every one of these rules.

After their initial discovery in *C. elegans*,[7,8] miRNAs have been described in essentially all life forms. Many miRNAs are highly conserved from primitive life forms to mammals. The let-7 family, in particular, was first described in *C. elegans* and is highly conserved throughout the animal kingdom including humans.[9] All known miRNAs have been curated at miRBase.org, the central repository and nomenclature organization of miRNAs.[10] The most recent revision (v21) had over 35,800 miRNAs across 223 species. The deepest discovery is for *Homo sapiens* where ~2,500 miRNAs have been reported. Recent deep sequencing studies of small miRNA have yielded thousands of additional putative miRNAs[5,11,12] resulting in the formation of miRCarta which lists 12,857 additional miRNA precursors.[13] Converse to this miRNA expansion has been the development of MirGeneDB 2.0, a hand-curated list of miRNAs that are derived from the canonical methods described above.[14] MirGeneDB 2.0 only lists 586 human miRNA genes. It is likely that many miRNA-like small RNA sequences exist, and a better definition of these entities, to help remove them from the canonical miRNA grouping, is needed. Some classes of miRNAs are ancient, present in most animals, while others are restricted to certain species. There are ~280 miRNAs shared across mammalian species, indicating deep functional conservation.[15]

Mature miRNAs have distinct and important naming conventions. A complete miRNA name includes a three character species code (ex. hsa for Homo sapiens, mmu for mus musculus), followed by a dash, followed by the letters miR, followed by another dash, followed by a

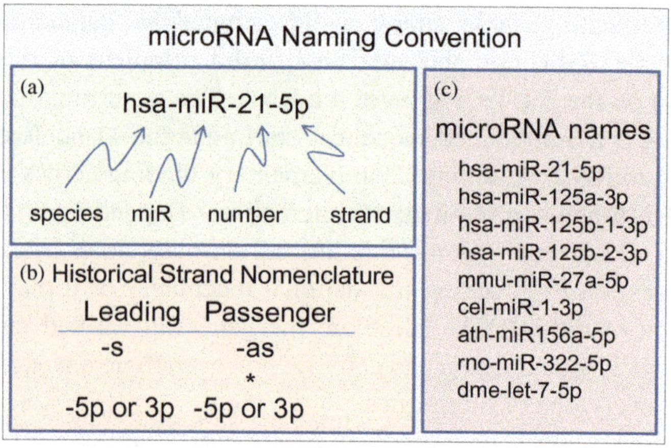

Figure 1. Mature microRNA Naming Convention. (a) All microRNAs have the same general style. The first three letters indicate the species. Then a generic miR is included, followed by a number and the strand direction. (b) Historically the abundant (leading) strand was considered the sense (-s) strand while the passenger strand was the antisense (-as) strand. This nomenclature was dropped and the leading strand had no strand information while the passenger strand was given a star (*). Now microRNAs are named based on whether they are from the 5p or 3p end of the pre-miRNA hairpin regardless of which is the abundant strand. (c) microRNAs are found in all species and share general naming conventions. The additional -1 and -2 in hsa-miR-125b-3p is the result of a duplication of the miR-125 locus and a sequence difference on the -3p microRNA, but not on the 5p microRNA.

number. Sometimes these numbers can have letters following them if they are miRNAs that share the same seed sequence and otherwise appear to be related through genetic loci duplication. After this number, miRNAs generally have another dash followed by a 5p or 3p designating the part of the pre-miRNA strand the miRNA comes from. These represent the leading and passenger strands, which have also been described as the sense (-s) and antisense (-as) strands or have a * notation signifying the passenger strand (Figure 1). The two exceptions to this general rule are the let-7 family and lin-4, both described in *C. elegans* as the *lethal-7* gene and the regulator of lin-14.[7] These distinct miRNA names (ex. hsa-let-7a-5p, cel-lin-4-3p) remain, and appear, across multiple species.[9] In general use, once a species has been established in a study, the three-character species code is removed. Also, most investigators do not report strand information (5p and 3p) when they are referring to the dominant strand. For the purposes of this chapter, all miRNAs will be human unless otherwise stated and the -5p and -3p convention will only be used as needed.

miRNAs exist in two different types of families. One family is based on genomic localization in which polycistronic clusters contain multiple miRNAs. One such group is the miR-17~92 family residing together on chromosome 13 and containing 6 miRNAs. These miRNAs all have different seed regions suggesting they regulate different genes but have some shared regulatory control.[16] Also, this cluster is partially duplicated on the X chromosome where some of these miRNAs have identical sequences to those on chromosome 13. Another polycistronic cluster, found on chromosome 19, contains 42 miRNAs that are highly expressed exclusively in placenta.[17] This cluster demonstrates shared transcriptional regulation of the pri-miRNA. Another familial clustering type is based on function. Some miRNAs have identical seed sequences providing functional redundancy of gene regulation. For example, miRNAs hsa-miR-215-5p and hsa-miR-192-5p (names from miRBase), found on separate chromosomes, each have the same TGACCTAT seed region. In MirGeneDB, these two are assigned the names Mir-192-P1 and Mir-192-P2 reflecting the shared seed sequence.[15] Often in knockout experiments there is mild to no phenotype, when all members sharing this redundancy have not been removed.

Another feature of miRNAs is that they exist in isomiR families. While there is a canonical sequence for each miRNA, as listed in miRBase and MirGeneDB 2.0, there are other variants to this sequence. These are mostly length variants that match the genomic sequence where more or fewer additional 3′ nucleotide or nucleotides may exist due to inexact dicer editing of the pre-miRNA. Additionally, new nucleotides, primarily adenosine and uracils, can be added to the 3′ end of the miRNA. The percent of each of these features is thought to vary by miRNA and two examples are given (Figure 2). Some recent research in the area of introns suggests useful discriminatory functionality in tissue studies.[18]

	IsomiRs of let-7a-5p	% of sequences
Canonical	TGAGGTAGTAGGTTGTATAGTT	67%
Shortened canonical	TGAGGTAGTAGGTTGTATA--	22%
Extended canonical	TGAGGTAGTAGGTTGTATAGTTTA	3%
3′ extension and edits	TGAGGTAGTAGGTTGTATAGTTNNNN	7%
5′ extension and edits	NNGAGGTAGTAGGTTGTATAGTT	0.4%
Internal edits	TG NGGTAGTAGGTTGTATNGTN	0.1%

	IsomiRs of miR-92a-3p	% of sequences
Canonical	TATTGCACTTGTCCCGGCCTGT	41%
Shortened canonical	TATTGCACTTGTCCCGGCC---	21%
Extended	TATTGCACTTGTCCCGGCCTGTTG	7%
3′ extension and edits	TATTGCACTTGTCCCGGCCTGTNNNN	31
5′ extension and edits	NNATTGCACTTGTCCCGGCCTGT	0.1%
Internal edits	TATTGCACTTGTCCCGGCNTGT	0.1%

Figure 2. Every microRNA is made of a family of sequences from the same loci, only some of which are canonical matches to miRBase.org. Common alterations are shortened sequences or 3′ extensions and edits.

II. microRNAs in Cells and Tissues

After the discovery of this broad class of RNAs, there was a spike of interest in localizing miRNAs to tissues. Multiple studies in humans and mice using array-based, hybridization-based, or early small RNA-seq based-investigation of five or more tissues.[19–22] They established some specific and ubiquitous expression patterns of miRNAs. For example, miR-122 was localized to the liver. This data was useful, but led to some confusion.[23] Within these tissues are a variety of cell types and each cell type could have its own miRNA expression pattern. Cell types, such as red blood cells, endothelial cells, and fibroblasts, can be found in most, if not all tissues. A miRNA, specific or enriched to one of these cell types, could be confused as a ubiquitously expressed miRNA across a variety of tissues. A classic example is the research concerning miRNAs miR-451a and miR-144. These two miRNAs are in a bicistronic cluster and are only expressed in red blood cells. From tissue-only data, they were thought to be widely expressed in a variety of cell types. Even worse, different levels of tissue washing (to remove red blood cells) led to altered expression of these miRNAs in normal and diseased tissues, leading to an implication of these miRNAs in disease.[24] Numerous studies of these miRNAs now exist in cancer and other diseases, for which they are unlikely to have any role.[25,26]

This confusion could only be resolved with cell-level data. In the summer of 2017, this data was finally generated. Three separate studies significantly enhanced our knowledge of cell specific miRNA localization using deep small RNA-seq across hundreds of samples.[5,11,27] The summation of this data can be found as a barChart track at the UCSC Genome Browser for which any miRNA can be evaluated for its expression level from over 75 unique primary cell types and 50 cancer cell lines.[28]

III. Obtaining Tissue and Cell Samples for microRNA Studies

The understanding of miRNA's role in human disease often begins with human tissues and cells. A typical study may compare a collection of normal and diseased tissues to identify dysregulation of miRNAs. It is important, when making these tissue and cell collections that one considers some important variables.

It is an underappreciated and exceedingly difficult task to match "control" and "diseased" tissues for most studies. The main reason is that the cellular composition of tissues will vary highly between different samples. As we have established that miRNAs vary by cell type, this will have an outsized influence on differences of miRNA levels in tissues. Two examples nicely illuminate this challenge. A group of researchers compared normal colon and inflamed colonic tissues.[29] They identified miR-150 to be overexpressed in the inflamed tissues and interpreted this as an upregulation of miR-150 in the colonic epithelial cells. However, miR-150 is only expressed in lymphocytes and it was the significant increase of lymphocytes in the diseased sample that drove this change, not dysregulation of miR-150. It is very challenging to find an appropriate tissue control for an inflamed disease tissue. In a second study, a group investigated miRNAs that were dysregulated in gastrointestinal stromal tumors (GISTs). They compared tumor tissue with adjacent normal colon tissue. A GIST is comprised of malignant mesenchymal cells akin to fibroblasts. It has no epithelial cell component. Conversely, normal colon tissue is a mix of epithelial cells, fibroblasts, smooth muscle cells, rarer inflammatory cells, and other cells. Therefore, it was not surprising they identified miRNAs exclusive to epithelial cells, such as the miR-200 family as being markedly reduced in GIST.[30] Again, the striking difference of the cellular composition of the tissues makes this an unfair comparison to explain any of the pathogenesis of the disease process. Therefore, picking appropriate matched tissue is a major challenge of expression studies.

Diseases most amenable to appropriate pairing between normal and disease include those in which the cellular composition is less variable, with the pathogenesis occurring with the normally present cells. Examples include minimal change disease of the kidney and hypertrophic cardiomyopathy in the heart. For other diseases, it may not be feasible to make a reasonable comparison and other methods should be employed.

One option is to isolate a particular cell type involved in the pathogenesis by a separation method and assay it specifically. Examples of separation methods include laser capture microdissection, magnetic bead separation, and flow-based isolation. In that way, a comparison of epithelial cells to epithelial cells in inflammatory bowel disease, for example, can be more of an "apples" to "apples" approach.

A second method is to deconvolute the total tissue expression level into the intrinsic cell types. Methods such as CIBERSORT rely on individual cell data to predict the composition of tissues.[31] They can inform on the percent composition of tissues to "normalize" the levels of each cell type to make more rational comparisons. The recent generation of cell-specific miRNA expression data should aid in this method.[5,28]

Within blood, comparisons may be more amenable to comparisons. Leukemias and other blood-based diseases may be more similar to control cells such that miRNA changes better reflect the disease process. However, a whole blood sample or buffy coat (total leukocytes of blood), may have marked skewing of leukocyte components between neutrophils and leukemic cells. Thus again, an isolation method such as flow-sorting should be considered.

IV. Sample Preparations of microRNAs

Regardless of the tissue used, several aspects for miRNA isolation should be recognized. Although it is always a good idea to use the highest quality tissue as possible, miRNAs may allow a greater tolerance of tissue quality. RNA quality is usually measured based on an RNA Integrity Number (RIN) scored between 0 and 10.[32] It is rare to get RNA scores above 8 for human tissues, due to devitalization time during surgeries or post-mortem intervals. While most groups discourage the use of RIN values <6 for genes in traditional RNA-seq, lower RIN values (down to RIN 4) may be reasonable for miRNA studies.[33] While total miRNA levels may be diminished, the relative levels of the miRNAs in degraded tissues are more generally maintained. The reason for this is two-fold. One is that shorter RNA species are less likely to fragment relative to longer mRNAs. Secondly, miRNAs are found in Ago2 and that localization seems to protect miRNAs from RNAses. However, not all studies agree.[34] A number of companies have developed mini-column-based approaches to isolate RNA. Several examples include miRNeasy (Qiagen), AnaPrep (Biochain), mirVana (ThermoFisher), mirPremier (Millipore Sigma) and NucleoSpin (Takara). Some isolation kits are more specific for specialty approaches such as formalin-fixed tissues (miRNeasy FFPE Kit) or blood-based isolation (PAXgene Blood). In particular, miRNAs obtained from formalin-fixed tissues are reported to be more robustly quantified than mRNAs.[33,35]

V. Measurement Systems for microRNAs

Once miRNAs have been isolated from whatever source and to whatever extent, they need to be assayed for expression levels. A number of methods exist to do this, and these methods can be divided by their ability to assay numerous miRNAs at once, or to assay one miRNA at a time.

There are multiple systems designed to evaluate large quantities of miRNAs that can essentially be broken down into hybridization, RNA sequencing, and quantitative PCR methods. The hybridization methods bind mature miRNAs to probes arrayed across solid-state systems and provide a fluorescent intensity signal for each location. Small RNA sequencing (RNA-seq) provides millions of reads of sequence data for each sample that can be processed to identify the miRNA component of the sample. Quantitative PCR (qPCR) uses primers to amplify the miRNAs of interest. Each system has its benefits and limitations.[36]

a. Hybridization systems

Hybridization systems include Affymetrix arrays, Agilent Technologies arrays, Nanostring nCounter system, and Exiqon arrays. The latest version of the Affymetrix array (GeneChip® miRNA 4.0 Array) contains 30,424 mature miRNAs (from 203 organisms including all human miRNAs in miRBase v20) and additional snoRNA and scaRNAs. It requires 130–1,000 ng of total RNA. Agilent Technologies' human miRNA Slide (Release 21.0) contains probes for 2,549 miRNAs and requires 100 ng of total RNA. The NanoString nCounter human miRNA v3 system contains 800 human miRNAs and is based on miRBase v21. It requires 100 ng of total RNA. The benefits of hybridization systems are the relatively low amount of starting RNA needed and the lower cost. The concern of these systems is cross-hybridization and mishybridization. Cross-hybridization is more likely to occur with the Affymetrix system containing 30,424 probes. Many of these are likely to vary by only a couple of nucleotides due to the strong conservation across species. It is also not clear what percentage of non-canonical isomiR miRNAs are captured by each probe. As seen in Figure 2, the percent of canonical reads can vary by miRNA.

b. Sequencing systems

Historically, the two major miRNA RNA-seq platforms are the Illumina TruSeq Small RNA system and the ThermoFisher Ion Torrent system. The Illumina system is coupled to a TruSeq Small RNA sample preparation kit, but kits by other manufacturers (New England Biolabs and Bioo Scientific) are also used. More recently, low-abundance small RNA-seq kits have appeared allowing as little at 1–10 ng of RNA to serve as starting material versus the 1 μg needed for older methods.[37] Ion Torrent™ sequencing systems work with Ion Total RNA-Seq Kits to prepare samples that can be sequenced on a range of systems including the Ion PGM™ System or The Ion Proton™ system. The outputs of these systems are FASTQ files containing raw sequencing read data. This data must be processed to identify all of the miRNAs in the file. While this used to be a tedious or specialized task, new software systems have made this much more accessible to scientists. In particular, Chimira (web-based) and miRge 2.0 (locally run) both align sequences to miRNA libraries, are easy to use, and provide fast and robust results.[38–40] Systems based on alignment to the total genome tend to be less robust than those aligning to dedicated libraries.

The different sequencing platforms each have a range of depth of coverage. The Illumina miSeq system can provide 30 million reads and the HiSeq systems can provide up to 300 million reads. The Ion PGM can provide up to 5.5 million reads and the Ion Proton provides 80 million reads. Additional sequencers such as the iSeq 100 have lower costs, but only provide up to 4 million reads. For the purpose of capturing and counting miRNAs, how many reads are needed? We, and others, have determined that having 1 million miRNA reads is sufficient to capture all miRNAs with a frequency of at least 10 reads per million miRNA reads (RPM).[39] However, other manuscripts have published with many fewer reads per sample (40,000+).[27] For well-constructed miRNA libraries, using traditional library preparation methods, the percent miRNAs are between 35% and 85%. For the current iterations of small input library preparations, the percent miRNAs have been closer to 10%. Thus, one needs 3–10 million reads in a sequencing run, depending on the library preparation. For most of these systems, multiplexing samples together can increase the cost-effectiveness of the system. The benefits of RNA-seq are that one can identify all miRNAs (including novel miRNAs) and accurately account for all isomiRs. The limitations are the relatively higher cost (although with multiplexing this is decreasing) and library preparation biases.

c. Quantitative PCR systems

There are numerous platforms to quantify miRNAs built around a qPCR structure. The two most well known are the TaqMan Cards/Open Array (Life Technologies) and miScript (Qiagen). The TaqMan® OpenArray® Human miRNA panel profiles 754 human miRNAs with only 100 ng of total RNA starting material. The TaqMan® Array miRNA Cards cover 740 miRNAs based on miRBase v20 and need as little as 1 ng of total RNA. Qiagen provides a number of different miScript profilers based on different disease processes, each containing a different mix of miRNAs. One product is designed specifically for serum and plasma and contains 84 detectable miRNAs. miScript uses a standard SYBR Green-based RT-PCR with ΔΔCT normalization against internal controls.

d. Platform comparisons

The large number of miRNA profiling systems has made cross-study comparisons difficult. It was unknown how well the systems performed relative to each other. To better understand this, the miRQC consortium evaluated 12 miRNA expression platforms.[41] Systems were compared for reproducibility, sensitivity, accuracy, specificity, and concordance of differential expression. Although no winners were declared, in general, the qPCR systems outperformed the other methods. In particular, they were better than the hybridization systems in accuracy and sensitivity. Among the sequencing systems, the Illumina system outperformed the Ion Torrent system, particularly in specificity and accuracy (Table 1).

VI. *In situ* Studies of microRNAs

In situ hybridization for RNA molecules is a standard molecular biology technique. miRNAs have also been evaluated by this method, although this is significantly more challenging than for longer RNAs. To understand why, let us begin with an mRNA approach. A current, popular RNA *in situ* hybridization (RISH) method, termed RNAScope, has very high sensitivity to detect even single RNA molecules.[42] The method takes advantage of using two ~20 bp independent Z probes that hybridize to the target sequence in tandem. Then a separate signal amplifying probe binds to both, upon which a fluorescence or chromogenic signal is created. The advantage of having an mRNA (or lncRNA) as a target is that a unique ~40 bp segment can usually be located with the 1 + kb of sequence to offer specificity to the binding strategy. Further, these can cross intronic regions to maximize RNA binding over DNA binding. In comparison, a miRNA is only ~20 bp and within certain families, such as the let-7 family, there can be high sequence conservation between members. This can cause non-specific, or off-target, hybridization. Additionally, the most abundant sequence of a miRNA family is typically only ~45% of all reads of a given

Table 1. miRNA quantitation techniques.

General description	Advantages	Disadvantages	Representative systems /vendors	RNA required
miRNA microarray				
Hybridization technique, massively parallel	Assay the entire known miRNAome at once; uses robust global normalization; measure more lowly expressed miRNAs	Cross-hybridization and hybridization efficiency challenges; no coverage of isomiRs; fluorescence noise can be misinterpreted as signal	GeneChip miRNA 4.0 Array/ Affymetrix miRNA Microarray/Agilent miRCURY LNA microRNA array/ Qiagen nCounter/Nanostring	ng-µg
Quantitative PCR (qPCR)				
Amplifies one miRNA at a time using standard PCR technique	Can be tailored from one to hundreds of miRNAs for individualized studies; sensitive	PCR efficiency variation between primers; normalization options	TaqMan/ABI TaqMan OpenArray/ABI miRCURY LNA miRNA PCR System/ Qiagen	pg-ng
Droplet digital PCR (ddPCR)				
Uses a water–oil emulsion technology to quantify PCR across 20,000 droplets	Absolute and simplified quantification	Not massively parallel	QX200/Bio-Rad	pg-ng
Small RNA sequencing (RNA-seq)				
Identify all RNA species below a size threshold (~30 bp) using sequencing by synthesis	Complete capture of all miRNAs; obtain other small RNA types; newer library preparation methods can use nanograms of RNA	Library preparation kit ligation differences; normalization challenges; requires millions of reads to robustly measure rare miRNAs	HiSeq; miSeq; iSeq; Nextseq/ Illumina Ion Torrent/ThermoFisher	ng-µg

Source: Adapted from PMID: 22510765.

miRNA, due to the isomiR family, but this percentage is highly variable.[5] A probe may or may not capture the isomiRs. Thus, with only a single probe for a given miRNA, roughly half of the signal may be missed. A potential additional challenge is that most miRNA, in a cell, is bound to Ago2 within a RISC complex. It is unknown how successful these probes can be to access the miRNA within this complex in formalin-fixed or frozen tissues. This is why only robustly expressed miRNAs are typically detected appropriately, as attempts to identify miRNAs with low signals usually results in high background staining. As a result, miRNA ISH is fraught with the potential for error.

With that said, the best commercially available ISH method utilizes locked nucleic acids (LNAs) (miRCURY by Qiagen). This method has been used in many publications. The Mourelatos laboratory has demonstrated convincing images for three miRNAs (miR-124a, miR-125b, and miR-9) in oligodendrogliomas using LNA-ISH.[43] The Mendell laboratory also showed clear miR-143 and miR-145 staining of mesenchymal cells in the absence of epithelial cells.[44] We demonstrated endothelial cell staining of miR-126, albeit with a high background showing positivity in some non-endothelial cell types.[45] The Maitra lab used this method to highlight miR-155 co-localizing with miR-21 in pancreatic adenocarcinoma.[46] miR-155 is well known to be expressed in immune cells and is not thought to be expressed in epithelial cells, and thus this may represent background or mishybridization signal.

At this time, visualization of the miRNA signal in tissues remains challenging without a clear means to improve on current methods due to the short nature of the mature miRNA.

VII. microRNAs Implicated by Tissue Studies

Due to the concerns raised above regarding the changing cellular composition of tissues in tissue expression studies, the number of robust miRNA expression alteration papers is limited. We will highlight some examples here.

The Bitzer laboratory investigated the role of miR-21 in glomerular injury.[47] Most of the work involved transgenic mouse models where they demonstrated a role for TGF-β1 to regulate miR-21 levels influencing podocyte activities. To confirm these findings in human subjects, they separately microdissected the glomeruli and tubulointerstitial fractions of 48 renal biopsies. This is important as it reduced the complexity of the tissues, minimizing the contribution of altered cell composition to miRNA levels. Within the glomeruli, they found significantly higher levels of miR-21 in individuals with macroalbuminuria (increased renal injury) compared to normalbuminuria and microalbulminura. In contrast, no change in the tubulointerstital cells was found. miR-21 is a ubiquitously expressed miRNA found abundantly in nearly all cells. In this study, altered miR-21 levels were specifically noted in the glomerulus. This association would have been difficult to make if the entire tissue been evaluated at once as the contributions of all other cells would have overwhelmed the altered miR-21 signal coming from a cell (or cells) in the glomerulus.

The Dorn laboratory investigated changes in miRNA and mRNA levels in human hearts that developed heart failure and recovery from heart failure with mechanical support.[48] This comparison of three tissues was slightly complicated by two of the tissues (non-failing and failing hearts) being obtained from one center and the recovered heart tissues being obtained from a second center. Nonetheless, non-inflammatory and, in particular, dilated cardiomyopathy hearts have generally robustly consistent cell ratios. This group found that a number of miRNAs, including myomiRs (muscle-enriched/specific miRNAs), were elevated in failing heart tissues. The higher myomiRs included miR-1, miR-133a, miR-133b, and miR-499. Other miRNAs that were higher included miR-126, most highly expressed in endothelial cells, and miR-143 which is highest in smooth muscle cells and fibroblasts. These upregulated miRNA signals generally abated in the recovery tissues. In contrast, many mRNAs were up- and down-expressed in heart failure. Fewer of these mRNA signals returned toward baseline expression in the recovered tissues.

Understanding the concerns of complex tissue on miRNA expression levels, the Lupold laboratory used expression microdissection (xMD) to isolate epithelium from stroma in formalin-fixed prostate cancer specimens.[49,50] Within a wider project, they demonstrated the cell-type specificity of several miRNAs. Specifically, they demonstrated miR-1 and miR-143 are specific to stroma while miR-141 was specific to epithelium.

Consistent with other work on the ubiquity of miR-21, this miRNA was found abundantly in both cell types. This data challenged some reports suggesting miR-1 was an epithelial-expressed tumor suppressor.[51] It nicely demonstrated how using an isolation step in human tissues can improve signal specificity.

Tissue studies can also be complemented by blood-based biomarker studies. The Nattel laboratory investigated miRNAs that play a role in atrial fibrillation.[52] In this complex study, the team investigated miR-29b in a mouse model, a canine model, and in canine fibroblasts. On the human side of the study, they obtained right atrial appendages of the heart from 36 individuals and blood plasma from 95 individuals who either had or did not have atrial fibrillation. They noted that in the atrial appendages, miR-29b was decreased by 40–54% in patients with different forms of fibrillation. In plasma, miR-29b levels were significantly reduced as well. This connection between miRNA alterations in tissues and plasma brings up some points worthy of consideration for blood-borne miRNAs that will be briefly covered here.

VIII. microRNA Subsets in Bodily Fluids

miRNAs, in bodily fluids, are found in basically three states. Some miRNAs are bound to Argonaute2 (Ago2) of the RISC complex which is freely carried within serum or plasma and protects the small RNA from RNAse activities.[53] Some miRNAs are bound to lipoprotein complexes.[54] Other miRNAs are found within extracellular vesicles. These include exosomes, apoptotic bodies, and other microvesicles. Of these three compartments, it is unclear where most miRNAs reside. One early report suggested they are predominately in the free Ago2-bound fraction.[53] A second report suggested they are concentrated in exosomes.[55]

The localization of a miRNA to each carrier state is based on different biologic activities. Ago2-bound free miRNAs are mostly the product of cell lysis, either through necrosis or through apoptosis.[56] Thus, diseases such as myocardial infarction, hepatonecrosis, or tumor necrosis result in elevated levels of free miRNAs that have exclusivity to these organs as described above. The extracellular vesicle-encapsulated miRNAs are the product of packaging of miRNAs from viable cells. The literature is

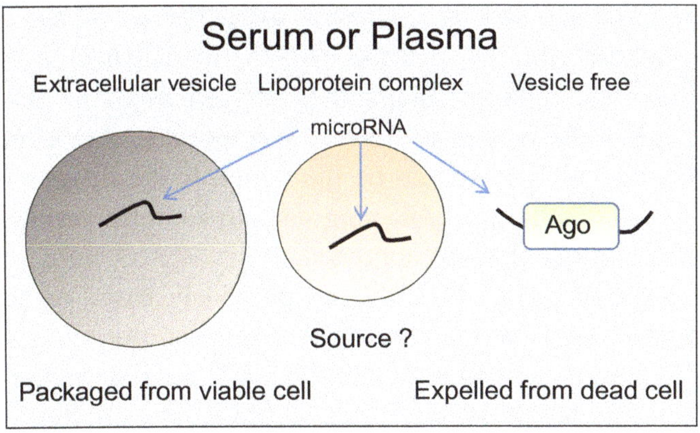

Figure 3. Blood-based microRNAs exist in essentially three forms. One is in microvesicles that contain microRNAs that may or may not be bound to Ago. The second form is in a lipoprotein complex (i.e., HDL). The third is free in the serum or plasma-bound to Ago. Microvesicle microRNAs are packaged from viable cells (unless they are in apoptotic vesicles), whereas most free microRNAs are likely the result of spillage of cellular contents from dead and dying cells. The source of lipoprotein microRNAs is unclear but may come from both viable and dead cells. (Structures are not drawn to scale).

unclear as to whether this is or is not a selective process. It also likely depends on the type of extracellular vesicle.[57] Either way, these miRNAs indicate the functional activity of a given cell. Higher or lower levels of miRNAs within extracellular vesicles indicates cellular function perturbation and altered signaling, but not cell necrosis. The source of lipoprotein complex miRNAs is less clear, but is likely an amalgam of both processes. (Figure 3).[54,58]

IX. Isolation Methods for Fluid-based microRNAs

Isolation methods to obtain miRNAs for serum and plasma significantly impact on which types of miRNAs are isolated. This has been a critically underappreciated aspect of miRNA biomarker studies and better approaches should be taken. Serum and plasma differ by the presence of blood clotting factors. Serum is depleted of clotting factors while plasma contains clotting factors. Within either of these fluids, extracellular vesicles, lipoproteins, and free miRNAs exist. For a miRNA biomarker study, the decision should be made early as to which type of miRNA would be important in understanding the pathogenesis of the disease being queried. If cell death-related miRNAs are important, then capturing the free Ago-bound miRNAs should be the target. If the disease process is unrelated to cell death, then collecting the extracellular vesicle-encapsulated miRNAs may be the best approach.

The expression levels of miRNAs vary widely based on the general isolation method used. In a key example of this concept, one sample was processed by different methods, comparing a $600 \times g$ spin with a $3{,}400 \times g$ spin and then a further $1{,}940 \times g$ spin followed by passage through a 0.22-μm filter.[59] Each step had variable effects on the miRNAs. Between the least and most selective purification methods, some miRNAs had expression levels that were more than >1,000 fold different (miR-142-3p and let-7a) while other miRNAs (miR-122, miR-451a) were less than 4-fold variable.

X. Subsetting Extracellular Vesicular microRNAs

Evaluating the extracellular vesicle fraction of serum or plasma for their miRNAs has not been routinely performed for biomarker studies. One reason is that the isolation of extracellular vesicles is fraught with a general lack of consistency mainly stemming from a lack of exacting definitions for different vesicles (exosomes vs. microparticles) and agreement on isolation methods.[60,61] In particular, the different names of extracellular vesicles, particularly "exosome," are used interchangeably, despite very different meanings. One method to define these particles is size/method-dependent. Ultracentrifugation at $10{,}000$–$20{,}000 \times g$ can collect a pellet that contains microvesicles/microparticles/ectosomes. Ultracentrifugation at higher speeds ($100{,}000$ or higher $\times g$) will pellet exosomes. These methods only achieve enrichment, not purity.[62] Protein complexes, lipoprotein particles, and other contaminants will be included.

XI. Origin of microRNAs in Blood-based Fluids

The Nattel study described above demonstrated lower levels of miR-29b in both tissue and plasma that correlated with atrial fibrillation.[52] It is important to consider the origin of the plasma miR-29b. This miRNA is highly expressed in macrophages and lymphocytes, with generally lower expression across a range of other cell types. Three data sets of small RNA-seq data of the heart (Sequence Read Archive numbers ERR038425-7) and additional data from iPSC-derived cardiomyocytes (SRR1542718) report very low, overall expression in the heart and cardiomyocyte.[5,28] It is possible to probable that the group accurately measured a reduced miR-29b in the tissue, based on a change in expression from a non-cardiomyocyte. An inflammatory cell present in the heart (macrophage or lymphocyte) could either reduce its expression of miR-29b or, perhaps, be reduced in number in atrial fibrillation tissues relative to other cell types (i.e. increasing fibroblasts), resulting in a lower value.[63] Squaring that tissue finding to the reduced miR-29b plasma data is more challenging.

Do the atrial cells responsible for the reduction of miR-29b levels contribute less miR-29b into plasma, thus reducing total plasma levels? Considering all of the inflammatory cells, and in fact, all of the other cells of the body that could shed/lose/actively export miR-29b into blood, the inflammatory cells of the atrium would only represent a tiny contributor to global miR-29b levels. It is hard to reconcile how the plasma signal could so strongly match this data. One idea is that perhaps subjects with atrial fibrillation have an intrinsically reduced miR-29b expression level that leaves them predisposed to this disease. However, that did not match the canine model data.[52] Regardless, this exercise demonstrates the complexity of the contributors to blood-based miRNA signals.

A significant amount of the circulating miRNAs in blood, that can be detected in serum or plasma, are from circulating blood cells such as red blood cells and platelets. At least one group has strongly championed using platelet-poor plasma as a starting point to remove this contamination (predominately miR-223).[64] The remainder of the miRNAs can be obtained from any cell type of the body. Thus, a serum or plasma sample is essentially a total-body biopsy. Therefore, capturing a modestly differentially expressed signal of a miRNA from a rare cell of the body, particularly of a miRNA that is ubiquitously expressed is somewhere between challenging and a foolhardy exercise. What works best, for serum or plasma, is to identify a miRNA that has specificity to a single cell type and that is shed upon injury of that cell type in disease. This is why the myomiR class of cardiac and skeletal muscle-specific miRNAs are successful for detecting cardiac injury such as myocardial infarction.[56]

The isolation of the total extracellular microvesicle fraction does reduce some of the complexity of the sample. However, these microvesicles in the circulation also come from a wide variety of cells. This includes hematologic cells (red blood cells, platelets), endothelial cells, epithelial cells, neural cells, myocytes, and others. Thus, this sample also does not have high specificity for any given cell type. Additionally, it is known that in certain disease states, some cells will increase their production of extracellular microvesicles. Thus, if a miRNA is enriched in a certain cell type that produces 2% of all exosomes (ex. miR-92a in lymphocytes) and that cell type increases their exosome production to 4%, there may be higher levels of miR-92a in the exosome fraction, but the cellular expression level would be unchanged.

An advantage of comparing microvesicle (rather than plasma/serum) miRNA signals to tissue studies is the potential for cell-type specificity. New methods are being developed to isolate subsets of these microvesicles that come directly from a particular cell type.[65] This intrinsically makes sense as it can allow one to focus directly on the miRNAs of the particular cell type. So, in the atrial fibrillation example above, isolating microvesicles derived from lymphocytes, macrophages, or whatever cell altered its miR-29b expression level in atrial tissues could lead to a more valid blood-based replication. This would be a more robust biomarker for the disease. One drawback of this approach, though, is capturing sufficient numbers of microvesicles to have enough material to assay in downstream systems. For example, isolating macrophage miRNA biomarkers may represent only 10–15% of all exosomes, which is already a subset of serum or plasma miRNAs. The total pool to assay is greatly diminished.

XII. microRNA Tissue Testing in Clinical Scenarios

One of the best opportunities to utilize tissue-based miRNAs is in the setting of fine needle aspiration. Fine needle aspiration is often performed to extract a small number of cells from a mass lesion to rule in/out a possible malignancy. Historically, these cells were reviewed by cytopathologists who can make a firm diagnosis of malignancy vs benign process based on cytology in only 60–90% of all cases, leaving opportunities for molecular methods.

One specific fine needle aspiration-challenging tumor that has been targeted is the thyroid. It is possible that other molecular tools (genes, DNA mutations, proteins, or miRNAs) could improve the prognostic capabilities of the fine needle procedure and better optimize the surgical approach to the patient. Keutgen *et al.* explored six miRNAs by qPCR in two groups of indeterminate samples of needle biopsies of thyroid. The group found that four of these miRNAs could predict with 100% sensitivity and 86% specificity malignancy from benign lesions.[66] If validated, this would be particularly useful and meaningful as a biomarker.

This line of reasoning was picked up by Rosetta Genomics for further study.[67] Instead of using the four miRNAs of the Keutgen work, they started anew with initially investigating over 2,000 miRNAs. From this initial discovery study, they selected 96 miRNAs to evaluate in 82 fine needle aspirate smears. This yielded 24 miRNAs in a final set to test against 375 fine needle aspirate smears and then validated them with an additional 189 smears. The simple goal was to identify miRNAs that could distinguish between malignant and benign thyroid masses. The overall test performance of the miRNA panel showed that it had a negative predictive value of 91% (95% CI 84–96%) along with a sensitivity of 85% (CI 74–93%) and a specificity of 72% (CI 63–79%). These values were outperformed by a group of three reviewing pathologists who, when they agreed on the diagnosis, had superior test statistics: negative predictive value 99% (CI 94–100%); sensitivity: 98% (CI 87–100%); specificity: 78% (CI 69–85%). What is important in the scope of this work is that miRNAs as tissue-based biomarkers are being seriously evaluated by industry.

XIII. Future Roles of Tissues in Establishing the Roles of microRNA in Disease and Diagnosis

It is clear from the first generation of tissue-based experiments with miRNAs that there are both opportunities and challenges. The greatest challenges include determining what is or is not a miRNA, consistent measurement techniques, and obtaining appropriate tissues for comparison of disease and normal states. The good news is that new efforts are solving these problems. The realization that not everything that is 21 bp long is a miRNA has led to a highly selective list of miRNAs in MirGeneDB2.0.[14] Technical advances keep improving our ability to assay miRNAs from smaller and smaller amounts of starting material. One microgram was required for many early sequencing approaches, but now 10 ng is acceptable. This can allow for us to use fewer, more selective cells from microdissection approaches or even single-cell RNA-seq methods.[68,69] Additionally, newer sequencing methods will reduce the bias of older approaches to more accurately determine the miRNAs involved in disease.[37] The last part of the puzzle is making the correct comparison between diseased and normal tissues based on cell composition. As we highlighted, microdissection techniques or tissue deconvolution algorithms may be needed for many tissues to correctly identify aberrant miRNA expression levels.

With these improved approaches, there is no limit to what may be discovered regarding miRNA dysregulation in disease. These aberrantly expressed miRNAs will in turn serve as therapeutic targets,[70] disease biomarkers[71] and guide future research into disease pathogenesis.

Acknowledgments

MKH is grateful for helpful comments from Avi Rosenberg. MKH is supported by grant 1R01HL137811 from the NIH and an American Heart Association Grant-in-Aid 17GRNT33670405.

References

1. Bartel DP, MicroRNAs: target recognition and regulatory functions. *Cell* 2009; **136**: 215–233.
2. Mendell JT, Olson EN. MicroRNAs in stress signaling and human disease. *Cell* 2012; **148**: 1172–1187.
3. Bartel DP. MicroRNAs: genomics, biogenesis, mechanism, and function. *Cell* 2004; **116**: 281–297.
4. Chang TC, Pertea M, Lee S, Salzberg SL, Mendell JT. Genome-wide annotation of microRNA primary transcript structures reveals novel regulatory mechanisms. *Genome Res* 2015; **25**: 1401–1409.
5. McCall MN, Kim MS, Adil M, Patil AH, Lu Y, Mitchell CJ, Leal-Rojas P, Xu J, Kumar M, Dawson VL. *et al.* Toward the human cellular microRNAome. *Genome Res* 2017; **27**: 1769–1781.

6. Vainberg Slutskin I, Weingarten-Gabbay S, Nir R, Weinberger A, Segal E. Unraveling the determinants of microRNA mediated regulation using a massively parallel reporter assay. *Nature Communications* 2018; **9**: 529.
7. Lee RC, Feinbaum RL, Ambros V. The C. elegans heterochronic gene lin-4 encodes small RNAs with antisense complementarity to lin-14. *Cell* 1993; **75**: 843–854.
8. Wightman B, Ha I, Ruvkun G. Posttranscriptional regulation of the heterochronic gene lin-14 by lin-4 mediates temporal pattern formation in C. elegans. *Cell* 1993; **75**: 855–862.
9. Roush S, Slack FJ. The let-7 family of microRNAs. *Trends in cell biology* 2008; **18**: 505–516.
10. Kozomara A, Griffiths-Jones S. miRBase: annotating high confidence microRNAs using deep sequencing data. *Nucleic Acids Res* 2014; **42**: D68–D73.
11. de Rie D, Abugessaisa I, Alam T, Arner E, Arner P, Ashoor H, Astrom G, Babina M, Bertin N, Burroughs AM, *et al.* An integrated expression atlas of miRNAs and their promoters in human and mouse. *Nature biotechnology* 2017; **35**: 872–878.
12. Londin E, Loher P, Telonis AG, Quann K, Clark P, Jing Y, Hatzimichael E, Kirino Y, Honda S, Lally M. *et al.* Analysis of 13 cell types reveals evidence for the expression of numerous novel primate- and tissue-specific microRNAs. *Proc Natl Acad Sci U S A* 2015; **112**: E1106–E1115.
13. Backes C, Fehlmann T, Kern F, Kehl T, Lenhof HP, Meese E. Keller A. miRCarta: a central repository for collecting miRNA candidates. *Nucleic Acids Res.* 2017
14. Fromm B, Domanska D, Hackenberg M, Mathelier A, Hoye E, Johansen M, Hovig E, Flatmark K, and Peterson KJ. (2018). MirGeneDB2.0: The curated microRNA Gene Database. BioRxiv, https://doi.org/10.1101/258749.
15. Fromm B, Billipp T, Peck LE, Johansen M, Tarver JE, King BL, Newcomb JM, Sempere LF, Flatmark K, Hovig E, *et al.* A uniform system for the annotation of vertebrate microRNA genes and the evolution of the human microRNAome. *Ann Revi Genetics* 2015; **49**: 213–242.
16. Chaulk SG, Thede GL, Kent OA, Xu Z, Gesner EM, Veldhoen RA, Khanna SK, Goping IS, MacMillan AM, Mendell JT, *et al.* Role of pri-miRNA tertiary structure in miR-17~92 miRNA biogenesis. *RNA Biology* 2011; **8**: 1105–1114.
17. Ouyang Y, Mouillet JF, Coyne CB, and Sadovsky Y, Review: placenta-specific microRNAs in exosomes — good things come in nano-packages. *Placenta* 2014; **35** (Suppl): S69–S73.
18. Telonis AG, Magee R, Loher P, Chervoneva I, Londin E, Rigoutsos I. Knowledge about the presence or absence of miRNA isoforms (isomiRs) can successfully discriminate amongst 32 TCGA cancer types. *Nucleic Acids Res* 2017; **45**: 2973–2985.
19. Landgraf P, Rusu M, Sheridan R, Sewer A, Iovino N, Aravin A, Pfeffer S, Rice A, Kamphorst AO, Landthaler M, *et al.* A mammalian microRNA expression atlas based on small RNA library sequencing. *Cell* 2007; **129**: 1401–1414.
20. Lagos-Quintana M, Rauhut R, Yalcin A, Meyer J, Lendeckel W, Tuschl T. Identification of tissue-specific microRNAs from mouse. *Current Biol: CB* 2002; **12**: 735–739.

21. Baskerville S, Bartel DP. Microarray profiling of microRNAs reveals frequent coexpression with neighboring miRNAs and host genes. *RNA* 2005; **11**: 241–247.
22. Ludwig N, Leidinger P, Becker K, Backes C, Fehlmann T, Pallasch C, Rheinheimer S, Meder B, Stahler C, Meese E. *et al.* Distribution of miRNA expression across human tissues. *Nucleic Acids Res* 2016: **44**: 3865–3877.
23. Kent OA, McCall MN, Cornish TC, Halushka MK. Lessons from miR-143/145: the importance of cell-type localization of miRNAs. *Nucleic Acids Res* 2014; **42**: 7528–7538.
24. Xu Z, Ramachandran S, Gunasekaran M, Zhou F, Trulock E, Kreisel D, Hachem R, and Mohanakumar T. MicroRNA-144 dysregulates the transforming growth factor-beta signaling cascade and contributes to the development of bronchiolitis obliterans syndrome after human lung transplantation. *The J Heart Lung Transpl: The Offi Publication of the International Society for Heart Transplantation* 2015; **34**: 1154–1162.
25. Halushka MK. MicroRNA-144 is unlikely to play a role in bronchiolitis obliterans syndrome. *The J Heart Lung Transpl: The Offi Publication of the International Society for Heart Transplantation* 2016; **35**, 543.
26. Witwer KW, Halushka MK. Toward the promise of microRNAs — Enhancing reproducibility and rigor in microRNA research. *RNA Biology* 2016; **13**: 1103–1116.
27. Juzenas S, Venkatesh G, Hubenthal M, Hoeppner MP, Du ZG, Paulsen M, Rosenstiel P, Senger P, Hofmann-Apitius M, Keller A. *et al.* A comprehensive, cell specific microRNA catalogue of human peripheral blood. *Nucleic Acids Res* 2017; **45**: 9290–9301.
28. Halushka MK, Fromm B, Peterson KJ, McCall MN. Big Strides in Cellular MicroRNA expression. *Trends Genetics: TIG* 2018; **34**: 165–167.
29. Bian Z, Li L, Cui J, Zhang H, Liu Y, Zhang CY, and Zen K. Role of miR-150-targeting c-Myb in colonic epithelial disruption during dextran sulphate sodium-induced murine experimental colitis and human ulcerative colitis. *J Pathol* 2011; **225**: 544–553.
30. Gyvyte U, Juzenas S, Salteniene V, Kupcinskas J, Poskiene L, Kucinskas L, Jarmalaite S, Stuopelyte K, Steponaitiene R, Hemmrich-Stanisak G. *et al.* MiRNA profiling of gastrointestinal stromal tumors by next-generation sequencing. *Oncotarget* 2017; **8**: 37225–37238.
31. Newman AM, Liu CL, Green MR, Gentles AJ, Feng W, Xu Y, Hoang CD, Diehn M, Alizadeh AA. Robust enumeration of cell subsets from tissue expression profiles. *Nature methods* 2015; **12**: 453–457.
32. Schroeder A, Mueller O, Stocker S, Salowsky R, Leiber M, Gassmann M, Lightfoot S, Menzel W, Granzow M, Ragg T. The RIN: An RNA integrity number for assigning integrity values to RNA measurements. *BMC molecular biology* 2006; **7**: 3.
33. Hall JS, Taylor J, Valentine HR, Irlam JJ, Eustace A, Hoskin PJ, Miller CJ, West CM, Enhanced stability of microRNA expression facilitates classification of FFPE

tumour samples exhibiting near total mRNA degradation. *Br J Cancer* 2012; **107**: 684–694.

34. Ibberson D, Benes V, Muckenthaler MU, Castoldi M. RNA degradation compromises the reliability of microRNA expression profiling. *BMC Biotechnology* 2009; **9**: 102.
35. Liu A, Tetzlaff MT, Vanbelle P, Elder D, Feldman M, Tobias JW, Sepulveda AR, and Xu X. MicroRNA expression profiling outperforms mRNA expression profiling in formalin-fixed paraffin-embedded tissues. *International journal of clinical and experimental pathology* 2009; **2**: 519–527.
36. Pritchard CC, Cheng HH, Tewari M. MicroRNA profiling: approaches and considerations. *Nat Rev Genet* 2012; **13**: 358–369.
37. Dard-Dascot C, Naquin D. d'Aubenton-Carafa Y. Alix K. Thermes C. van Dijk E. Systematic comparison of small RNA library preparation protocols for next-generation sequencing. *BMC Genomics* 2018; **19**: 118.
38. Vitsios DM, Enright AJ. Chimira: analysis of small RNA sequencing data and microRNA modifications. *Bioinformatics* 2015; **31**: 3365–3367.
39. Baras AS, Mitchell CJ, Myers JR, Gupta S, Weng LC, Ashton JM, Cornish TC, Pandey A, Halushka MK. miRge — A Multiplexed method of processing small RNA-Seq data to determine microRNA entropy. *PLoS One* 2015; **10:** e0143066.
40. Lu Y, Baras AS, Halushka MK. (2018). miRge 2.0: An updated tool to comprehensively analyze microRNA sequencing data. BioRxiv 10.1101/250779.
41. Mestdagh P, Hartmann N, Baeriswyl L, Andreasen D, Bernard N, Chen C, Cheo D, D'Andrade P, DeMayo M, Dennis L. *et al.* Evaluation of quantitative miRNA expression platforms in the microRNA quality control (miRQC) study. *Nat Met* 2014; **11**: 809–815.
42. Guedes LB, Morais CL, Almutairi F, Haffner MC, Zheng Q, Isaacs JT. Antonarakis ES, Lu C, Tsai H, Luo J. *et al.* Analytic validation of RNA *in situ* hybridization (RISH) for AR and AR-V7 expression in human prostate cancer. *Clin Cancer Res* 2016; **22**: 4651–4663.
43. Nelson PT, Baldwin DA, Kloosterman WP, Kauppinen S, Plasterk RH, Mourelatos Z. RAKE and LNA-ISH reveal microRNA expression and localization in archival human brain. *RNA* 2006; **12**: 187–191.
44. Chivukula RR, Shi G, Acharya A, Mills EW, Zeitels LR, Anandam JL, Abdelnaby AA, Balch GC, Mansour JC, Yopp AC, *et al.* An essential mesenchymal function for miR-143/145 in intestinal epithelial regeneration. *Cell* 2014; **157**: 1104–1116.
45. McCall MN, Kent OA, Yu J, Fox-Talbot K, Zaiman AL, Halushka MK. MicroRNA profiling of diverse endothelial cell types. *BMC Med Genomics* 2011; **4**: 78.
46. Habbe N, Koorstra JB, Mendell JT, Offerhaus GJ, Ryu JK, Feldmann G, Mullendore ME, Goggins MG, Hong SM, Maitra A. MicroRNA miR-155 is a biomarker of early pancreatic neoplasia. *Cancer Biology & Therapy* 2009; **8**: 340–346.

47. Lai JY, Luo J, O'Connor C, Jing X, Nair V, Ju W, Randolph A, Ben-Dov IZ, Matar RN, Briskin D. *et al.* MicroRNA-21 in glomerular injury. *J Am Soc Nephrol: JASN* 2015; **26**: 805–816.
48. Matkovich SJ, Van Booven DJ, Youker KA, Torre-Amione G, Diwan A, Eschenbacher WH, Dorn LE, Watson MA, Margulies KB, Dorn GW. 2nd. Reciprocal regulation of myocardial microRNAs and messenger RNA in human cardiomyopathy and reversal of the microRNA signature by biomechanical support. *Circulation* 2009; **119**: 1263–1271.
49. Rosenberg AZ, Armani MD, Fetsch PA, Xi L, Pham TT, Raffeld M, Chen Y, O'Flaherty N, Stussman R, Blackler AR, *et al.* High-Throughput Microdissection for Next-Generation Sequencing. *PLoS One* 2016; **11**: e0151775.
50. Kumar B, Rosenberg AZ, Choi SM, Fox Talbot K, de Marzo AM, Nonn L, Brennen WN, Marchionni L Halushka MK, Lupold SE. (2018). Cell-type specific expression of oncogenic and tumor suppressive microRNAs in the human prostate and prostate cancer. Sci Rep. 2018 May 8; **8**(1): 7189.
51. Chang YS, Chen WY, Yin JJ, Sheppard-Tillman H, Huang J, Liu YN. EGF Receptor Promotes Prostate Cancer Bone Metastasis by Downregulating miR-1 and Activating TWIST1. *Cancer Res* 2015; **75**: 3077–3086.
52. Dawson K, Wakili R, Ordog B, Clauss S, Chen Y, Iwasaki Y, Voigt N, Qi XY, Sinner MF, Dobrev D, *et al.* MicroRNA29: a mechanistic contributor and potential biomarker in atrial fibrillation. *Circulation* 2013; **127**: 1466–1475, 1461–1428.
53. Arroyo JD, Chevillet JR, Kroh EM, Ruf IK, Pritchard CC, Gibson DF, Mitchell PS, Bennett CF, Pogosova-Agadjanyan EL, Stirewalt DL, *et al.* Argonaute2 complexes carry a population of circulating microRNAs independent of vesicles in human plasma. *Proc Natl Acad Sci U S A* 2011; **108**: 5003–5008.
54. Vickers KC, Palmisano BT, Shoucri BM, Shamburek RD, Remaley AT, MicroRNAs are transported in plasma and delivered to recipient cells by high-density lipoproteins. *Nature Cell Biology* 2011; **13**: 423–433.
55. Gallo A, Tandon M, Alevizos I, Illei GG. The majority of microRNAs detectable in serum and saliva is concentrated in exosomes. *PLoS One* 2012; **7**: e30679.
56. Haider BA, Baras AS, McCall MN, Hertel JA, Cornish TC, and Halushka MK. A critical evaluation of microRNA biomarkers in non-neoplastic disease. *PLoS One* 2014; **9**: e89565.
57. Das S, Halushka MK. Extracellular vesicle microRNA transfer in cardiovascular disease. *Cardiovasc Pathol* 2015; **24**: 199–206.
58. Michell DL, Vickers KC. Lipoprotein carriers of microRNAs. *Biochimica et biophysica acta* 2016; **1861**: 2069–2074.
59. Cheng HH, Yi HS, Kim Y, Kroh EM, Chien JW, Eaton KD, Goodman MT, Tait JF, Tewari M, Pritchard CC. Plasma processing conditions substantially influence circulating microRNA biomarker levels. *PLoS One* 2013; **8**: e64795.
60. Lotvall J, Hill AF, Hochberg F, Buzas EI, Di Vizio D, Gardiner C, Gho YS Kurochkin IV, Mathivanan S, Quesenberry P, *et al.* Minimal experimental requirements for definition

of extracellular vesicles and their functions: a position statement from the International Society for Extracellular Vesicles. *J Extracellular Vesicles* 2014; **3**: 26913.

61. Witwer KW, Soekmadji C, Hill AF, Wauben MH, Buzas EI, Di Vizio D, Falcon-Perez JM, Gardiner C, Hochberg F, Kurochkin IV, *et al.* Updating the MISEV minimal requirements for extracellular vesicle studies: building bridges to reproducibility. *Journal of Extracellular Vesicles* 2017; **6**: 1396823.
62. Witwer KW, Buzas EI, Bemis LT, Bora A, Lasser C, Lotvall J, Nolte-'t Hoen EN, Piper MG, Sivaraman S, Skog J. *et al.* Standardization of sample collection, isolation and analysis methods in extracellular vesicle research. *J Extracellular Vesicles* 2013; **2**.
63. McCall MN, Illei PB, Halushka MK. Complex sources of variation in tissue expression data: Analysis of the GTEx lung transcriptome. *Am J Human Genetics* 2016; **99**: 624–635.
64. Sunderland N, Skroblin P, Barwari T, Huntley RP, Lu R, Joshi A, Lovering RC, Mayr M. MicroRNA biomarkers and platelet reactivity: The clot thickens. *Circ Res* 2017; **120**: 418–435.
65. Zhao Z, Yang Y, Zeng Y, He M. A microfluidic ExoSearch chip for multiplexed exosome detection towards blood-based ovarian cancer diagnosis. *Lab on a Chip* 2016; **16**: 489–496.
66. Keutgen XM, Filicori F, Crowley MJ, Wang Y, Scognamiglio T, Hoda R, Buitrago D, Cooper D, Zeiger MA, Zarnegar R. *et al.* A panel of four miRNAs accurately differentiates malignant from benign indeterminate thyroid lesions on fine needle aspiration. *Clin Cancer Res* 2012; **18**: 2032–2038.
67. Lithwick-Yanai G, Dromi N, Shtabsky A, Morgenstern S, Strenov Y, Feinmesser M, Kravtsov V, Leon ME, Hajduch M, Ali SZ, *et al.* Multicentre validation of a microRNA-based assay for diagnosing indeterminate thyroid nodules utilising fine needle aspirate smears. *Journal of Clinical Pathology* 2017; **70**: 500–507.
68. Faridani OR, Abdullayev I, Hagemann-Jensen M, Schell JP, Lanner F, Sandberg R. Single-cell sequencing of the small-RNA transcriptome. *Nature Biotechnology* 2016; **34**: 1264–1266.
69. Alberti C, Manzenreither RA, Sowemimo I, Burkard TR, Wang J, Mahofsky K, Ameres SL, Cochella L. Cell-type specific sequencing of microRNAs from complex animal tissues. *Nat Meth* 2018.
70. Janssen HL, Reesink HW, Lawitz EJ, Zeuzem S, Rodriguez-Torres M, Patel K, van der Meer AJ, Patick AK, Chen A, Zhou Y. *et al.* Treatment of HCV infection by targeting microRNA. *New Engl J Med* 2013; **368**: 1685–1694.
71. Barana A, Matamoros M, Dolz-Gaiton P, Perez-Hernandez M, Amoros I, Nunez M, Sacristan S, Pedraz A, Pinto A, Fernandez-Aviles F, *et al.* Chronic atrial fibrillation increases microRNA-21 in human atrial myocytes decreasing L-type calcium current. *Circul Arrhyth Electrophysiol* 2014; **7**: 861–868.

Informatics, Implementation, and Genomic Cohorts

12 Example of a Scalable and Adaptable Approach for NGS Analyses Leveraging High-Performance Computing

Cody Ashby,* Michael A. Bauer,* Erich A. Peterson,* and Donald J. Johann, Jr.*,†

*Winthrop P. Rockefeller Cancer Institute, University of Arkansas for Medical Sciences, Little Rock, AR
†don.johann@gmail.com

Contents

I. Introduction

Next-generation sequencing (NGS) is situated to contribute toward advanced understanding of health and disease in an unprecedented manner. In oncology, these advances are spanning the gamut from basic science to guideline-based clinical management of many cancers. National guidelines (e.g., NCCN version 2.2018 for non-small lung cancer) strongly advise broader molecular profiling (i.e., precision medicine) to identify driver mutations to ensure patients receive the most appropriate treatment. Under specific conditions, the Centers for Medicare and Medicaid Services (CMS) has proposed that the evidence is sufficient to cover NGS as a diagnostic laboratory test when ordered by a physician and performed in a CLIA-certified lab. NGS-based tumor profiling is soon to become the standard of care for advanced cancer patients, and liquid biopsy assays, which profile cell-free DNA, are an application of deep sequencing. Along with enabling a revolution in the fields of molecular diagnostics and drug discovery and development, the finding of novel classes of biomolecules (e.g., new types of non-coding RNAs) is being facilitated by NGS.

However, there exists a meta-problem in life sciences research. Big data from NGS is outpacing the capabilities of information technology. Maintaining state-of-the-art informatics needs can lead to an ever-shortening cycle of purchasing new high-performance computing (HPC) systems, which may lead to fiscal strife, especially for organizations new to the field. To this end, an information architecture and suite of tools for the deployment and retrieval of study data were developed. The NGS pipeline software was engineered from open-source best practices using an adaptable methodology along with a Docker container approach, so that the same code is executed regardless of the HPC facility to which a study is deployed.

Described in the present chapter is a comprehensive methodology for NGS analysis, which leverages in-house HPC, the scalability of a commercial cloud provider, and the computational resources present at an affiliated research university. NGS analyses have been performed at three HPC facilities using the same pipeline code, and software configuration management tasks have been greatly reduced. Also gained are improvements in reliability with an effective approach toward scalability. Finally, the promotion of fiscal responsibility is achieved, through the use of greater university-wide and commercial resources.

II. Background

NGS is rapidly changing the manner in which biomedical research is performed and clinical medicine is practiced. This is especially true for medical oncology. Former President Obama's *Precision Medicine Initiative* (PMI) calls for a near-term focus on cancers and a longer-term aim to generate knowledge applicable to a whole range of health and disease.[1] The recently announced *Cancer Moonshot* furthers the PMI by galvanizing research efforts against cancer, managed by a prominent national task force, and led by the former Vice President Biden.[2]

Why now? These goals are within reach due to: (i) advances in genomics over the past 10 years, (ii) increasing use of electronic health records, (iii) technical advances in health devices residing in smart phones and the fact a majority of US adults own one, (iv) advances in data science especially concerning big data, and (v) the changing role of patient partnerships especially crowd sourcing and citizen science, where people want to participate with feedback. The National Academy of Sciences is recommending a new taxonomic classification of disease, especially cancer, which requires definition by the many levels of molecular information (genome, epigenome, microbiome, etc.) and integrated with clinical data, into a *knowledge network that learns*.[3] Since the cost of a whole-exome sequencing (WES) study is ~$,1000 (the price of a CT scan), NGS assays are expected to proliferate and be reimbursed, so there is an urgent need for innovative and cost effective computational approaches.

NGS informatics related to complex pipeline deployment and management of data has produced many advances. For instance, Galaxy [4] is an open-sourced web-based platform that facilitates the analysis of NGS experiments. It is a powerful and flexible approach designed for use by non-programmers. Galaxy allows users to run pipelines without in-depth knowledge of any single tool, which is a significant innovation. However, it's sometimes hard to see the relationship between the GUI and command line parameter options used inside a pipeline. In addition, some users may need access to more programmatic or custom options from within the pipeline, which would require programming abilities. In addition, GUI options do not gracefully handle batch NGS analyses with multiple stages and inputs. Furthermore, some Galaxy pipeline processing steps can be hard to integrate with non-Galaxy-based tools and an integrated data cleaning (e.g., ETL) along with database archiving of pipeline-based study results is absent.

Omics Pipe[5] is an open-source python package that can be installed on a local machine, a compute cluster, or a cloud computing platform such as amazon web services (AWS). Its primary purpose is collecting community-supported "best practices" pipelines in an attempt to increase the overall reproducibility of results, and it claims to be an easily extensible approach to pipeline deployment and management. However, there are some configuration requirements. If not using the AWS distribution, all third-party tools must be installed before using a given pipeline. Other important features not addressed concern the management of related experiments tied to pipeline deployment as well as annotating, archiving, and presenting the data results returned by the pipeline in a useful manner.

openBIS is a powerful tool that has proven itself useful in addressing problems of big data management.[6] It provides generic approaches toward the management of high-throughput biological data. However, upon careful study of the requirements associated with this particular project, it was decided that a more specialized approach was needed. In-house developers with significant expertise in database design created a custom schema, which has allowed more in-depth data handling of multiple molecular profiling modalities and the establishment of interfaces for the interaction with third-party tools, such as IGV. The development of third-party tool interfaces and linkages may be more challenging with externally developed software.

Other innovative approaches include Taverna,[7] which is a workflow management system used in a variety of scientific fields including bioinformatics. It is a desktop authoring system which allows a user to integrate a wide variety of software components into a computable solution. These features make it very useful for non-programmers. Other rapid prototyping approaches that may be applied to NGS pipeline construction include Bpipe[8] and Snakemake.[9] However, the developer is required to specify the pipeline components and any specific interfacing steps that may be required.

III. New System — Technical Approach

There are a wide variety of analysis tools and pipelines with different deployment modalities and parameters,[3,10–12] and a custom solution addressing all requirements and constraints is illustrated here. The main challenges were to satisfy the following two requirements: first, to have the ability to utilize: (i) in-house, (ii) local affiliated university, or (iii) commercially provided HPC resources. Second, was to minimize software configuration management, and yet have robust pipelines which are platform independent. Illustrated in this chapter is a novel methodology for the processing of NGS experimental data, which utilizes an *adaptable approach* for the computational pipeline architecture. This methodology, which utilizes an information architecture, through the Molecular Profiling Database (MPDB, discussed later), is independent of any particular HPC system. Open source tools and the best practices from the Broad Institute and The McDonnell Genome Institute at Washington University were used in the pipeline construction. Pipeline design goals were: (i) to be fast but flexible, (ii) flexible but correct, and (iii) able to generate reports that are easily understood by clinicians/scientists.

This overall approach greatly reduces configuration management since there is only one version of the pipeline software that runs on different HPC systems. It also improves reliability, aids scalability, and promotes fiscal responsibility through the use of greater university-wide and commercial cloud provider resources.

a. System design

Figure 1 illustrates the adaptable pipeline system diagram. Figure 1A shows the role of the MPDB in the coordination, deployment, and retrieval of all HPC activities at different locations. There is more detail on MPDB later in the chapter. Figure 1B illustrates the software architecture for the in-house HPC system at UAMS, and consists of the WES pipeline (used for illustrative purposes) that interacts with the resident operating system, Red Hat Enterprise.

Figure 1C displays the software architectures utilized for HPC systems accessed through the Google Cloud and Figure 1D shows the same for The University of Arkansas (UA) at Fayetteville HPC Center. In these cases, the software stack contains the same version of the WES pipeline that interacts with the Docker container, which interacts with a version of

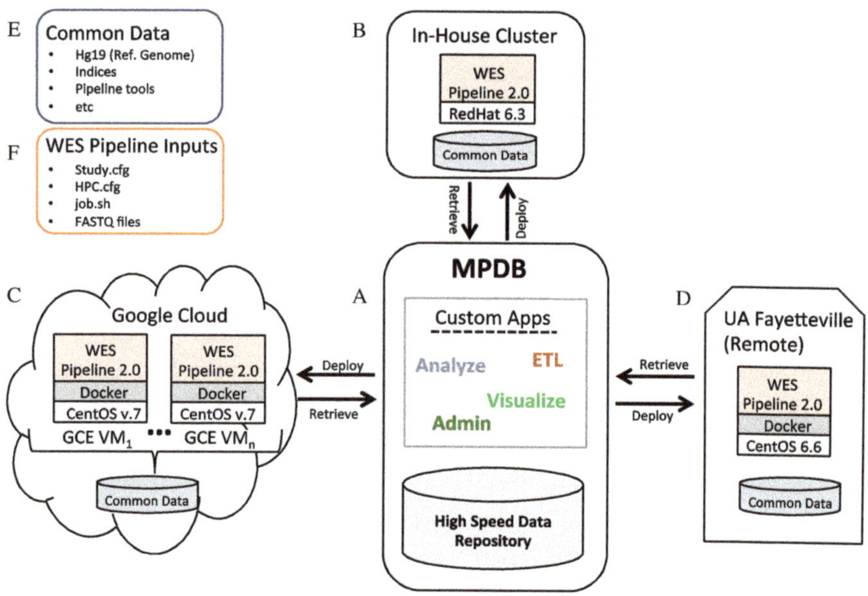

Figure 1. Adaptable pipeline system diagram.

CentOS®. Importantly, all three HPC configurations use identical versions of the WES pipeline. This design greatly reduces configuration management and improves system reliability and experimental data fidelity. As shown in Figure 1E, all instances use a common data disk containing the human genome reference assembly and associated indices along with tools used by the pipeline. Figure 1F shows the WES pipeline with the associated and automatically generated input files. The pipeline requires these files for operation, and they are discussed later in the chapter.

b. Molecular profiling database (MPDB)

It is possible to accomplish small molecular profiling studies without the need for a commercial-grade database management system (DBMS). However, as the number of studies and modalities increase, an extensible and scalable database schema along with interfacing tools becomes necessary. A locally installed and architected solution was developed for the storage, retrieval, annotation, and transformation for molecular profiling studies called the MPDB. Figure 2 illustrates the information architecture.

Following pipeline execution and retrieval of data, a user imports the raw output files into MPDB. The MPDB Administrative (Admin) website interface is a graphical means for a user to schedule the importation of output files into "raw" tables within the database. Figures 2A and 2B show this step. A raw table exists for the processing of each archived output file. These raw tables closely mimic file formats. For example, the typical final step of a WES pipeline is variant calling and the production of a variant call format (VCF) file. Within MPDB, a table exists to capture all data within this VCF file in its original format. There are consistency checks on records against their respective file formats along with data cleaning steps (e.g., range checks, data type checks) during importation.

After the initial importation of raw data, more extensive validation routines (e.g., custom aggregation queries) may optionally be exercised. Once satisfied with the fidelity of the imported data, the user toggles a switch and the raw data undergoes a further and more comprehensive processing. The imported records undergo extensive extract, transform, load (ETL) operations, as data migrates from raw to "processed" tables. The exact operations during this process depend on the particular

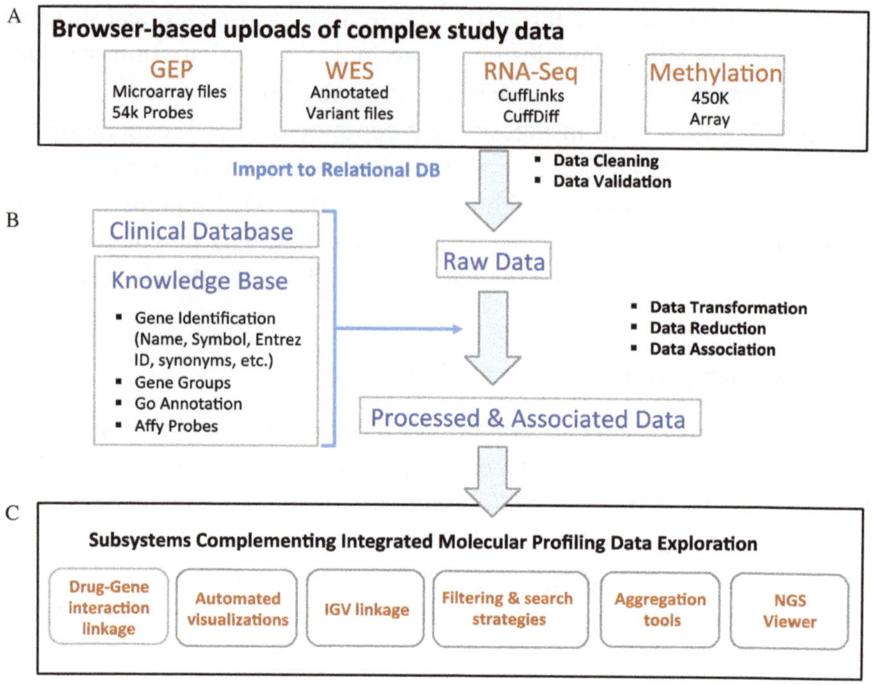

Figure 2. Information architecture: MPDB.

molecular profiling modality used in the imported study. Common operations are the association of Entrez, Ensembl, and Gene Ontology (GO) identifiers to each record.

Following successful data import, additional in-house and third-party tools may be exercised, which aid in further analyses via integration, aggregation, visualization, association, and validation activities. Figure 2C illustrates this capability. The NGS Viewer[13] is an in-house developed software suite, which contains internally developed and custom interfaces to well-regarded third-party tools for support of supplementary analyses. Rapid data integration for viewing "mash-ups" of heterogeneous modalities is an important feature for integrative genomic inquiries. Additionally, data filtering schemes coupled with subsequent sequential search abilities involving integrated data modalities (e.g., WES and RNA-seq) with optional graphic rendering via D3[14] is provided. Another example involving a third-party tool is the direct linkage between NGS Viewer and the integrative genomics viewer (IGV).[15] This automated functionality allows one to validate NGS study results by visualizing experimental data regions, against genomic references and the graphic annotations provided by IGV, by simply clicking on a single hyperlink within the NGS Viewer.

c. MPDB admin

In order to facilitate the automation of multiple NGS experimental studies, a custom website that interfaces to the MPDB was developed. This interface allows input of NGS experimental and pertinent patient metadata relating to a study. The current version has the capability to collect relevant molecular profiling modality data, the library prep kit used, sample collection dates, histopathology, as well as other relevant clinical and descriptive information. These items are used to define parameters for an in-house or deployed NGS pipeline-based experiment.

d. Docker

Virtual machines (VMs) are popular in bioinformatics, especially for complex applications utilizing suites of tools or pipelines, which may require a specific software architecture or environment.[12] There are several advantages to VMs in the realm of replication of experimental results and pre-configuration of complex environments. In comparison, bare metal HPC clusters have a variety of configuration options in both operating system and hardware. However, as reported in this IBM study, deploying applications in a VM results in reduced performance due to extra levels of abstraction (e.g., emulating virtual hardware).[16]

New advances in container-based virtualization simplify the deployment of applications, allow for isolation, and improved performance. Docker is an open-source container management utility that allows developers to build, ship, and run distributed applications by a portable and lightweight runtime and packaging tool.[17] This allows for the advantages of VMs while improving run time by only abstracting the Linux kernel. An IBM study stated that "container performance is equal or better performance than VMs in almost all cases."[16] Although Docker utilization is relatively new, it has already shown value with regard to reproducibility of results.[18,19]

The significance of using Docker is maximizing platform independence while minimizing software configuration. The Docker image utilized by the WES pipeline establishes an interface into a well-known version of Linux (CentOS 7) and provides for the packaging of software tool dependencies (e.g., compilers, language processors, etc.). The Docker image was created on a local desktop computer from the CentOS 7 base image pulled from the Docker repository. All dependencies were installed into the image and as an example include: gcc, python, latex, java, R, etc. None of the pipeline proper tools (e.g., BWA, GATK, etc.) are installed directly in the Docker image. Rather, they are accessed following the read-only mounting of a common shared disk drive (see Figure 1, Common Disk). This maximizes compatibility of the tools due to a predefined run environment provided by the Docker container and eases future upgrades.

e. WES pipeline

Although microarray and RNA-seq are currently supported by MPDB, to date, WES has received the majority of work emphasis. An adaptable pipeline for WES, built as a python package and using open source tools and best practices was developed. Along with FASTQ files, the pipeline package uses three additional automatically generated files as input: (i) study configuration file (*Study.cfg*), (ii) HPC configuration file (*HPC.cfg*), and (iii) linux shell script file (*job.sh*). Figure 1F illustrates the pipeline with associated input files. Generation of these three files is automatic, based upon the operator's experiment setup performed via MPDB Admin (Figure 1A). The study configuration file sets up locations, experimental analysis options, and tagging parameters for data retrieval as well as verification and validation procedures. The HPC configuration and job.sh files model the attributes of a given HPC system. For instance, the HPC configuration file contains the paths to reference genomes, the paths to pipeline tools, and machine-specific parameters such as the number of cores available to the pipeline. The job.sh file for the in-house UAMS cluster simply invokes the WES pipeline script by submitting it to the Sun Grid Engine. On the

Google Cloud, job.sh is slightly more involved due to (i) required data transfers of FASTQ and configuration files and the need for checksum validation, (ii) sending emails signalling job completion, (iii) compressing data for download, and (iv) shutting down the compute instance once the job is complete. The job.sh script is relatively small (e.g., less than 30 lines of code). It is HPC system dependent by design and serves as an abstraction to the pipeline proper. Job.sh along with the HPC configuration file are the only two files that differ between HPC platforms.

The pipeline takes a top-down approach and is composed of several submodules that are themselves extensible. The submodules are: initialization, alignment, variant calling, and utility. Figure 3 shows an overview and steps of the pipeline. The initialization module controls the various early steps taken before executing alignment. The current implementation runs FASTQC[20] on the tumor and germline FASTQ files to check quality. Before proceeding to the next step, a manual verification is undertaken to ensure that the FASTQ quality is within an acceptable range. The alignment module

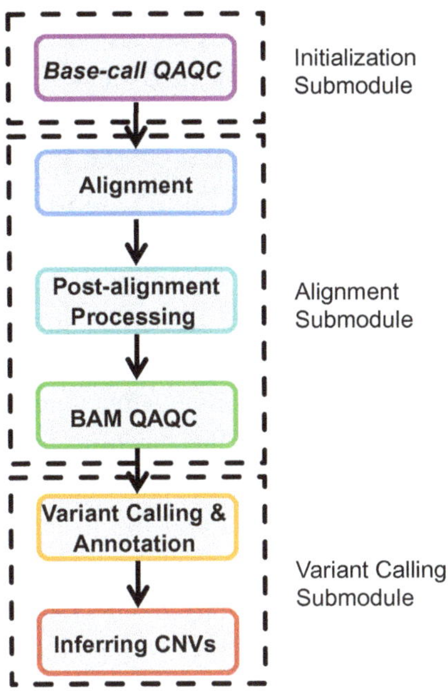

Figure 3. WES pipeline.

handles the procedure of aligning FASTQ files to the Hg19 (GRCh37) reference genome. It utilizes either hybrid approach of BWA[21] and STAMPY,[22] or just BWA-MEM. Next, duplicate read removal is handled by Picard Tools.[23] The Genome Analysis Toolkit (GATK),[24] is used for sequence recalibration and local realignment. *Note*: Base recalibration is optional and is controlled by a parameter in the study configuration file. The tumor and germline are run in parallel (simultaneously) during the above steps.

The final module, variant calling, utilizes Strelka[25] and/or MuTect2, and SnpEff[26] performs VCF annotation. Copy numbers are inferred with ExomeCNV[27] or Sequenza.[28] The utility submodule contains functions for parallel execution, logging of pipeline data, along with directory and path creation. A master script encapsulates and controls the execution of all submodules. The modular design of this software provides for updates and additions with relative ease.

Following the user/operator experiment setup procedure, deployment to a HPC system is fully automatic. MPDB abstracts an experimental run queue along with dashboard-like viewing, which provides status tracking of: (i) currently queued (waiting), (ii) running, and (iii) completed jobs. A Grid Engine serves as the queue in the case of UAMS deployed HPC jobs. The combination of an MPDB table and custom scripts handle NGS experimental uploads and functions as the queue for remote jobs, such as those deployed to the Google Cloud Engine (GCE).

Secure data uploads to non-UAMS HPC systems are managed via a custom script and utilize the following procedure. First, input files are check-summed using MD5 on the client side, next, uploaded through BBCP,[29] and then verified with server side MD5 summation and comparison before pipeline execution begins. Upon job completion in the cloud, the operator receives a notification e-mail. After sending the e-mail, the job script halts the instance, and keeps the "results disk" for later retrieval. After retrieval, verification, and validation checks, salient output files (e.g., VCF, copy number) are downloaded and the user transfers and imports them into MPDB. An in-house high-speed data repository stores the other output files (e.g., BAM, log files) for future analyses and reference.

An investigative clinical genomics report generating application, developed in-house, creates an Excel-based summary report of the study and is intended for use by clinician/scientists. This report is especially useful for molecular tumor board conferences.[30] The report contains clinically relevant information for mutations, e.g., gene symbol, hyperlinks to Entrez Gene and Ensembl Gene website entries, DNA mutation and amino acid effects, likely effect on protein function (modifier), depth and allelic frequency information for both tumor and germline, and the inferred copy number. In addition, results can be filtered based on an investigator's domain-specific list of biologically relevant genes of interest. A custom interface to COSMIC[31] serves to identify driver gene mutations.

Adaptability is the major strength of this approach. Any suitably powerful (at least a quad-core CPU with 48GB of memory) HPC environment will work as long as it meets three requirements: (i) it has an HPC configuration file, (ii) Docker is installed or all pipeline-related software is configured, and (iii) the pipeline and genome references are available.

f. Installation and setup procedure

The minimal setup procedure for a new HPC environment is defined as follows: (i) upload copies of the tools, references and Docker image to the new HPC environment, (ii) install Docker on the HPC and import the WES Pipeline Docker image, (iii) upload input data (FASTQ files, job.sh, HPC configuration file, and study configuration file), (iv) create a Docker container from the Docker image and execute the pipeline script through job.sh (providing the input data as parameters).

g. Cloud computing

Google Compute Engine (GCE) has the capabilities for snapshots. A snapshot allows the capture of the contents of a disk at an instant in time. First, a base CentOS 7 system was created with the Docker image installed and an e-mail server (for notification e-mails). Next, a snapshot was taken using the GCE API. The snapshot is now finalized and can be utilized. All new GCE VMs are created using this snapshot, which provides them with CentOS 7 and the Docker image. All dependencies (e.g., compilers, interpreters, R, etc.) have been incorporated into the Docker image. Tools and reference genomes are accessed by mounting in a read-only fashion a persistent 2.5 TB drive (see Figure 1, *Common Data* disk).

Once a snapshot has been established and a GCE VM is created, all cloud-based initiation steps are fully automated through MPDB Admin. It employs a series of associated scripts that interact with the GCE API. Once a job has finished on a VM residing in the cloud, the data is brought back down to the UAMS cluster. It is then manually checked by the user, (e.g., VCF file size and general contents), as a verification step, before being imported into MPDB. Portability is realized through the Docker image (which is easily transportable to new HPC environments) and the HPC configuration file. An overall-system level diagram of this process is shown in Figure 1.

h. Regression test sets

There are five internal regression tests consisting of paired tumor/germline data. The *Full Presentation* (tumor/germline samples from a de-identified patient newly diagnosed with cancer) was chosen due to the fact that it was studied at length on numerous occasions. The other four regression test sets are modifications and are as follows: *Full Chromosome 1*, *Full Chromosome 17*, *Chromosome 1 — 10%* (only ten percent of reads), and *Chromosome 17 — 10%* (only ten percent of reads). These four test sets were constructed due to the speed at which the pipeline could be fully executed, as well as the degree of known variants each data set provided.

For the reduced sets, Samtools[32] was used to produce a pair of FASTQ files containing only chromosome 1 or 17, and a custom script captured the top ten percent. MPDB holds the expected output for all of these studies. Whenever there is a major or minor change to the pipeline, the appropriate test set is exercised to verify functionality (i.e., software regression testing) by comparing observed versus expected results. The *Chromosome 17 — 10%* regression test is especially useful since it also exercises the entire pipeline and completes in less than 20 minutes. The *Full Presentation* regression test is considered a more in-depth evaluation, exercises every part of the pipeline, and completes in approximately two days.

IV. Significance

There is a meta-problem in life sciences research involving high-throughput technologies. The capabilities of information technology cannot keep pace with big data from NGS. A new and expensive HPC system may become incapable of processing state-of-the-art NGS modalities within an 18–24 month period.[2,16,30] To address this issue, a comprehensive methodology was developed for NGS analysis that is able to leverage in-house HPC, the scalability of a commercial provider, and the computational resources present at an affiliated research university. Figure 1 illustrates the system design. The computational approach (pipeline) is adaptable, executing the same code regardless of the HPC facility.

Figure 4 further illustrates details of the NGS workflow that is HPC independent. This approach was initially implemented for WES and now other modalities have been developed. The deployment and running of a NGS pipeline study first involves a dialog through a web interface that communicates to a database system (Figure 4A) via custom administrative tools (Figure 4B). Figure 4C contains the inputs for

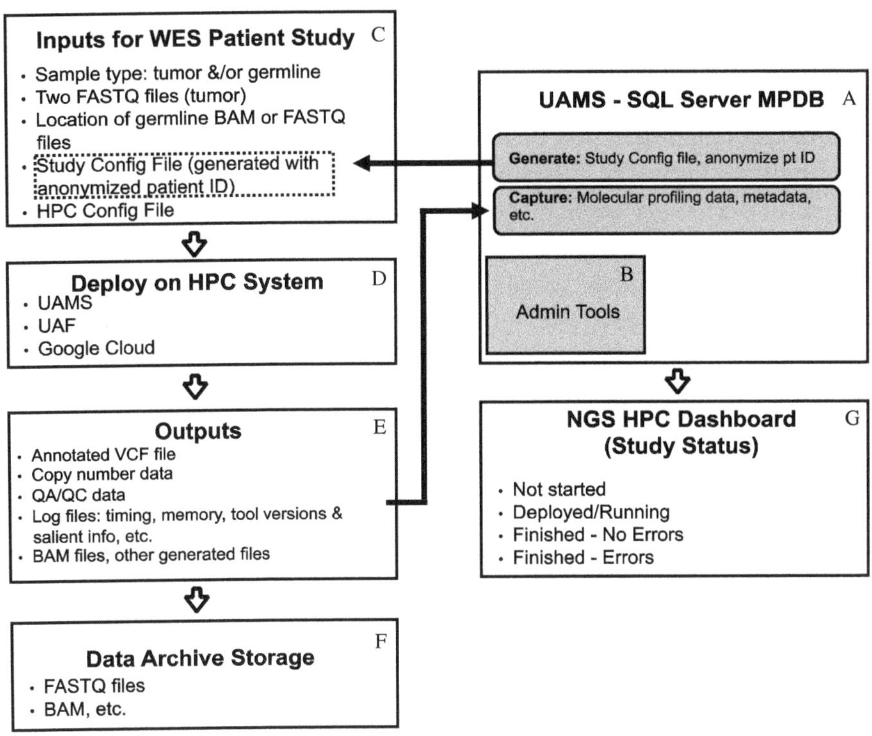

Figure 4. NGS HPC workflow (HPC independent).

a WES patient study. These include specifying a tumor plus/minus germline, along with the location of the necessary FASTQ files or an existing BAM file (e.g., existing patient germline). Study information is automatically captured including an anonymous patient ID, with the generation of a Study Configuration file. Additional steps include specification of HPC system deployment (Figure 4D) along with automated preparation of system entries for NGS study results (Figure 4E), and data archiving (Figure 4F). The NGS HPC Dashboard displays study status as shown in Figure 4G.

a. Deployment on google compute engine

When running remote jobs, data transfer is a necessary concern. The overage cost of uploading the tools needed by pipelines, management of snapshots each time a tool updates can be significant. The solution employed was to create a UAMS Genomics 2.5 TB persistent disk that contains all tools needed to run the WES pipeline. This persistent disk, seen as "Common Data" in Figure 1C, will be mounted by multiple running instances in read only mode. The impact on performance concerning the UAMS Genomics disk was carefully accessed when connected to one and then multiple instances. Systems employing Linux containers having I/O intensive workloads need to be prudently examined. This was accomplished by creating a single instance, then four simultaneous instances and finally ten simultaneous instances, with all instances running the *Full Presentation* regression test set through the pipeline. Google compute engine scalability and stress testing is shown in Figure 5 and profiles the I/O testing impact across all stages of the pipeline, which was found to be minimal.

Google compute engine is a compelling option from both a fiscal and scalability standpoint. Table 1 shows a cost breakdown of running a typical WES job on a Google compute instance. The average run time is ~33 hours with standard deviation of six hours, and the total compute cost is ~$40. All cost and performance assessments are based on an instance with 16 virtual CPUs and 104 GB of RAM (i.e., Google Highmem-16 configuration). Deployment to the HPC at the University of Arkansas Fayetteville was done in an analogous fashion.

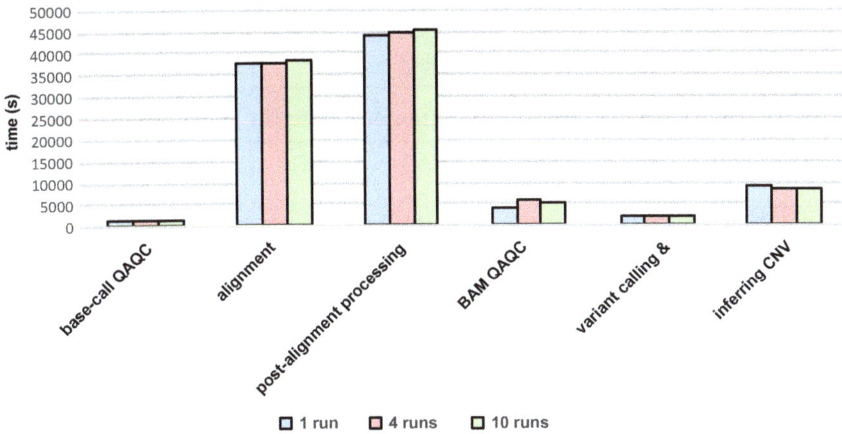

Figure 5. Google compute engine scalability stress testing.

Table 1: Timing and cost analysis. > 100 WES samples.

	Breakdown of cost			
Breakdown of time hours	CPU	Disk	Network	Total
	\$29.75	\$6.29	~\$5.00	\$41.16

b. Internal testing and capabilities

The MPDB contains a multitude of experiments from the WES pipeline in a tumor/normal configuration and can serve functions for QA/QC report generation, data mining, and providing abstractions to third-party tools through a view mechanism. Table 2 shows a breakdown of variant calling statistics collected from ~100 WES studies and presented in aggregate. The columns are broken into three categories: (i) type, (ii) effect, and (iii) effect impact. Type is simply whether the variant is an SNV or an small InDel. Effect is whether the mutation is of the class frameshift, nonsense, or missense. Effect impact is a metric that tells us the putative impact of the variant on protein function. All of these annotation steps are defined through SnpEff.[26] Note that if there are multiple variants that hit the same position, there may be some overlap in these columns.

Table 2: Variant calling statistics, >100 WES samples.

	Type		Effect			Effect impact		
	SNVs	InDels	Frameshifts	Missense	Nonsense	High	Low	Moderate
Average	459	68	38	265	16	27	196	119
Min	160	10	0	74	0	5	40	42
Max	1975	122	99	982	133	90	1085	452
Std Dev	329	28	20	202	24	14	182	79

Generally, it is expected that patient relapse (cancer recurrence) data will contain more mutations than presentation (initial diagnosis). However, this is not always the case and may be due to factors related to a particular treatment regimen and/or the underlying cancer biology. Nevertheless, it is a good metric to access when looking at data from presentation–relapse pairs, and straightforward to construct this report. Table 3 contains a breakdown of 27 pairs of presentation–relapse data. The columns show (i) an anonymized pair ID, (ii) the number of variant caller-passed mutations in the presentation, (iii) the number of variant caller-passed mutations in the relapse, and (iv) arithmetic difference. Pairs that bear potential investigation are colored light red, and pairs that are a potential pass are colored blue. These statistics along with the timing performance presented in Table 2 offer an internal QA/QC role, by providing metrics to assess whether study results fall within acceptable ranges.

V. Future Work

Future work will involve RNA-seq and establishing pipelines for other rapidly evolving NGS modalities. Recently developed is a RNS-seq mutation calling pipeline that uses the following workflow: STAR,[33] HTSeq Count,[34] and Haplotype Caller.[24] This pipeline was deployed from within a Docker container on GCE as well as on the UAMS cluster. An additional RNA-seq pipeline that uses Bpipe[8] as the pipeline manager has also been developed, and implements the following software components: Tophat,[35] RNASeQC,[36] Picard Tools,[23] Samtools flagstat,[32] and Cufflinks.[37] Bpipe and similar tools, such as Omics Pipe,[5] will enable rapid construction of pipelines and with the adaptability of the established methodology, future

Table 3: Paired presentation-relapse comparisons.

ID	Passed presentation	Passed relapse	Difference
1	481	110	−371
12	257	77	−180
21	211	125	−86
22	139	116	−23
11	82	69	−13
13	97	87	−10
23	110	104	−6
2	113	108	−5
7	75	75	0
6	91	96	5
18	142	153	11
17	123	146	23
26	119	148	29
27	135	167	32
16	58	101	43
15	99	150	51
8	47	99	52
3	91	144	53
9	129	183	54
5	112	177	65
10	314	387	73
25	64	142	78
24	119	198	79
20	58	163	105
4	120	234	114
19	107	262	155
14	198	446	248

pipeline construction for ChIP-seq, miRNA-seq, and methylation studies may simply be a matter of adding the metadata capabilities associated with a study into the MPDB Admin module. The end goal is to support these modalities as thoroughly as WES.

VI. Conclusions

Described and discussed in this chapter is a novel methodology for the processing of NGS experimental data that utilizes an *adaptable approach* for computational pipeline architecture. This methodology satisfies the following two requirements of the *main challenge* associated with this work: first, to have the ability to utilize (i) in-house, (ii) local affiliated university, or (iii) commercially provided HPC resources. Second, to keep software configuration management to a minimum, and provide a platform independent version of a pipeline architecture. It is likely that this scenario exists at other academic centers, thus this approach has significant value to the greater bioinformatics community. An information architecture independent of any particular HPC system was utilized. Open source tools and the best practices from the Broad Institute and The McDonnell Genome Institute at Washington University were used. Design goals for the pipeline were: (i) fast but flexible, (ii) flexible but correct, and (iii) had the ability to generate reports that are easily understood by clinicians/scientists. These aims have been demonstrated via timing data, the construction of a variety of software regression test sets, and the design and development of a programmatically controlled genomics clinical report generator, which renders an Excel output file of salient NGS study data along with hyperlinks to relevant Entrez and Ensembl website entries.

Deployment on three different HPC systems with verification and validation of results from over 100 WES samples has been achieved and QA/QC statistics and other types of aggregation-based or search and sorting analyses are easily generated. This approach has also facilitated inter-university collaboration through the University of Arkansas system by virtue of minimal setup for a new environment. For instance, when deploying the code-base to University of Arkansas Fayetteville for the first time, the entire pipeline including transfer of tools and FASTQ data was operational in less than an hour. The number of man-hours associated with deploying and running the pipeline are greatly reduced due to full automation and deployment capabilities through MPDB Admin module. This automation also improves reliability by minimizing the chance of manual errors. Reliability is also improved by maintaining only one version of the pipeline regardless of the HPC platform. In addition, personnel with minimal programming experience are able to run the pipeline due to the support gained through the information architecture.

In essence, the foundation for a NGS data warehouse has also been established. Importantly, this overall approach greatly reduces configuration management since there is only one version of the pipeline software that runs on many different HPC systems. It also improves reliability, aids scalability, and promotes fiscal responsibility (a complete study on Google costs ~$40) through the use of greater university-wide and commercial cloud provider resources. Ultimately, as the system grows and matures it will likely have a central role in cancer biology discovery and improving patient outcomes.

Competing Interests

The authors declare that they have no competing interests.

Authors' Contributions

DJJ conceived and designed the system. CA, MAB, EAP, and DJJ designed the software. CA, MAB, and EAP implemented the software. DJJ and CA wrote the chapter. All authors read and approved the chapter.

References

1. Collins FS, Varmus H. A new initiative on precision medicine. *New Engl J Med* 2015; **372**(9): 793–795.
2. Lowy DR, Collins FS. Aiming high — changing the trajectory for cancer. *New Engl J Med* 2016; **374**(20): 1901–1904.
3. NationalAcademies: Toward precision medicine: building a knowledge network for biomedical research and a new taxonomy of disease. US. National Academies Press, 2011.
4. Goecks J, Nekrutenko A, Taylor J. Galaxy: A comprehensive approach for supporting accessible, reproducible, and transparent computational research in the life sciences. *Genome Biol* 2010; **11**(8): R86.
5. Fisch KM, Meißner T, Gioia L, Ducom J-C, Carland T, Loguercio S, Su AI: Omics pipe: A computational framework for reproducible multi-omics data analysis. *BioRxiv* 2014: 008383.

6. Bauch A, Adamczyk I, Buczek P, Elmer F-J, Enimanev K, Glyzewski P, Kohler M, Pylak T, Quandt A, Ramakrishnan C. openBIS: A flexible framework for managing and analyzing complex data in biology research. *BMC Bioinformatics* 2011; **12**(1): 468.
7. Taverna [http://www.taverna.org.uk].
8. Sadedin SP, Pope B, Oshlack A. Bpipe: A tool for running and managing bioinformatics pipelines. *Bioinformatics* 2012; **28**(11): 1525–1526.
9. Köster J, Rahmann S. Snakemake — A scalable bioinformatics workflow engine. *Bioinformatics* 2012; **28**(19): 2520–2522.
10. Fischer M, Snajder R, Pabinger S, Dander A, Schossig A, Zschocke J, Trajanoski Z, Stocker G. SIMPLEX: cloud-enabled pipeline for the comprehensive analysis of exome sequencing data. *PLoS One* 2012; **7**(8): e41948.
11. Torres-García W, Zheng S, Sivachenko A, Vegesna R, Wang Q, Yao R, Berger MF, Weinstein JN, Getz G, Verhaak RG. PRADA: Pipeline for RNA sequencing data analysis. *Bioinformatics* 2014; **30**(15): 2224–2226.
12. Nocq J, Celton M, Gendron P, Lemieux S, Wilhelm BT. Harnessing virtual machines to simplify next-generation DNA sequencing analysis. *Bioinformatics* 2013; **29**(17): 2075–2083.
13. Bauer MA, Chavan SS, Peterson EA, Heuck CJ, Johann DJ. Leveraging the new with the old: Providing a framework for the integration of historic microarray studies with next generation sequencing. *BMC Bioinformatics* 2014; **15**(Suppl 11): S3.
14. Data Driven Documents [http://d3js.org].
15. Robinson JT, Thorvaldsdóttir H, Winckler W, Guttman M, Lander ES, Getz G, Mesirov JP. Integrative genomics viewer. *Nature Biotechnology* 2011; **29**(1): 24–26.
16. Felter W, Ferreira A, Rajamony R, Rubio J. An updated performance comparison of virtual machines and linux containers. In *2015 IEEE International Symposium On: Performance Analysis of Systems and Software (ISPASS),* IEEE, 2015; pp. 171–172.
17. Docker [http://www.docker.com].
18. Boettiger C. An introduction to Docker for reproducible research, with examples from the R environment, 2014, arXiv preprint arXiv:14100846.
19. Chamberlain R, Invenshure LLC, Schommer J. Using Docker to support reproducible research. 2014.
20. Andrews S. FastQC: A quality control tool for high throughput sequence data. *Reference Source* 2010.
21. Li H, Durbin R. Fast and accurate short read alignment with Burrows — Wheeler transform. *Bioinformatics* 2009; **25**(14):1754–1760.
22. Lunter G, Goodson M. Stampy: A statistical algorithm for sensitive and fast mapping of Illumina sequence reads. *Genome Res* 2011; **21**(6): 936–939.
23. Picard Tools [http://broadinstitute.github.io/picard/].
24. McKenna A, Hanna M, Banks E, Sivachenko A, Cibulskis K, Kernytsky A, Garimella K, Altshuler D, Gabriel S, Daly M *et al.* The Genome Analysis Toolkit: a MapReduce

framework for analyzing next-generation DNA sequencing data. *Genome Res* 2010; **20**(9): 1297–1303.

25. Saunders CT, Wong WSW, Swamy S, Becq J, Murray LJ, Cheetham RK. Strelka: accurate somatic small-variant calling from sequenced tumor — normal sample pairs. *Bioinformatics* 2012; **28**(14): 1811–1817.
26. Cingolani P, Platts A, Wang LL, Coon M, Nguyen T, Wang L, Land SJ, Lu X, Ruden DM. A program for annotating and predicting the effects of single nucleotide polymorphisms, SnpEff: SNPs in the genome of Drosophila melanogaster strain w1118; iso-2; iso-3. *Fly* 2012; **6**(2): 80–92.
27. Sathirapongsasuti JF, Lee H, Horst BAJ, Brunner G, Cochran AJ, Binder S, Quackenbush J, Nelson SF. Exome sequencing-based copy-number variation and loss of heterozygosity detection: ExomeCNV. *Bioinformatics* 2011; **27**(19): 2648–2654.
28. Sathirapongsasuti JF. Pushing the boundaries of somatic copy-number variation detection: Advances and challenges. *Ann Oncol* 2015; **26**(1): 11–12.
29. BBCP Tool [http://www.slac.stanford.edu/~abh/bbcp/].
30. Beltran H, Eng K, Mosquera JM, Sigaras A, Romanel A, Rennert H, Kossai M, Pauli C, Faltas B, Fontugne J. Whole-exome sequencing of metastatic cancer and biomarkers of treatment response. *JAMA Oncol* 2015; **1**(4): 466–474.
31. Bamford S, Dawson E, Forbes S, Clements J, Pettett R, Dogan A, Flanagan A, Teague J, Futreal PA, Stratton M. The COSMIC (Catalogue of Somatic Mutations in Cancer) database and website. *Brit J Cancer* 2004; **91**(2): 355–358.
32. Li H, Handsaker B, Wysoker A, Fennell T, Ruan J, Homer N, Marth G, Abecasis G, Durbin R, others. The sequence alignment/map format and SAMtools. *Bioinformatics* 2009; **25**(16): 2078–2079.
33. Dobin A, Davis CA, Schlesinger F, Drenkow J, Zaleski C, Jha S, Batut P, Chaisson M, Gingeras TR. STAR: ultrafast universal RNA-seq aligner. *Bioinformatics* 2013; **29**(1): 15–21.
34. Anders S, Pyl PT, Huber W. HTSeq–A Python framework to work with high-throughput sequencing data. *Bioinformatics* 2014: btu638.
35. Trapnell C, Pachter L, Salzberg SL. TopHat: discovering splice junctions with RNA-Seq. *Bioinformatics* 2009; **25**(9): 1105–1111.
36. DeLuca DS, Levin JZ, Sivachenko A, Fennell T, Nazaire M-D, Williams C, Reich M, Winckler W, Getz G. RNA-SeQC: RNA-seq metrics for quality control and process optimization. *Bioinformatics* 2012; **28**(11): 1530–1532.
37. Trapnell C, Roberts A, Goff L, Pertea G, Kim D, Kelley DR, Pimentel H, Salzberg SL, Rinn JL, Pachter L. Differential gene and transcript expression analysis of RNA-seq experiments with TopHat and Cufflinks. *Nat Protocols* 2012; 7(3): 562–578.

13 Kidney Diseases in the Age of Molecular Diagnostics: An Integrated Histopathologic and Molecular Approach

Marco Delsante*, Jeffrey B. Kopp†, and Avi Z Rosenberg*

*Johns Hopkins University, Baltimore, MD, 21218, USA
†NIDDK, NIH, Bethesda, MD, 20892, USA

Contents

I. Introduction

Renal diseases are a leading cause of morbidity and mortality worldwide.[1] The clinical workup of a patient with unexplained renal functional abnormalities such as elevated creatinine or excessive urine protein includes the clinical context, personal and familial history, renal imaging, and serological and urinary studies. Despite this thorough evaluation, ultrasound-guided kidney needle biopsy remains the gold standard for the diagnosis of renal parenchymal diseases.[2] Though renal biopsies only sample a small percentage of the kidney and sampling error can be an issue on occasion, as most renal parenchymal diseases affect kidney tissue homogeneously, it is nevertheless sufficient for many diagnoses.[3] The routine pathological workup includes (1) histologic stains (hematoxylin and eosin — H&E, periodic acid Schiff (PAS), and silver and Masson trichrome stains), (2) immunofluorescence studies including antibodies directed against immunoglobulin heavy and light chains, complement components, albumin, and fibrinogen, and (3) ultrastructural studies by transmission electron microscopy of glomerular and extraglomerular structures. Together, this multimodal approach along with the clinical information can help achieve a contextual diagnosis with vital prognostic information.

Despite this standard approach, a specific diagnosis may be elusive or not as specific to be most helpful for patient management. In an era of precision/personalized medicine, diagnostic studies need to be extended to provide (a) a more specific diagnosis and (b) guidance for therapeutic choices. This is the promise of focused or global molecular studies in renal biopsy analysis. Precision or personalized medicine has been defined as "an emerging approach for disease treatment and prevention that takes into account individual variability in genes, environment, and lifestyle for each person".[4] Coupling integrated molecular studies (targeted or global/-omic approaches) with kidney biopsy diagnosis by standard renal histopathologic tissue workup is a promising path forward in achieving these goals. While the term "omic" was originally derived from the study of the genome, it has subsequently been addended to the global study of many other bioanalytes relevant to molecular diagnostics including, but not limited to the following: genes (genomics), gene transcripts (transcriptomics), polypeptides (proteomics), and metabolites (metabolomics). These molecular/-omic approaches have the potential to provide additional dimensions towards achieving a precision/personalized diagnosis. In this chapter, we will review molecular diagnostic tools applied to nephrology and specific nephrological diseases for which omic sciences play a relevant role in the diagnostic workup.

II. Molecular Diagnostic Techniques in Kidney Disease

Selection of the appropriate molecular diagnostic tool depends on the bioanalyte to be assessed and the sensitivity and specificity required for reliable and accurate reporting. Analytes to be discussed in the coming section include nucleic acids (DNA, RNAs), proteins, and metabolites.

a. DNA studies

When molecular diagnostics were first introduced, the principal technology for DNA genotyping was targeted sequencing for a known mutation. Traditional Sanger sequencing using sequencing primers directed against a particular genetic locus allows for the sequential insertion of nucleotides to derive the patient's genotype for comparison to a consensus sequence. While this approach has enjoyed broad deployment, next-generation sequencing technologies (NGS) are rapidly displacing it. NGS and its associated multiplexing technologies have greatly expanded the capacity for both discovery efforts and diagnostics. NGS relies on generating a cDNA library that can be either genome-wide library, e.g., whole exome approaches, or targeted panels using specific primer pairs to amplify genomic loci of interest, e.g., panels from Ion Torrent (ThermoFisher Scientific, Waltham MA) and Illumina (San Francisco, CA) and/or multiplexed library generation with a Fluidigm (South San Francisco, CA) microfluidics PCR method.[5] These burgeoning technologies have generated genomic output more rapidly than was previously possible and have become central to clinical genomics and diagnostics. Importantly, these approaches are amenable to DNA derived from a broad variety of matrices ranging from blood, fresh tissue, and, particularly relevant to the histopathologist, formalin-fixed paraffin-embedded tissues (FFPE). In addition to relatively routine clinical use in understanding the genetic basis of individuals renal parenchymal disease in clinical practice, many research consortia studying renal parenchymal disease have adopted NGS as a central methodology to study target genes and/or the whole exome.

b. RNA studies

Transcriptomics are a promising approach to the discovery of disease markers, disease-related processes, and implications of disease-causing or associated genes. Compared to DNA, RNA is a much more challenging analyte to assess because of it labiality and sensitivity to RNAses. Many techniques most readily allow for the analysis of tissue-derived RNAs from fresh and frozen tissue and include *in situ* hybridization and expression analysis (microarray and RNAseq). More recently, hybridization-based technologies that use short RNA fragments such as Nanostring (Nanostring Technologies, Seattle, WA) and ligation *in situ* hybridization (LISH) have emerged, making it possible to characterize the fragmented RNAs characteristic of FFPE material.[6] Finally, using these same technologies, non-coding RNA species such as miRNAs and lncRNA can be analyzed in tissues and in circulating exosomes. These approaches have accelerated efforts to understand the relevance of the diverse RNA species to pathobiology and molecular diagnostics. We will provide examples wherein RNA based-technologies have demonstrated a relevance to the diagnosis and prognosis of renal diseases.

c. Protein studies

In contrast to the methods for nucleic acids, the methods for protein and peptide analysis in tissues are rather limited. For example, routine renal biopsy immunofluorescence staining is performed for qualitative and semiquantitative assessment of immunoglobulins (heavy and light chains), IgG subclass staining (IgG1–IgG4), complement components (C3 and C1q), as well as albumin and fibrinogen. In the setting of renal transplantation, C3d and C4d staining have proven important in the assessment of rejection. Studies of collagen IV chains are also useful in the determination and confirmation of thin basement membrane diseases/Alport syndrome. Most recently, a role for phospholipase A2 receptor (PLA2R) staining in membranous glomerulopathy has emerged and proven useful as well in helping to distinguish the primary form of membranous nephropathy from those arising from secondary etiologies. As can be appreciated, these methods are all built, for the most part, on single to at most dual staining for the particular targets. By contrast, a proteomic approach is being employed in the diagnosis of fibrillary deposits (specifically amyloid) in the kidney and other tissues. Standard workflows for the diagnosis and subtyping of amyloids includes immunostaining for amyloid proteins such as serum amyloid A and transthyretin in conjunction with Congo red staining and ultrastructural analysis.

While additional useful immunohistochemical approaches are likely to emerge, there is at present limited multiplex capacity. By contrast, over the last five years peptide mass spectroscopy (MS) has been leveraged to identify amyloid subtypes in an unbiased manner. For example, when a positive Congo red stain is observed, the involved tissue can be isolated by microdissection and analyzed by shotgun mass spectrometry identifying the particular subtype of amyloid and the associated molecular retinue.[7] Using this approach, 25 discrete amyloids as well as a diagnostic component of fibrillary glomerulonephritis have been identified. Additional applications for proteomics includes the utility to study exosomes, for example in the setting of allograft rejection.[8] Further advances in targeted mass spectrometry (i.e., multiplex reaction monitoring) hold great promise to advance the field of serum and tissue proteomics in the setting of kidney diseases.

d. Metabolomics

A final set of analytes of interest in the kidney are by-products of metabolism. Metabolomics is the comprehensive characterization of the small-molecule metabolites, typically <1,000 Da. These molecules and their interactions within a biological system define the metabolome. Study of the metabolome is a promising approach to further characterize disease processes and identify disease markers. Common methods to study metabolites in biological systems include nuclear magnetic resonance spectrometry. Mass spectrometry techniques (including shotgun, targeted, and imaging approaches) hold promise to gain insight into metabolic changes at the cellular and tissue levels and to identify aberrations in health and disease. Thus, a recent study identified an association between serum metabolic profiles and different etiologies of chronic kidney disease (CKD).[9] However, with regard to metabolomics, this application of technology remains under development and thus it is not currently part of the diagnostic workup in kidney disease.

III. Molecular Diagnosis in Clinical Nephrology Practice

a. Glomerular diseases

i. *Idiopathic nephrotic syndrome and focal segmental glomerulosclerosis*

Idiopathic nephrotic syndromes may present across the age spectrum, from congenital nephrotic syndromes to diseases of the elderly such as dysproteinemias. Nephrotic syndrome is characterized by the classic pattern of proteinuria, hypoalbuminemia, edema, and dyslipidemia. There is a broad range of histopathological features associated with nephrotic syndromes, from minimal change disease (MCD), focal segmental segmental glomerulosclerosis (FSGS) to membranous nephropathy, with relative prevalence influenced of the affected individual's age group.

First-line remittive therapy for MCD and FSGS is typically a glucocorticoid. For those with frequent relapse, cyclophosphamide may be used. For those who are steroid-resistant, calcineurin inhibitor is often used, followed by mycophenolate mofetil.

Congenital nephrotic syndromes present with severe proteinuria within the first three months of life and have a cumulative incidence of 0.5/100.000 live births.[10] Non-steroidal anti-inflammatory agents and angiotensin pathway inhibitors are first-line therapy for these individuals, but parsing out the precise pathogenetic process underlying the particular congenital nephrotic syndrome compared to less responsive proteinuric processes is important. Formerly, congenital nephrotic syndromes were referred to as "of the Finnish type", this reflected their increased prevalence in those of Finnish descent (1/8,000 live births). The discovery of mutations in the nephrin gene (*NPHS1*)[11] enabled genetic, structural, and functional insights into mechanisms for podocyte dysfunction in proteinuria in these patients. Further efforts discovered additional mutations in another podocyte protein, *NPHS2* (podocin), which were found to be causative in another form of steroid-resistant nephrotic syndrome. To date, more than 50 genes have been implicated in steroid-resistant nephrotic syndrome, mostly in infants and children. These conditions are sometimes part of a syndrome, and inheritance patterns include autosomal recessive, autosomal dominant, sex-linked, and mitochondrial (Table 1).

The etiology of steroid-resistant nephrotic syndrome (SRNS) varies with age, and the prevalence of monogenic forms decreases with age. In the pediatric population, the clinical workup generally does not include a renal biopsy at the time of presentation. Rather, children with nephrotic syndrome are empirically treated with steroids and the disease further classified based on the response to steroid therapy (steroid sensitive, steroid dependent, frequently relapsing, secondary resistant, and resistant). Patients with steroid-dependent nephrotic syndrome SSNS or SRNS will typically receive immunosuppressive therapies, such as cyclophosphamide,

Table 1. Nephrotic syndrome genes

Autosomal dominant nephrotic syndrome	Autosomal recessive nephrotic syndrome	Syndromic nephrotic syndrome
• NPHS1 • NPHS2 • NPHS3 • ACTN4 • TRPC6 • PAX2 • ARHGAP24 • ANLN • APOL1	• CRB2 • TTC21B • COQ2 • COQ6 • PDSS2 • ADCK4 • NUP93 • NUP95 • XPO5 • EMP2 • PTPRO • GLEPP1 • KANK1 • KANK2 • KANK4	• WT1 (Denys-Drash and Frasier Syndrome) • LAMB2 (Pierson Syndrome) • SMARCAL1 (Schimke Syndrome) • LMX1B (Nail-Patella Syndrome) • INF2 (Charcot-Marie-Tooth Syndrome) • WDR73 (Galloway-Monwat Syndrome) • SCARB2/LIMP2 (Myoclonus-Renal failure Syndrome)

rituximab, or calcineurin inhibitors (cyclosporine or tacrolimus). Renal biopsy is most commonly performed in patients with SRNS and less frequently in SSNS, with FSGS being the most frequent histopathological pattern in the former group. In these patients, genetic studies focused on the genes listed in Table 1 to identify hereditary/congenital types of nephrotic syndrome, which are more typically resistant immunosuppressant therapy, saving needless exposure to steroids; response to other immunosuppressive modalities is also limited in this setting.[12]

FSGS most typically presents as nephrotic syndrome with progressive loss of renal function, although isolated nephrotic proteinuria and sub-nephrotic proteinuria are also seen. Six mechanistic groups of FSGS can be identified. Three forms are most common: primary, adaptive, and APOL1-associated FSGS. Three forms are less common: high-penetrance genetic, virus-associated, and medication-associated FSGS.

Primary FSGS is believed to be caused by a plasma factor. It has long been postulated and experimentally demonstrated that the plasma of some FSGS patients have a factor responsible for driving the glomerular permeability in the primary presentation and post-transplant FSGS recurrence setting.[13–15] Candidate molecules include cardiotrophin-like cytokine factor

1 and serum-urine type plasminogen-activator receptor (suPAR), although the evidence for both remain preliminary. Further, plasma antibodies may be markers for recurrence risk. Most recently, a panel of seven antibodies (CD40, PTPRO, CGB5, FAS, P2RY11, SNRPB2, and APOL2) was demonstrated to predict 92% of cases of post transplant recurrent FSGS.[16]

1. *APOL1-related nephropathies*

Coding variants in the gene encoding apolipoprotein L1 (*APOL1*), termed G1 and G2 risk alleles, have been associated with glomerulopathies in individuals of sub-Saharan Africa descent. Glomerular disease risk is largely confined to individuals with 2 risk alleles (RA), which is termed high-risk (HR) status. The APOL1 risk alleles initially evolved in sub-Saharan Africa to evade trypanosomal defense mechanisms, which in turn had evolved to evade killing by the APOL1 common variant G0, and thus confer resistance to trypanosome-transmitted infections. APOL1-related nephropathies comprise an expanding spectrum of renal histopathologic entities. The influence of *APOL1* HR status was first identified for its association with proteinuric glomerulopathies in the setting of FSGS and HIV associated nephropathy (HIVAN).[17,18] Subsequently *APOL1* HR status was associated with collapsing glomerulopathy in the settings of systemic lupus erythematosus (SLE),[19,20] PLA2R-positive membranous glomerulopathy[21] and exogenous interferon administration.[22] Interferons are potent drivers of APOL1 gene expression, and this suggests why APOL1 nephropathy arises in the setting of viral infection, most notably HIV, in lupus nephropathy, and with the administration of therapeutic interferon. The common thread is increased interferon release, which drives APOL1 gene transcription. *APOL1* RA expression mediates podocyte toxicity and podocytopathic alterations in collapsing glomerulopathy and other forms of FSGS.[22] More broadly, Kopp *et al.*, recently reported that *APOL1* HR status correlates with more glomerulosclerosis (segmental and global) and tubulointersitial chronicity[23] in a pediatric FSGS cohort.

Less is known about the role of *APOL1* variants in other histopathologic entities. For example, further work will be needed to elaborate the impact of *APOL1* RA status on renal allograft outcomes where donor (but not recipient) *APOL1* RA status correlates with accelerated loss of renal allografts[24,25] but did not impact HIV reinfection of allograft in HIV-positive recipients.[26] Finally, and most broadly, a weaker signal has been noted for *APOL1* HR status in type 2 diabetes-associated end-stage kidney disease (ESKD).[27] Additional cohort renal biopsies (particularly in the setting of diabetes) with genome-wide data will be needed to better understand the role of *APOL1* RA status in more complex genetic states. A current controversy concerns the utility of screening for *APOL1* RA in at-risk cohorts in order to inform the risk of end organ damage and risk to the kidney donor and recipient of an allograft with *APOL1* HR status; this is the subject of study in the APOL1 Long-term Kidney Transplantation Outcomes Network (APOLLO) network.

Glomerular proteinuria is largely due to podocyte dysfunction, together with contributions from endothelial dysfunction in some conditions, and so methods of assessing podocyte numbers are imperative. Recent evidence shows that progressive podocyte loss underlies renal allograft failure.[28] While assessment of podocyte density using tissue sections is a possibility using staining for podocyte markers such as Wilms tumor 1 (WT1) or transducin-like enhancer protein 4 (TLE4),[29] this relies on the availability of tissue, which is not routinely available for longitudinal assessment of glomerular disease. Wiggins has proposed an alternative podometric approach involving the measurement of urine podocyte mRNAs, such as the ratio between urine podocin/nephrin and aquaporin mRNAs[30–32] with normalization to urine creatinine. These assays can be performed robustly and longitudinally across a spectrum of disease states. Additional work with large cohorts are still needed to demonstrate the clinical utility of these for predicting disease activity and risk of progression.

ii. *Basement membrane-related disorders*

Collagen IV-related diseases are a complex set of rare genetic disorders first described in the early 20th century with renal manifestations (progressive kidney disease) and extrarenal manifestations (sensorineural hearing loss, eye defects, and leiomyomatosis).[33–35] The primary renal presentation of these disorders is microscopic hematuria. There are two conditions that overlap clinically, the more common thin basement membrane nephropathy (TBMN) and the rare disease, Alport syndrome. TBMN is the most common cause of hematuria (prevalence 1%) in the setting of primary glomerular disease.[35] By contrast, the prevalence of Alport syndrome ranges from 1:5,000 (Utah and southern Idaho) to 1:53,000 (Finland) (http://www.orpha.net).[36,37] The current basis for the diagnosis of Alport syndrome relies on a combination of clinical history, family history, histological examination of skin and/or kidney biopsy, and confirmation by genetic testing. Alport syndrome typically shows variable but definite rates of progression depending on the genetic form, and this contrasts to the renal-limited and more indolent though occasionally progressive TBMN.[38,39] Three genetic forms of Alport syndrome are known, the most common of which is X-linked (80%), followed by autosomal recessive (15%) and autosomal dominant (5%) subtypes. Patients with renal presentation of Alport syndrome show early hematuria with progressive renal insufficiency.

It is now established that collagen IV defects account for most cases of AS, with X-linked variant caused by alterations in the COL4A5 gene,[40] autosomal recessive caused by alterations in both copies of either COL4A3 or COL4A4[41,42] gene, and autosomal dominant in one copy of either COL4A3 or COL4A4.[43,44] Because of the complexity of these genetic variants and the hundreds of genetic alterations observed, diagnosis involves a multi-modal approach to include clinical evaluation, renal and skin biopsies evaluated with ultrastructural and immunohistochemical analysis, and DNA analysis. The array of mutations in these diseases are many and include point mutations, insertions, deletions and inversions (rarely). As a results, an array of molecular diagnostic modalities is used. First, in a cases where ultrastructural examination shows thin and/or lamellated basement membranes characteristic of Alport disease, immunofluorescence studies can be performed on frozen biopsy material for collagen 4 alpha chains (usually alpha 1, alpha 3, and alpha 5) (Wieslab) or a cocktail of alpha 2 and alpha 5 performed using 2-color immunofluorescence (Shigei Medical Research Institute).[45] When COL4A3 or COL4A5 chains are lost, absent staining of the glomerular basement membranes is noted. However, this technique relies on tissue availability (kidney or skin), indirect immunofluorescence methods, and complete

Table 2. Integrated histopathologic and molecular analysis of primary kidney diseases

Name	Pathogenesis	Omic platform	Histologic lesions	Histopathology
Focal segmental glomerulosclerosis	Genetic or acquired podocyte injury	Genomics	• Segmental glomerular consolidation (picture) • Variable podocyte hypertrophy • Variable (usually extensive) podocyte foot process effacement on EM	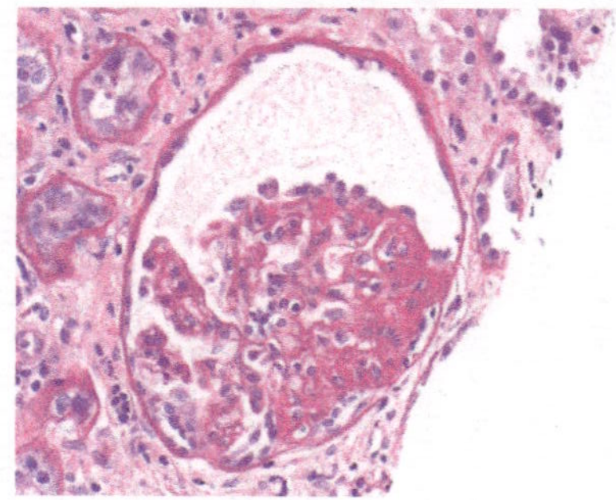
Alport disease	Genetic (autosomal or X-linked) mutation of type IV collagen	Genomics/ Proteomics	• Mostly normal looking glomeruli on LM • Lack of specific Collagen IV subunits stain on IF • GBM irregularities on EM (picture)	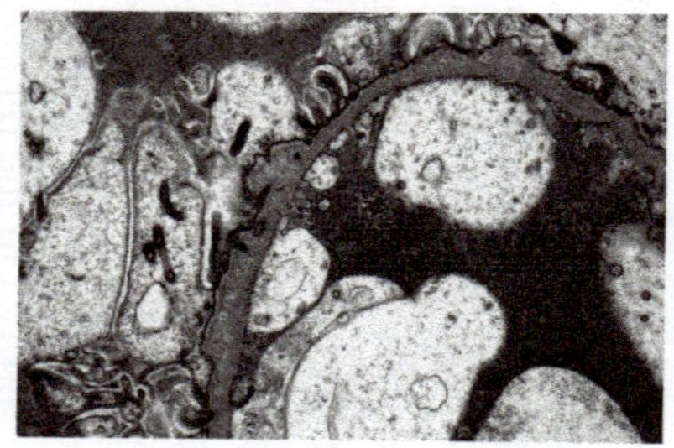

Atypical hemolytic uremic syndrome (aHUS)	Alternative complement pathway dysregulation	Genomics/ Proteomics	• Variable lesions on LM, including glomerular thrombi, fibrinoid necrosis, and arteriolar onion skinning (picture) • Non-specific IF and EM findings	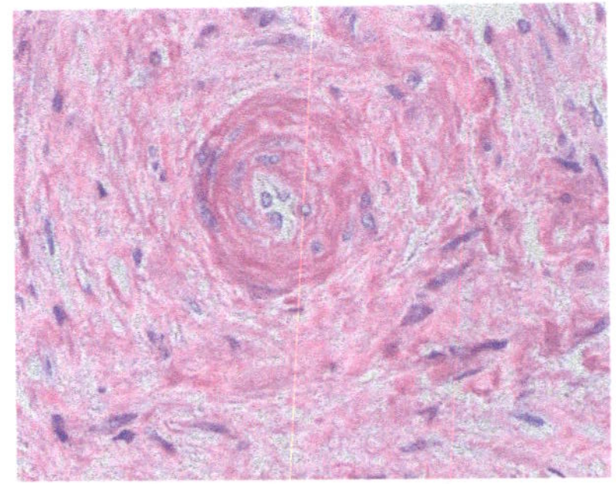
C3-glomerulopathies	Alternative complement pathway dysregulation	Genomics/ Proteomics	• Variable LM lesion, usually MPGN pattern • Dominant/exclusive C3 deposition on IF (picture) • Variable subendothelial/ mesangial electron dense deposits on EM (including specific pattern of dense deposit disease)	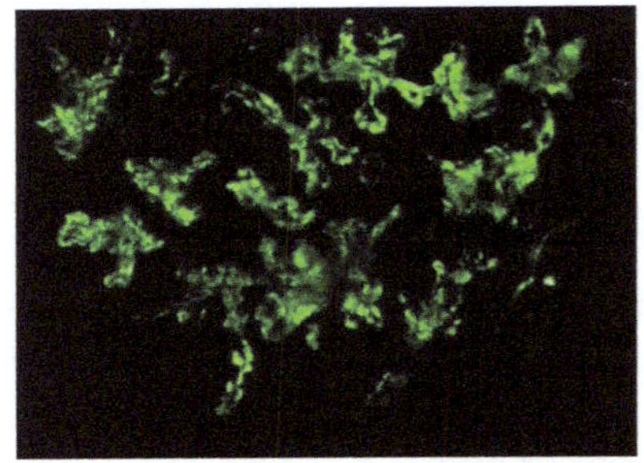

(*Continued*)

Table 2. (*Continued*)

Name	Pathogenesis	Omic platform	Histologic lesions	Histopathology
IgA Nephropathy	Abnormal glomerular IgA accumulation	Proteomics	• Mostly mesangioproliferative pattern; MPGN and crescents can be observed • IgA and C3 mesangial deposits on IF (picture) • Electron-dense mesangial deposits on EM	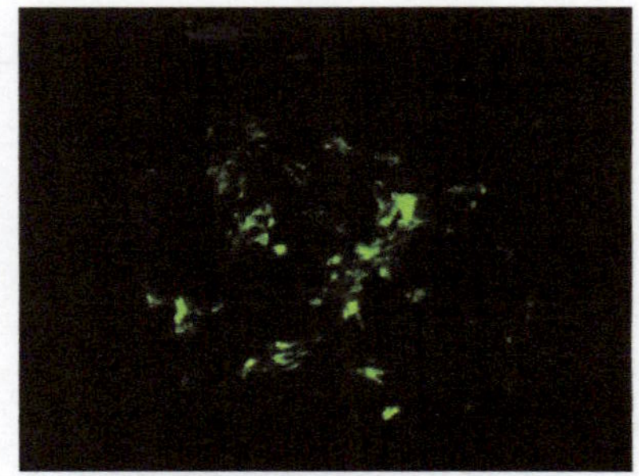
Membranous Glomerulopathy	Glomerular subepithelial deposits of complement and IgG against glomerular antigens	Proteomics	• GBM thickening on light microscopy with variable degree of GBM projections on silver methenamine stain (spikes; picture, right panel) • IgG and C3 granular GBM deposits by IF (picture, left panel) • Electron-dense subepithelial deposits on EM	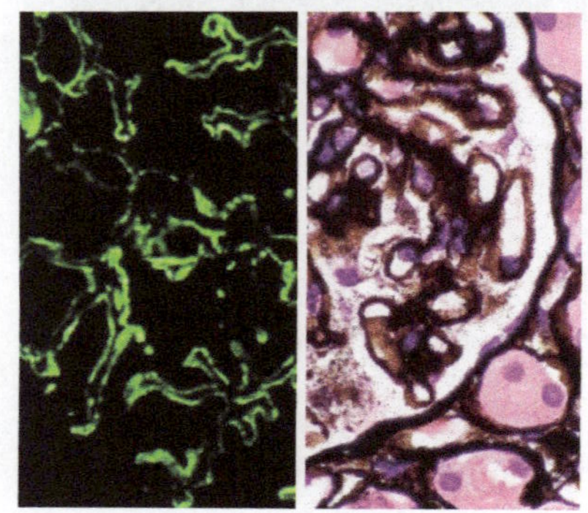

Amyloidosis	Extracellular glomerular beta fibril deposition	Proteomics	• Variable pattern by LM, mostly involving mesangial expansion. Amyloid deposition can be seen in vessels and interstitium. • Apple green birefringence on Congo red stain under polarized light (picture) • IF can show monoclonal light chain deposition (AL type amyloidosis) • Organized electron dense deposits (7–12nm fibrils)	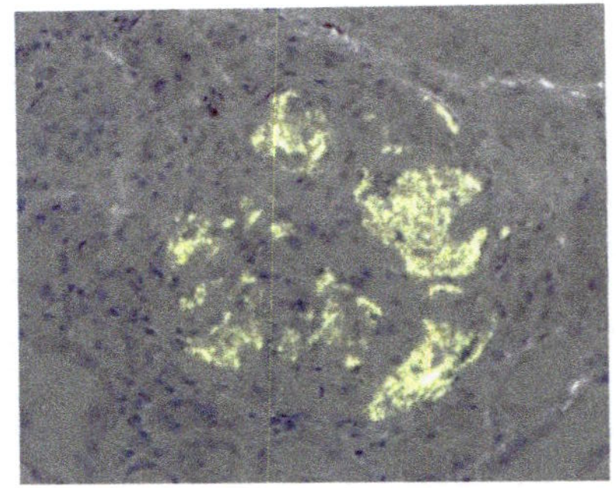

Note: LM: light microscopy; IF: immunofluorescence; EM: electron microscopy; GBM: glomerular basement membranes.

loss of staining for diagnostic certainty. Martin *et al.* described a technique using PCR and direct sequencing to explore mutations in COL4A5 intronic flanking sequences and 51 exons with a mutation detection rate of 82%.[46] In order to have adequate coverage of the various mutations in COL4A3, COL4A4, and COL4A5, it has been necessary to date to use multiple techniques to cover the various mutational mechanisms and would take several months by traditional methods. Recently, reports of using next-generation sequencing have demonstrated the robustness to cover COL4 chains for Alport's testing with great efficiency.[47–51]

Advanced DNA analysis has been helpful at unmasking the mutational complexity over the large collagen IV genes (48–53 exons, >5,000 base pairs) but makes genetic diagnosis, albeit specific and sensitive (>95%), a challenge. For adequate diagnostic utility, testing strategies must include splice sites and 5′ untranslated regions for adequate coverage of the mutational spectrum. As a result of the complexity of mutational possibilities, current approaches to mutational analysis includes direct sequencing (exons and flanking introns), reverse transcriptase PCR of mRNA derived from cultured skin fibroblasts (splicing site variants), multiplex ligation-dependent probe amplification (submicroscopic deletions/duplications), conventional cytogenetic analysis (structural chromosome rearrangements), and next-generation sequencing, which is a promising new technique providing a single platform to assess the whole genomic context of the collagen IV genes.

The pathomechanism underlying glomerular injury in Alport syndrome and progression is similarly complex.[52] Collagen IV protomers undergo a switch in glomerular basement membrane (GBM) composition from embryogenesis to adulthood from collagen IV a1,a1,a2 protomers to collagen IV a3,a4,a5 protomers with two protomers forming a hexamer. With aberrant adult GBM collagen IV composition in the setting of Alport syndrome, weaker GBMs susceptible to recurrent fracturing and repairing result. The embryonic GBM collagen IV is very susceptible to matrix metalloproteinase (MMP) activity, in particular MMP12 which is highly expressed by macrophages of mice with AS.[53] With recurrent injury, additional chemokines including transforming growth factor beta (TGFβ) and tumor necrosis factor (TNF) are thought to contribute to the progression of Alport syndrome.

The prognostic significant of staining for COL4A3 and COL4A5 for X-linked Alport syndrome has been explored, and in fact loss of persistent expression of COL4A3 in the context of COL4A5 loss is a positive prognostic factor.[54] Furthermore, the value of co-staining has also been suggested in Alport syndrome diagnosis.[55] Proteomics can be useful to evaluate other putative biomarkers upregulated in the context of Alport syndrome models, and these include integrin a1, integrin a3, and vimentin.[56] Additional biomarkers were reported in limited cohorts, such as the 23 putative biomarkers reported by Baum *et al.*[57] and MMP12/CCR2.[53] In summary, the collagenopathies have systemic and renal manifestations, and a combination of genetic, histopathologic, and serum biomarkers will play a role in the future of diagnosis.

iii. *Complement-related disorders*

Inherited and acquired alternative complement pathway dysregulation may elicit acute or chronic kidney injury. These rare defects can cause a broad spectrum of clinical and pathological manifestations whose characterization and classification are still being defined. Atypical hemolytic uremic syndrome (aHUS) and C3 glomerulopathies are the most common renal manifestations.[58,59]

1. *Atypical hemolytic uremic syndrome (aHUS)*

aHUS is an acute/subacute disease clinically characterized by endothelial injury, leading to intravascular hemolysis, thrombocytopenia, small vessels thrombosis (thrombotic microangiopathy), and variable degrees of renal and extrarenal organ failure. The syndrome is frequently triggered by an infection. The endothelial injury is caused by abnormal activation of the complement alternative pathway, due to congenital or acquired anomalies. Acquired anomalies include formation of autoantibodies against alternative pathway regulatory protein factor H, and the inactivation of this protein leads to dysregulated complement activation and endothelial injury. On the other hand, congenital anomalies of genes involved in alternative pathway regulation can cause an identical phenotype: loss-of-function mutations in a negative regulatory gene (*CFH*, *CFI*, or *CD46*) or gain-of-function mutations in an effector gene (*CFB* or *C3*) have been described.[60] Moreover, several different variants of the same genes have been associated with increased risk of aHUS. The renal biopsy findings are indistinguishable from typical (infection-related) HUS and show vascular anomalies (including small vessels thrombosis and glomerular thrombotic/ischemic alterations) and variable degrees of tubular ischemic injury.[61] The family history can be positive for thrombotic microangiopathy (sometimes misdiagnosed as malignant hypertension) and/or unexplained renal disease. Genetic testing for pathogenetic mutations or risk variants is essential to guide therapeutic choices, which include the use of specific complement inhibitors, in some cases for the entire life of the patient.

2. *C3-glomerulopathies*

C3-glomerulopathies include dense deposit disease (DDD) and C3-glomerulonephritis (C3GN), both characterized clinically by urinary abnormalities (proteinuria and/or hematuria), renal dysfunction, and histologically by glomerular deposition of C3 as demonstrated by immunofluorescence. The two diseases are differentiated by the glomerular electron microscopy patterns: DDD shows dense intramembranous deposits, whereas in C3GN the deposits are typically in the subendothelial space and mesangium.[62] Both clinical and pathological manifestations are caused by alterative complement pathway dysregulation. In both syndromes, alternative pathway dysregulation can be due either to the presence of antibodies that stabilize the alternative pathway C3 convertase (these antibodies are called C3-nephritic factor, and are seen in up to 80% of patients with DDD).[63-65] or inherited or acquired defects in the function of the negative regulator protein Factor H (either Factor H gene loss of function mutation or C3 gene mutations causing resistance to factor H).[66–69] In both syndromes, familial cases have been described and genetic studies are indispensable to identify the protein responsible for alternative pathway dysregulation.

iv. *Others*

1. *IgA nephropathy*

IgA Nephropathy (IgAN) is glomerular disease characterized by the deposition of atypical (hypoglycosylated) IgA in the glomeruli.[70] While microhematuria is almost invariably present, patients can also show variable degrees of proteinuria and many develop progressive loss of renal function.

Epidemiologic studies show marked ethnic differences, being frequent in Asian population, relatively common among European-descent populations, and rare in Africa;[71,72] familial cases have been reported.[73] Although no molecular diagnostic is currently part of the routine diagnostic workup for IgA nephropathy, recent genome-wide association studies (GWAS) have identified several risk loci linked to the development of the disease, including major histocomaptibility complex (MHC) region and genes involved in complement pathway regulation (factor H).[74–76] With regard to the MHC region, the frequency of these risk variants among continental populations mirrors the prevalence of IgA nephropathy in those populations (Asia > Europe > Africa).[74]

Proteomic techniques have been employed to study the characteristics of mesangial IgA deposits: these IgA molecules are mostly IgA1 with poor O-glycosylation and sialylation alterations.[77,78]

Despite these molecular and genetic insights, the pathogenesis of IgA nephropathy remains unknown, and the diagnosis is based on the typical histologic findings in renal biopsy.

2. *Idiopathic membranous nephropathy*

Membranous nephropathy is a leading cause of nephrotic syndrome in adults. Histologically, the disease shows the presence of immune deposits (immunoglobulins, mainly IgG, and complement) in the glomerular basement membranes.[79] Membranous nephropathy can be idiopathic or secondary to drugs or other systemic diseases, such as chronic viral infections, collagenopathies, or neoplasms. In approximately 75% of the cases of idiopathic membranous nephropathy, IgG are directed against a glomerular protein, the phospholipase A2 receptor (PLA2r).[80,81] Protein composition of glomerular deposits is studied, in clinical practice, through immunofluorescence or immunohistochemistry, to show deposition of IgG (mostly IgG4 and IgG1 subclasses), complement C3 fraction, and increased expression of PLA2r antigen. ELISA kit are available for isolation of PLA2r antibodies in serum.

Several GWAS have shown association between idiopathic membranous nephropathy and *PLA2R1* gene variants and histocompatibility leucocyte antigen (HLA)-DQA1 allele, but, as for IgAN, genetic testing is not currently part of the clinical workup.[82]

3. *Renal amyloidosis and paraprotein-related disorders*

Amyloidosis is an extensive family of diseases characterized by extracellular accumulation of fibrils assembling with a beta-sheet structure; at least 25 proteins have been associated with this pathologic process.[83] Most commonly, the fibrils derive from immunoglobulin light (AL) or heavy (AH) chains or amyloid protein A (AA). Less common variants include familial transthyretin, amyloid beta protein (associated with Alzheimer disease), and several rare forms.[84] Renal amyloidosis, most frequently observed in the elderly, is clinically characterized by variable degree of proteinuria and renal insufficiency, and progression to end stage kidney disease (ESKD) is not uncommon.[85,86]

The renal biopsy shows amyloid deposits in glomeruli (mesangium and glomerular basement membranes) and less frequently in blood vessels and tubulointerstitium. Deposits are identified through characteristic staining behavior (PAS and silver negativity, positivity of Congo Red stain with apple green birefringence on polarized light) and ultrastructural studies, which will show 7–12 nm thick, randomly arranged fibrils. Until recently, deposit composition was exclusively studied by immunofluorescence and immunohistochemistry, to identify the different amyloid components (e.g., light chains, amyloid A, etc.). However, the sensitivity and specificity of these approaches are not sufficient for a precise diagnosis.[87,88] Proteomics have been used in the last few years to assess amyloid composition. Particular renal parenchymal components (e.g., glomeruli) can be isolated through laser microdissection and analyzed using mass spectrometry.[89–91]

b. Extraglomerular diseases

The renal tubulointerstitial compartment is invariably involved in the late stages of glomerular disease but can be the primary site of renal injury in several pathologic conditions. Some of these diseases are genetically determined, and so omic sciences offer new opportunities in the diagnostic approach. Polycystic kidney disease represents one of the most frequent hereditary cause of ESKD worldwide. Newborn screening for genetic disorders is widely used in developed countries. In the USA, 54 conditions are tested for, but the list of conditions does not include any disease discussed in this chapter. Other countries test for fewer conditions.

i. *Nephrolithiasis/nephrocalcinosis*

A recent retrospective study from Halbritter *et al.* involving subjects from Europe and the United States showed that monogenic mutations in 14 genes account for 15% of nephrolithiasis/nephrocalcinosis cases.[92] The most frequently involved are cystinuria genes (SLC7A9 and CSNU1, autosomal dominant and SLC3A1, autosomal recessive), observed in up to 1–2% of the kidney stone cases, with higher percentages in children (5%).[93] The clinical use of genetic testing for these disorders is still debated and, in some settings, may be limited to pediatric and familial cases.

ii. *Fabry disease*

Anderson–Fabry disease is an X-linked inborn error of metabolism, involving deficit of the enzyme alpha-galactosidase A (alpha-Gal A), which results in accumulation of globotriaosylceramide within lysosomes in a wide variety of cells, including renal tubular and glomerular cells. This process leads to variable degree of renal dysfunction and slow progression toward ESKD. The disease is rare and, due to the X-linked transmission, is mainly observed in males, but the presence of pathogenic mutations is probably largely underestimated in the general population (genetic screening studies suggest presence of mutation in 1:22,000–40,000) among European descent populations.[94,95]

The alpha-gal A encoding gene (*GLA*) is located on the long arm of X chromosome, and hundreds of different disease-causing mutations have been identified. The diagnosis is based on measurement of alpha-Gal A activity in serum, with high sensitivity and specificity in males with typical manifestations but much lower performance in females and males with atypical disease. Therefore, genetic testing of *GLA* is recommended in patients with marginal levels of alpha-Gal A activity and, more broadly, in at least one member of each involved family.

iii. *Polycystic kidney disease and congenital anomalies of kidney and urinary tract (CAKUT)*

Autosomal dominant polycystic kidney disease (ADPKD) is a relatively common disease, affecting 1:400–1,000 live births among European descent populations.[96] Approximately 80% of patients have a mutation on chromosome 6 (*PKD1* gene), whereas 14% show alteration on chromosome 4 (*PKD2*). No mutation is identified in the remaining 6% of the patients. Clinically, the disease is characterized by the gradual development of multiple cysts in kidney parenchyma, leading to compression of the functioning kidney and variably progressive renal dysfunction. Cysts are also present in other solid organs such as liver and pancreas, but functional impairment is less frequent is these sites. Patients may show vascular aneurysms in several regions, including brain arteries. PKD2 patients have less severe disease with later onset and slower progression to ESKD.

In patients with family history of ADPKD, the diagnosis is based on imaging, but genetic studies are the necessary in specific cases (i.e., pediatric patients with equivocal ultrasound findings) and allow to identify the specific mutation. The method of choice is Sanger sequencing of all exons and splice junctions of PKD1 and PKD2 genes. Most of changes are missense mutations (more than 1,500 and 240 mutations have been identified for PKD1 and PKD2, respectively) [Autosomal dominant polycystic kidney disease mutation database: PKDB http://pkdb.mayo.edu/].

Homozygosity for PKD1 or PKD2 mutations results in lethal embryonic phenotype, but a recessive variant of polycystic disease (autosomal recessive polycystic kidney disease, ARPKD) is caused by mutation in *PKHD1* gene, which can lead to hepatic and lung cystic involvement.[97] Clinical presentation is variable and is overall more severe if the onset is during infancy. Genetic testing is not necessary when clinical criteria are fulfilled, but may be useful in cases with uncertain radiologic findings or for preimplantation and prenatal diagnosis. Available techniques include gene sequencing and targeted mutation analysis. However, false negative genetic testing results are possible because of the large size of the *PKHD1* gene and the large number of mutations associated with ARPKD.[98] Next-generation sequencing provides an accurate tool and may prove to be advantageous in the early diagnosis.[99]

Other cystic kidney disease, which may be included in the differential diagnosis with ADPKD and ARPKD, are tuberous sclerosis complex, glomerulocystic kidney disease, HNF1B-related cystic disease, and nephronophthisis (NPHP). The latter is an autosomal recessive genetically heterogenous disorder for which mutations have been identified in several genes that encode proteins involved in the function of primary cilia, basal bodies, and centrosomes (including NPHP1 to 13 and several other genes). Clinical presentation and age of onset are highly variable, but the disease typically progresses to ESKD.[100,101] In patients with high clinical suspicion for NPHP, screening for *NPHP1* deletion should be performed as these mutations are the most common cause of NPHP. Homozygous deletions can be detected by the lack of amplification by PCR of genomic DNA markers; this test permits a quick and accurate diagnosis of the disease in affected patients.[102] Testing for mutation in all the other NPHP genes is not widely available, but next-generation sequencing techniques will likely provide accurate tools for molecular diagnostic of NPHP.

Approximately 3–7 out of 1,000 births show congenital anomalies of the kidney and urinary tract (CAKUT), accounting for about 20% of birth defects.[103,104] These malformations account for 40–50% of pediatric and 7% of adult ESKD worldwide[105]. The clinical phenotype and the renal prognosis are extremely variable. The differentiation defects may occur along the full extent of the kidney and urinary tract. Currently, the diagnosis of CAKUT is based on prenatal or postnatal imaging studies demonstrating absent or small kidneys with or without additional urinary tract defects. The etiology of the majority of cases remains unknown. However, multiple lines of evidence suggest a strong genetic contribution to the pathogenesis of these birth defects. For example, they can be associated with many cytogenetic abnormalities and genetic syndromes, and up to 10% of patients with kidney malformations show mutations associated with genetic syndromes. Copy-number disorders have been associated with sporadic cases of CAKUT,[106] whereas familial and syndromic forms have shown mutations of several different genes.[107,108] Molecular diagnosis is not currently part of the standard diagnostic workup in children and adults with CAKUT, but in syndromic cases, alterations of frequently involved genes (such as *PAX* genes and *HNF1β*) should be investigated using traditional (Sanger) or next-generation sequencing.

c. Kidney transplantation

Once renal parenchymal diseases progress to ESKD, a decision whether or not to proceed with a kidney transplant must be made. While renal transplantation offers survival and quality of life benefits over dialysis, it is not without significant long-term challenges and consequences, including transplant rejection, infection, and cancer. Various conditions may make renal transplant inadvisable, including active cancer, severe obesity, substance abuse, dementia, and inability to adhere to a medication regimen. Omic sciences play a fundamental role in the different steps of kidney transplantation, form pre-transplant tests to long term follow-up.

i. *Pre-transplant tests*

Before proceeding with renal transplantation, immunological matching between donor and recipient must be assessed. First, the HLA profile of both donor and recipient is defined through PCR studies. Even though renal transplant can be performed in presence of any degree of HLA incompatibility (mismatch) between donor and recipient, the number of different HLA alleles in the couple will influence the graft survival and choice of immunosuppressive therapy and plays a variable role in different organ allocation algorithms.[109–111] Typically, HLA mismatching is assessed between HLA-A, B, and DR, but additional HLA alleles, including C, DP, DQ, and DW have been studied for their possible relevance in incidence of rejection and allograft survival.[112,113]

A patient on the transplant waiting list can have pre-formed antibodies against different HLA alleles. If the transplant is performed using an organ expressing one of these HLA molecules, the risk for hyperacute rejection (with intraoperative vascular thrombosis and immediate graft loss) is very high. Therefore, patients waiting for a transplant are screened for the presence of these antibodies both periodically and at the time of the transplant.[114] If a preformed donor-specific antibody against HLA molecule (HLA-donor specific antibodies, DSA) is identified, the transplant is generally waived, unless specific desensitization protocol is available. Proteomic methods used to screen for pre-formed HLA-DSA include the following:

1. *Cytotoxic cross-match*

This technique, developed at the end of the 1960s,[115] consists of incubating recipient's serum with donor's lymphocytes (showing high HLA expression) and complement factors. If HLA-DSA are present, donor's lymphocytes (the test is performed separately for B- and T-lymphocytes) lysis will be observed as a consequence of antibody-mediated complement activation. This test is highly specific for the presence of DSA, but false negatives are possible.

2. *Flow cytometry cross-match*

Flow cytometry cross-match increases the sensitivity of cytotoxic cross-match and is used to assess the presence on circulating DSA in recipient's serum. Donor lymphocyte and recipient serum are mixed and a fluorescein marked anti-human IgG antibody is added. If DSA are present, there will be a fluorescent signal on the lymphocyte surface that will be read by flow cytometry.[116]

3. *Donor specific cross-match by flow cytometry*

In 1998, a novel testing strategy was described that uses purified human leucocyte antigen (HLA) molecules coated on microspheres.[117] These microspheres are also coated with fluorochromes that can be detected by flow-cytometry-based techniques (Luminex, Austin, TX). Thus, recipient sera can be "cross-matched" with the HLA-coated microspheres. Fluorescence intensity will express a semi-quantitation of DSA levels, which correlates with the positivity for that specific serum to cytotoxicity of flow cytometry crossmatch (allowing to generate "virtual crossmatches").

The technique is extremely sensitive for the detection of DSA and has revolutionized the pre- and post-transplant management of kidney transplant recipients.[118]

ii. *Post-transplant tests*

Kidney allograft malfunction can be the consequence of both non-immunological and immunological factors, the latter presenting as variable phenotypes of transplant rejection. The gold standard for assessing the cause of allograft dysfunction is renal biopsy, which, when adequate, will allow to identify the presence of rejection and to categorize the immune injury based on a specific classification system (Banff Classification).[119] Increasing interest in this process is shown for omic sciences, which include serological and tissue tests involved in the diagnostic flow. Some of these tests (Luminex, described above) are currently part of the evaluation of a transplanted patient with kidney dysfunction: the presence of DSA through Luminex test should be ruled out every time a clinical suspect of antibody-mediated rejection is posed.

Other tests are being validated to assist histology in the diagnosis of renal allograft rejection and some are already part of the diagnostic algorithm.

1. *RNA expression in renal allograft biopsy*

Analysis of gene expression though messenger (m)RNA sequencing has been studied for the diagnosis of T-cell-mediated rejection (TCMR)[120,121] and antibody-mediated rejection (ABMR).[122,123] The test is based on the observation that different rejection subclasses lead to diverse tissue gene expression, with TCMR being characterized by activated-T-cell and antigen-presenting-cell transcripts overexpression (including CTLA4, IFNG, ADAMDEC1, and CXCL13), and ABMR showing elevated tissue levels of natural killer (NK) cells (GNLY), endothelial (DARC), and IFNG-induced (CXCL11 and PLA1A) transcripts. On the other hand, non-rejection causes of allograft dysfunction (such as acute kidney injury or interstitial fibrosis and tubular atrophy) show different molecular signatures, allowing a clinically useful distinction of different pathologic processes. Although highly promising, mRNA expression studies on allograft biopsies have some limitations, including elevated costs and lack of standardization among different centers.[122] To note, specific gene expression signature has been included in the criteria for diagnosis of rejection by Banff system.[124]

2. *Circulating donor cell-free DNA*

Donor-derived cell-free DNA (dd-DNA) can be detected in the blood of transplant recipients and has been proposed as a marker for non-invasive diagnosis of rejection. Several studies have shown increased levels of circulating dd-DNA during rejection compared to normal controls. However, the ability to distinguish different types of rejection is still not satisfactory and dd-DNA testing is not currently part of clinical workup of rejection.[125,126]

3. *Other tests*

Tissue sample are used to rule out viral infection involving the allograft through immunohistochemistry, such as antibodies against SV40 antigen to evaluate for BK virus infection, as the two polyoma viruses share homology, or in situ hybridization for polyoma RNA or DNA.

New markers are being actively investigated to allow non-invasive diagnosis/screening of rejection. These include plasma and urinary levels of CXCL9 for cell-mediated rejection[127] and CXCL10 for antibody-mediated rejection.[128]

IV. Conclusions and Future Directions

In part due to the application of omic sciences, progress has been made in the last 20 years toward precision medicine and new disease classification in renal biopsy interpretation, moving from a histopathological/descriptive approach to a more pathogenetic disease assessment. Specific nephrology branches such as pediatric nephrology and renal transplantation have already experienced substantial advances via molecular studies and proteomics. However, limitations to most of omic studies include the challenges of multicentric technique standardization and sometimes prohibitive costs. These limitations may be overcome by consortia developing and testing analytic protocols that use modern and affordable technologies.

Acknowledgement

This work supported in part by the Intramural Research Program, NIDDK, NIH (JBK).

References

1. Mills KT, Xu Y, Zhang W, *et al.* A systematic analysis of worldwide population-based data on the global burden of chronic kidney disease in 2010. *Kidney Int.* 2015; **88**(5): 950–957.
2. Dhaun N, Bellamy CO, Cattran DC, Kluth DC. Utility of renal biopsy in the clinical management of renal disease. *Kidney Int.* 2014; **85**(5): 1039–1048.
3. Scheckner B, Peyser A, Rube J, *et al.* Diagnostic yield of renal biopsies: A retrospective single center review. *BMC Nephrol.* 2009; **10**: 11.
4. Health. NIo. What is precision medicine? 2018; Available at https://ghr.nlm.nih.gov/primer/precisionmedicine/definition. Accessed 4/17/2018.
5. Gillies CE, Otto EA, Vega-Warner V, *et al.* tarSVM: Improving the accuracy of variant calls derived from microfluidic PCR-based targeted next generation sequencing using a support vector machine. *BMC Bioinformatics.* 2016; **17**(1): 233.
6. Credle JJ, Itoh CY, Yuan T, *et al.* Multiplexed analysis of fixed tissue RNA using Ligation in situ Hybridization. *Nucleic Acids Res.* 2017; **45**(14): e128.
7. Vrana JA, Gamez JD, Madden BJ, Theis JD, Bergen HR, Dogan A. Classification of amyloidosis by laser microdissection and mass spectrometry-based proteomic analysis in clinical biopsy specimens. *Blood.* 2009; **114**(24): 4957–4959.
8. Pisitkun T, Gandolfo MT, Das S, Knepper MA, Bagnasco SM. Application of systems biology principles to protein biomarker discovery: Urinary exosomal proteome in renal transplantation. *Proteomics Clin Appl.* 2012; **6**(5–6): 268–278.
9. Grams ME, Tin A, Rebholz CM, *et al.* Metabolomic alterations associated with cause of CKD. *Clin J Am Soc Nephrol.* 2017; **12**(11): 1787–1794.
10. Bérody S, Heidet L, Gribouval O, *et al.* Treatment and outcome of congenital nephrotic syndrome. *Nephrol Dial Transplant.* 2018.
11. Kestilä M, Lenkkeri U, Männikkö M, *et al.* Positionally cloned gene for a novel glomerular protein — nephrin — is mutated in congenital nephrotic syndrome. *Mol Cell.* 1998; **1**(4): 575–582.

12. Santín S, Bullich G, Tazón-Vega B, *et al.* Clinical utility of genetic testing in children and adults with steroid-resistant nephrotic syndrome. *Clin J Am Soc Nephrol.* 2011; **6**(5): 1139–1148.
13. Sharma M, Sharma R, McCarthy ET, Savin VJ. "The FSGS factor:" Enrichment and *in vivo* effect of activity from focal segmental glomerulosclerosis plasma. *J Am Soc Nephrol.* 1999; **10**(3): 552–561.
14. Sharma M, Sharma R, McCarthy ET, Savin VJ. The focal segmental glomerulosclerosis permeability factor: biochemical characteristics and biological effects. *Exp Biol Med (Maywood).* 2004; **229**(1): 85–98.
15. Moriconi L, Lenti C, Puccini R, *et al.* Proteinuria in focal segmental glomerulosclerosis: role of circulating factors and therapeutic approach. *Ren Fail.* 2001; **23**(3–4): 533–541.
16. Delville M, Sigdel TK, Wei C, *et al.* A circulating antibody panel for pretransplant prediction of FSGS recurrence after kidney transplantation. *Sci Transl Med.* 2014; **6**(256): 256ra136.
17. Fine DM, Wasser WG, Estrella MM, *et al.* APOL1 risk variants predict histopathology and progression to ESRD in HIV-related kidney disease. *J Am Soc Nephrol.* 2012; **23**(2): 343–350.
18. Genovese G, Friedman DJ, Ross MD, *et al.* Association of trypanolytic ApoL1 variants with kidney disease in African Americans. *Science.* 2010; **329**(5993): 841–845.
19. Kofman T, Narjoz C, Raimbourg Q, *et al.* Collapsing glomerulopathy associated lupus in a black female with homozygous APOL1 mutation. *Lupus.* 2012; **21**(13): 1459–1462.
20. Larsen CP, Beggs ML, Saeed M, Walker PD. Apolipoprotein L1 risk variants associate with systemic lupus erythematosus-associated collapsing glomerulopathy. *J Am Soc Nephrol.* 2013; **24**(5): 722–725.
21. Larsen CP, Beggs ML, Walker PD, Saeed M, Ambruzs JM, Messias NC. Histopathologic effect of APOL1 risk alleles in PLA2R-associated membranous glomerulopathy. *Am J Kidney Dis.* 2014; **64**(1): 161–163.
22. Nichols B, Jog P, Lee JH, *et al.* Innate immunity pathways regulate the nephropathy gene Apolipoprotein L1. *Kidney Int.* 2015; **87**(2): 332–342.
23. Kopp JB, Winkler CA, Zhao X, *et al.* Clinical Features and Histology of Apolipoprotein L1-Associated Nephropathy in the FSGS Clinical Trial. *J Am Soc Nephrol.* 2015; **26**(6): 1443–1448.
24. Reeves-Daniel AM, DePalma JA, Bleyer AJ, *et al.* The APOL1 gene and allograft survival after kidney transplantation. *Am J Transplant.* 2011; **11**(5): 1025–1030.
25. Lee BT, Kumar V, Williams TA, *et al.* The APOL1 genotype of African American kidney transplant recipients does not impact 5-year allograft survival. *Am J Transplant.* 2012; **12**(7): 1924–1928.
26. Canaud G, Dejucq-Rainsford N, Avettand-Fenoël V, *et al.* The kidney as a reservoir for HIV-1 after renal transplantation. *J Am Soc Nephrol.* 2014; **25**(2): 407–419.

27. Palmer ND, Ng MC, Hicks PJ, *et al.* Evaluation of candidate nephropathy susceptibility genes in a genome-wide association study of African American diabetic kidney disease. *PLoS One.* 2014; **9**(2): e88273.
28. Yang Y, Hodgin JB, Afshinnia F, *et al.* The two kidney to one kidney transition and transplant glomerulopathy: a podocyte perspective. *J Am Soc Nephrol.* 2015; **26**(6): 1450–1465.
29. Venkatareddy M, Wang S, Yang Y, *et al.* Estimating podocyte number and density using a single histologic section. *J Am Soc Nephrol.* 2014; **25**(5): 1118–1129.
30. Fukuda A, Wickman LT, Venkatareddy MP, *et al.* Urine podocin:nephrin mRNA ratio (PNR) as a podocyte stress biomarker. *Nephrol Dial Transplant.* 2012; **27**(11): 4079–4087.
31. Sato Y, Wharram BL, Lee SK, *et al.* Urine podocyte mRNAs mark progression of renal disease. *J Am Soc Nephrol.* 2009; **20**(5): 1041–1052.
32. Wickman L, Afshinnia F, Wang SQ, *et al.* Urine podocyte mRNAs, proteinuria, and progression in human glomerular diseases. *J Am Soc Nephrol.* 2013; **24**(12): 2081–2095.
33. Kashtan CE, Michael AF. Alport syndrome. *Kidney Int.* 1996; **50**(5): 1445–1463.
34. Kashtan CE. Familial hematuria due to type IV collagen mutations: Alport syndrome and thin basement membrane nephropathy. *Curr Opin Pediatr.* 2004; **16**(2): 177–181.
35. Grünfeld JP. The clinical spectrum of hereditary nephritis. *Kidney Int.* 1985; **27**(1): 83–92.
36. Hasstedt SJ, Atkin CL. X-linked inheritance of Alport syndrome: family P revisited. *Am J Hum Genet.* 1983; **35**(6): 1241–1251.
37. Pajari H, Kääriäinen H, Muhonen T, Koskimies O. Alport's syndrome in 78 patients: epidemiological and clinical study. *Acta Paediatr.* 1996; **85**(11): 1300–1306.
38. Voskarides K, Damianou L, Neocleous V, *et al.* COL4A3/COL4A4 mutations producing focal segmental glomerulosclerosis and renal failure in thin basement membrane nephropathy. *J Am Soc Nephrol.* 2007; **18**(11): 3004–3016.
39. Gast C, Pengelly RJ, Lyon M, *et al.* Collagen (COL4A) mutations are the most frequent mutations underlying adult focal segmental glomerulosclerosis. *Nephrol Dial Transplant.* 2016; **31**(6): 961–970.
40. Atkin CL, Hasstedt SJ, Menlove L, *et al.* Mapping of Alport syndrome to the long arm of the X chromosome. *Am J Hum Genet.* 1988; **42**(2): 249–255.
41. Mochizuki T, Lemmink HH, Mariyama M, *et al.* Identification of mutations in the alpha 3(IV) and alpha 4(IV) collagen genes in autosomal recessive Alport syndrome. *Nat Genet.* 1994; **8**(1): 77–81.
42. Lemmink HH, Mochizuki T, van den Heuvel LP, *et al.* Mutations in the type IV collagen alpha 3 (COL4A3) gene in autosomal recessive Alport syndrome. *Hum Mol Genet.* 1994; **3**(8): 1269–1273.
43. Zhou J, Barker DF, Hostikka SL, Gregory MC, Atkin CL, Tryggvason K. Single base mutation in alpha 5(IV) collagen chain gene converting a conserved cysteine to serine in Alport syndrome. *Genomics.* 1991; **9**(1): 10–18.

44. Jefferson JA, Lemmink HH, Hughes AE, *et al.* Autosomal dominant Alport syndrome linked to the type IV collage alpha 3 and alpha 4 genes (COL4A3 and COL4A4). *Nephrol Dial Transplant.* 1997; **12**(8): 1595–1599.
45. Said SM, Fidler ME, Valeri AM, *et al.* Negative Staining for COL4A5 Correlates With Worse Prognosis and More Severe Ultrastructural Alterations in Males With Alport Syndrome. *Kidney Int Rep.* 2017; **2**(1): 44–52.
46. Martin P, Heiskari N, Zhou J, *et al.* High mutation detection rate in the COL4A5 collagen gene in suspected Alport syndrome using PCR and direct DNA sequencing. *J Am Soc Nephrol.* 1998; **9**(12): 2291–2301.
47. Chatterjee R, Hoffman M, Cliften P, Seshan S, Liapis H, Jain S. Targeted exome sequencing integrated with clinicopathological information reveals novel and rare mutations in atypical, suspected and unknown cases of Alport syndrome or proteinuria. *PLoS One.* 2013; **8**(10): e76360.
48. Fallerini C, Dosa L, Tita R, *et al.* Unbiased next generation sequencing analysis confirms the existence of autosomal dominant Alport syndrome in a relevant fraction of cases. *Clin Genet.* 2014; **86**(3): 252–257.
49. Beicht S, Strobl-Wildemann G, Rath S, *et al.* Next generation sequencing as a useful tool in the diagnostics of mosaicism in Alport syndrome. *Gene.* 2013; **526**(2): 474–477.
50. Artuso R, Fallerini C, Dosa L, *et al.* Advances in Alport syndrome diagnosis using next-generation sequencing. *Eur J Hum Genet.* 2012; **20**(1): 50–57.
51. Morinière V, Dahan K, Hilbert P, *et al.* Improving mutation screening in familial hematuric nephropathies through next generation sequencing. *J Am Soc Nephrol.* 2014; **25**(12): 2740–2751.
52. Noone D, Licht C. An update on the pathomechanisms and future therapies of Alport syndrome. *Pediatr Nephrol.* 2013; **28**(7): 1025–1036.
53. Rao VH, Meehan DT, Delimont D, *et al.* Role for macrophage metalloelastase in glomerular basement membrane damage associated with alport syndrome. *Am J Pathol.* 2006; **169**(1): 32–46.
54. Massella L, Gangemi C, Giannakakis K, *et al.* Prognostic value of glomerular collagen IV immunofluorescence studies in male patients with X-linked Alport syndrome. *Clin J Am Soc Nephrol.* 2013; **8**(5): 749–755.
55. Jiang YS, Ehara T. Trypsin digestion on paraffin sections is a useful tool for diagnosis of alport syndrome. *Ultrastruct Pathol.* 2010; **34**(4): 199–206.
56. Steenhard BM, Vanacore R, Friedman D, *et al.* Upregulated expression of integrin α1 in mesangial cells and integrin α3 and vimentin in podocytes of Col4a3-null (Alport) mice. *PLoS One.* 2012; **7**(12): e50745.
57. Baum A, Pohl M, Kreusch S, *et al.* Searching biomarker candidates in serum using multidimensional native chromatography. II Method evaluation with Alport syndrome and severe inflammation. *J Chromatogr B Analyt Technol Biomed Life Sci.* 2008; **876**(1): 31–40.
58. De Vriese AS, Sethi S, Van Praet J, Nath KA, Fervenza FC. Kidney disease caused by dysregulation of the complement alternative pathway: An etiologic approach. *J Am Soc Nephrol.* 2015; **26**(12): 2917–2929.

59. Noris M, Remuzzi G. Atypical hemolytic-uremic syndrome. *N Engl J Med.* 2009; **361**(17): 1676–1687.
60. Waters AM, Licht C. aHUS caused by complement dysregulation: new therapies on the horizon. *Pediatr Nephrol.* 2011; **26**(1): 41–57.
61. Matsumae T, Takebayashi S, Naito S. The clinico-pathological characteristics and outcome in hemolytic-uremic syndrome of adults. *Clin Nephrol.* 1996; **45**(3): 153–162.
62. Barbour TD, Pickering MC, Terence Cook H. Dense deposit disease and C3 glomerulopathy. *Semin Nephrol.* 2013; **33**(6): 493–507.
63. Sibley RK, Kim Y. Dense intramembranous deposit disease: new pathologic features. *Kidney Int.* 1984; **25**(4): 660–670.
64. Swainson CP, Robson JS, Thomson D, MacDonald MK. Mesangiocapillary glomerulonephritis: A long-term study of 40 cases. *J Pathol.* 1983; **141**(4): 449–468.
65. Schwertz R, Rother U, Anders D, Gretz N, Schärer K, Kirschfink M. Complement analysis in children with idiopathic membranoproliferative glomerulonephritis: a long-term follow-up. *Pediatr Allergy Immunol.* 2001; **12**(3): 166–172.
66. Smith RJ, Alexander J, Barlow PN, *et al.* New approaches to the treatment of dense deposit disease. *J Am Soc Nephrol.* 2007; **18**(9): 2447–2456.
67. Levy M, Halbwachs-Mecarelli L, Gubler MC, *et al.* H deficiency in two brothers with atypical dense intramembranous deposit disease. *Kidney Int.* 1986; **30**(6): 949–956.
68. Licht C, Heinen S, Józsi M, *et al.* Deletion of Lys224 in regulatory domain 4 of Factor H reveals a novel pathomechanism for dense deposit disease (MPGN II). *Kidney Int.* 2006; **70**(1): 42–50.
69. Martínez-Barricarte R, Heurich M, Valdes-Cañedo F, *et al.* Human C3 mutation reveals a mechanism of dense deposit disease pathogenesis and provides insights into complement activation and regulation. *J Clin Invest.* 2010; **120**(10): 3702–3712.
70. Wyatt RJ, Julian BA. IgA nephropathy. *N Engl J Med.* 2013; **368**(25): 2402–2414.
71. Li LS, Liu ZH. Epidemiologic data of renal diseases from a single unit in China: analysis based on 13,519 renal biopsies. *Kidney Int.* 2004; **66**(3): 920–923.
72. Crowley-Nowick PA, Julian BA, Wyatt RJ, *et al.* IgA nephropathy in blacks: Studies of IgA2 allotypes and clinical course. *Kidney Int.* 1991; **39**(6): 1218–1224.
73. Fennelly NK, Kennedy C, Jenkinson AC, *et al.* Clinical Heterogeneity in Familial IgA Nephropathy. *Nephron.* 2018.
74. Gharavi AG, Kiryluk K, Choi M, *et al.* Genome-wide association study identifies susceptibility loci for IgA nephropathy. *Nat Genet.* 2011; **43**(4): 321–327.
75. Kiryluk K, Li Y, Scolari F, *et al.* Discovery of new risk loci for IgA nephropathy implicates genes involved in immunity against intestinal pathogens. *Nat Genet.* 2014; **46**(11): 1187–1196.
76. Kiryluk K, Li Y, Sanna-Cherchi S, *et al.* Geographic differences in genetic susceptibility to IgA nephropathy: GWAS replication study and geospatial risk analysis. *PLoS Genet.* 2012; **8**(6): e1002765.

77. Hiki Y, Odani H, Takahashi M, *et al.* Mass spectrometry proves under-O-glycosylation of glomerular IgA1 in IgA nephropathy. *Kidney Int.* 2001; **59**(3):1077–1085.
78. Allen AC, Bailey EM, Brenchley PE, Buck KS, Barratt J, Feehally J. Mesangial IgA1 in IgA nephropathy exhibits aberrant O-glycosylation: observations in three patients. *Kidney Int.* 2001; **60**(3): 969–973.
79. Fogo AB, Lusco MA, Najafian B, Alpers CE. AJKD Atlas of Renal Pathology: Membranous Nephropathy. *Am J Kidney Dis.* 2015; **66**(3): e15–e17.
80. Cattran DC, Brenchley PE. Membranous nephropathy: integrating basic science into improved clinical management. *Kidney Int.* 2017; **91**(3): 566–574.
81. Beck LH, Bonegio RG, Lambeau G, *et al.* M-type phospholipase A2 receptor as target antigen in idiopathic membranous nephropathy. *N Engl J Med.* 2009; **361**(1): 11–21.
82. Stanescu HC, Arcos-Burgos M, Medlar A, *et al.* Risk HLA-DQA1 and PLA(2)R1 alleles in idiopathic membranous nephropathy. *N Engl J Med.* 2011; **364**(7): 616–626.
83. Glenner GG. Amyloid deposits and amyloidosis. The beta-fibrilloses (first of two parts). *N Engl J Med.* 1980; **302**(23): 1283–1292.
84. Sipe JD, Benson MD, Buxbaum JN, *et al.* Nomenclature 2014: Amyloid fibril proteins and clinical classification of the amyloidosis. *Amyloid.* 2014; **21**(4): 221–224.
85. Palladini G, Merlini G. Current treatment of AL amyloidosis. *Haematologica.* 2009; **94**(8): 1044–1048.
86. Wechalekar AD, Hawkins PN, Gillmore JD. Perspectives in treatment of AL amyloidosis. *Br J Haematol.* 2008; **140**(4): 365–377.
87. Satoskar AA, Burdge K, Cowden DJ, Nadasdy GM, Hebert LA, Nadasdy T. Typing of amyloidosis in renal biopsies: diagnostic pitfalls. *Arch Pathol Lab Med.* 2007; **131**(6): 917–922.
88. Picken MM. Immunoglobulin light and heavy chain amyloidosis AL/AH: renal pathology and differential diagnosis. *Contrib Nephrol.* 2007; **153**: 135–155.
89. Sethi S, Vrana JA, Theis JD, *et al.* Laser microdissection and mass spectrometry-based proteomics aids the diagnosis and typing of renal amyloidosis. *Kidney Int.* 2012; **82**(2): 226–234.
90. Sethi S, Theis JD, Leung N, *et al.* Mass spectrometry-based proteomic diagnosis of renal immunoglobulin heavy chain amyloidosis. *Clin J Am Soc Nephrol.* 2010; **5**(12): 2180–2187.
91. Aoki M, Kang D, Katayama A, *et al.* Optimal conditions and the advantages of using laser microdissection and liquid chromatography tandem mass spectrometry for diagnosing renal amyloidosis. *Clin Exp Nephrol.* 2018.
92. Halbritter J, Baum M, Hynes AM, *et al.* Fourteen monogenic genes account for 15% of nephrolithiasis/nephrocalcinosis. *J Am Soc Nephrol.* 2015; **26**(3): 543–551.
93. Stapleton FB. Childhood stones. *Endocrinol Metab Clin North Am.* 2002; **31**(4): 1001–1015, ix.

94. Spada M, Pagliardini S, Yasuda M, *et al.* High incidence of later-onset fabry disease revealed by newborn screening. *Am J Hum Genet.* 2006; **79**(1): 31–40.
95. Hwu WL, Chien YH, Lee NC, *et al.* Newborn screening for Fabry disease in Taiwan reveals a high incidence of the later-onset GLA mutation c.936+919G>A (IVS4+919G>A). *Hum Mutat.* 2009; **30**(10): 1397–1405.
96. Levy M, Feingold J. Estimating prevalence in single-gene kidney diseases progressing to renal failure. *Kidney Int.* 2000; **58**(3): 925–943.
97. Ward CJ, Hogan MC, Rossetti S, *et al.* The gene mutated in autosomal recessive polycystic kidney disease encodes a large, receptor-like protein. *Nat Genet.* 2002; **30**(3): 259–269.
98. Hartung EA, Guay-Woodford LM. Autosomal recessive polycystic kidney disease: A hepatorenal fibrocystic disorder with pleiotropic effects. *Pediatrics.* 2014; **134**(3): e833–e845.
99. Edrees BM, Athar M, Al-Allaf FA, *et al.* Next-generation sequencing for molecular diagnosis of autosomal recessive polycystic kidney disease. *Gene.* 2016; **591**(1): 214–226.
100. Srivastava S, Sayer JA. Nephronophthisis. *J Pediatr Genet.* 2014; **3**(2): 103–114.
101. Srivastava S, Molinari E, Raman S, Sayer JA. Many genes-one disease? genetics of nephronophthisis (NPHP) and NPHP-associated disorders. *Front Pediatr.* 2017; **5**: 287.
102. Heninger E, Otto E, Imm A, Caridi G, Hildebrandt F. Improved strategy for molecular genetic diagnostics in juvenile nephronophthisis. *Am J Kidney Dis.* 2001; **37**(6): 1131–1139.
103. Melo BF, Aguiar MB, Bouzada MC, *et al.* Early risk factors for neonatal mortality in CAKUT: analysis of 524 affected newborns. *Pediatr Nephrol.* 2012; **27**(6): 965–972.
104. Queisser-Luft A, Stolz G, Wiesel A, Schlaefer K, Spranger J. Malformations in newborn: results based on 30,940 infants and fetuses from the Mainz congenital birth defect monitoring system (1990–1998). *Arch Gynecol Obstet.* 2002; **266**(3): 163–167.
105. Seikaly MG, Ho PL, Emmett L, Fine RN, Tejani A. Chronic renal insufficiency in children: the 2001 Annual Report of the NAPRTCS. *Pediatr Nephrol.* 2003; **18**(8): 796–804.
106. Sanna-Cherchi S, Kiryluk K, Burgess KE, *et al.* Copy-number disorders are a common cause of congenital kidney malformations. *Am J Hum Genet.* 2012; **91**(6): 987–997.
107. Sanna-Cherchi S, Sampogna RV, Papeta N, *et al.* Mutations in DSTYK and dominant urinary tract malformations. *N Engl J Med.* 2013; **369**(7): 621–629.
108. Lopez-Rivera E, Liu YP, Verbitsky M, *et al.* Genetic drivers of kidney defects in the DiGeorge syndrome. *N Engl J Med.* 2017; **376**(8): 742–754.
109. Delmonico FL, Harmon WE, Lorber MI, *et al.* A new allocation plan for renal transplantation. *Transplantation.* 1999; **67**(2): 303–309.

110. Lim WH, Chadban SJ, Clayton P, *et al.* Human leukocyte antigen mismatches associated with increased risk of rejection, graft failure, and death independent of initial immunosuppression in renal transplant recipients. *Clin Transplant.* 2012; **26**(4): E428–E437.
111. Goes N, Chandraker A. Human leukocyte antigen matching in renal transplantation: an update. *Curr Opin Nephrol Hypertens.* 2000; **9**(6): 683–687.
112. Lim WH, Chapman JR, Coates PT, *et al.* HLA-DQ Mismatches and rejection in kidney transplant recipients. *Clin J Am Soc Nephrol.* 2016; **11**(5): 875–883.
113. Tran TH, Döhler B, Heinold A, Scherer S, Ruhenstroth A, Opelz G. Deleterious impact of mismatching for human leukocyte antigen-C in presensitized recipients of kidney transplants. *Transplantation.* 2011; **92**(4): 419–425.
114. Althaf MM, El Kossi M, Jin JK, Sharma A, Halawa AM. Human leukocyte antigen typing and crossmatch: A comprehensive review. *World J Transplant.* 2017; **7**(6): 339–348.
115. Patel R, Terasaki PI. Significance of the positive crossmatch test in kidney transplantation. *N Engl J Med.* 1969; **280**(14): 735–739.
116. Martin S, Liggett H, Robson A, Connolly J, Johnson RW. The association between a positive T and B cell flow cytometry crossmatch and renal transplant failure. *Transpl Immunol.* 1993; **1**(4): 270–276.
117. Pei R, Wang G, Tarsitani C, *et al.* Simultaneous HLA class I and class II antibodies screening with flow cytometry. *Hum Immunol.* 1998; **59**(5): 313–322.
118. Focosi D. Advances in pretransplant donor-specific antibody testing in solid organ transplantation: From bench to bedside. *Int Rev Immunol.* 2016; **35**(4): 351–368.
119. Haas M, Loupy A, Lefaucheur C, *et al.* The Banff 2017 kidney meeting report: Revised diagnostic criteria for chronic active T cell-mediated rejection, antibody-mediated rejection, and prospects for integrative endpoints for next-generation clinical trials. *Am J Transplant.* 2018; **18**(2): 293–307.
120. Reeve J, Sellarés J, Mengel M, *et al.* Molecular diagnosis of T cell-mediated rejection in human kidney transplant biopsies. *Am J Transplant.* 2013; **13**(3): 645–655.
121. Reeve J, Chang J, Salazar ID, Lopez MM, Halloran PF. Using molecular phenotyping to guide improvements in the histologic diagnosis of T cell-mediated rejection. *Am J Transplant.* 2016; **16**(4): 1183–1192.
122. Halloran PF, Pereira AB, Chang J, *et al.* Microarray diagnosis of antibody-mediated rejection in kidney transplant biopsies: an international prospective study (INTERCOM). *Am J Transplant.* 2013; **13**(11): 2865–2874.
123. Sellarés J, Reeve J, Loupy A, *et al.* Molecular diagnosis of antibody-mediated rejection in human kidney transplants. *Am J Transplant.* 2013; **13**(4): 971–983.
124. Haas M, Sis B, Racusen LC, *et al.* Banff 2013 meeting report: inclusion of c4d-negative antibody-mediated rejection and antibody-associated arterial lesions. *Am J Transplant.* 2014; **14**(2): 272–283.
125. Bloom RD, Bromberg JS, Poggio ED, *et al.* Cell-free DNA and active rejection in kidney allografts. *J Am Soc Nephrol.* 2017; **28**(7): 2221–2232.

126. Grskovic M, Hiller DJ, Eubank LA, *et al.* Validation of a clinical-grade assay to measure donor-derived cell-free DNA in solid organ transplant recipients. *J Mol Diagn.* 2016; **18**(6): 890–902.
127. Gandolfini I, Harris C, Abecassis M, *et al.* Rapid biolayer interferometry measurements of urinary CXCL9 to detect cellular infiltrates noninvasively after kidney transplantation. *Kidney Int Rep.* 2017; **2**(6): 1186–1193.
128. Hirt-Minkowski P, Ho J, Gao A, *et al.* Prediction of long-term renal allograft outcome by early urinary CXCL10 chemokine levels. *Transplant Direct.* 2015; **1**(8): e31.

14 DICER1 Syndrome Testing: The Role of Molecular Pathology in Improving Diagnosis and Elucidating Pathogenesis

D. Ashley Hill,[*,†] Amanda Field,[*,†]
Mark A. Brenneman,[*,†] and Kris Ann Schultz[‡]

[*] Division of Anatomic Pathology, Children's National Health System, Washington, DC, USA
[†] ResourcePath LLC, Sterling, VA, USA
[‡] International Pleuropulmonary Blastoma Registry, International Ovarian and Testicular Stromal Tumor Registry, Cancer and Blood Disorders, Children's Minnesota, Minneapolis, MN, USA

Contents

I. DICER1 Syndrome and Pleuropulmonary Blastoma (PPB)

In this chapter, we will review the pathologic and molecular features of PPB and other DICER1 syndrome tumors. We will also discuss the role of molecular diagnostics in understanding the pathogenesis of PPB as well as the clinical management of children and families affected.

PPB (OMIM #601200) is a rare pediatric lung cancer that was first recognized as a unique pathological entity in 1988.[1] PPB has been diagnosed around the world and does not show any overt geographic or cultural predilection. Fewer than 200 cases of PPB are diagnosed each year in the United States.[2] PPB is analogous to other embryonal tumors such as Wilms tumor (WT) of the kidney and neuroblastoma of the adrenal gland where the neoplastic process resembles the embryonic stage in organogenesis.[2–4] Similar to WT and neuroblastoma, PPB almost exclusively affects children younger 6 years of age.[2,3] PPB is the pathognomonic manifestation of a childhood cancer predisposition syndrome that features increased risk for numerous benign and malignant neoplasms including Sertoli–Leydig cell tumor (SLCT) of the ovary, cystic nephroma (CN), renal sarcoma, or WT of the kidney, nodular hyperplasia and carcinoma of the thyroid gland, nasal chondromesenchymal hamartoma (NCMH), embryonal rhabdomyosarcoma (ERMS), neuroblastoma, ciliary body medulloepithelioma (CBME), pituitary blastoma, and pineoblastoma (Table 1).[5–40] These tumors share a basis in organ development and also disproportionately affect children and young women (specifically ovarian SLCT and ERMS of the uterine cervix in young women).

Table 1. DICER1 syndrome conditions by age group

0–7 years	Children and adolescents	Adolescents and young adults	Any age
PPB	Sertoli–Leydig tumor	Sertoli–Leydig tumor	Thyroid nodular hyperplasia
Cystic nephroma	Renal sarcoma	Gynandroblastoma	
Wilms tumor	Ocular medulloepithelioma	Renal sarcoma	Thyroid cancer
Pineoblastoma	Embryonal rhabdomyosarcoma of the uterine cervix	Embryonal rhabdomyosarcoma of the uterine cervix	Lung cysts
Pituitary blastoma			Nasal chondromesenchymal hamartoma
Ocular medulloepithelioma	Pineoblastoma	Pineoblastoma	

II. The Natural History of PPB

Much of what we have learned so far about the biology of PPB has been from morphologic observations of tumors at different time points during early childhood.[1–3,41] Based on our study of the earliest PPBs (from 31 weeks gestation), we believe that the tumor arises from an embryonic uncommitted lung mesenchymal cell. The embryonic lung mesenchyme originates from the mesoderm and is characterized by loosely packed primitive cells that are the source of progenitors for endothelial cells, fibroblasts, smooth muscle, and cartilage.[42] Lung mesenchyme plays an important role in the growth of the developing lung.[43] One of the key functions of the mesenchyme is signaling the endodermally derived epithelium to proliferate, branch, and differentiate into the various types of lining cells.[44] Similarly, one of the key roles of the epithelium (and mesothelium) is to drive proliferation and regulate differentiation of the mesenchyme.[45]

The architecture of PPB early in development is characterized by multilocular cysts within the lung parenchyma (Type I PPB) (Figure 1).[41] The cyst walls are lined by benign-appearing alveolar-type epithelium. The cyst walls contain variable number of primitive mesenchymal cells, some of which show differentiation along either skeletal muscle or cartilaginous lines. Typically, these early cases present in infants and young children (median: 8 months of age) who come to medical attention for difficulty breathing due to a large space-occupying cyst in the

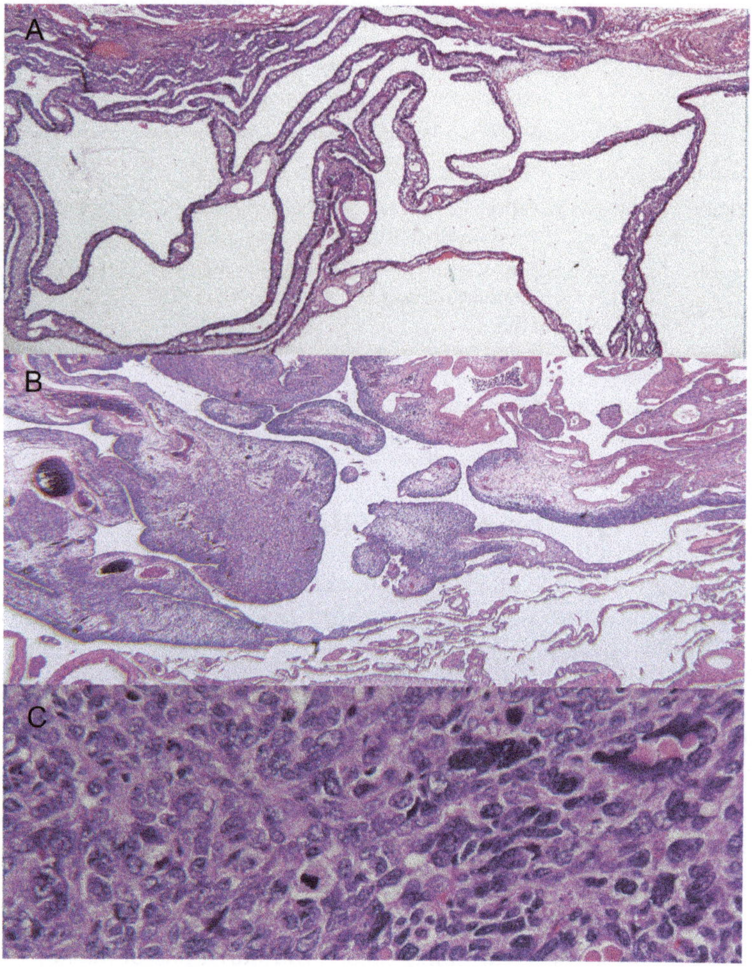

Figure 1. The Morphologic Natural History of PPB. (A) Type I PPB in a 1-month-old infant shows expansion of the alveolar septa with cellular mesenchyme as the earliest feature of PPB. The airspaces are dilated. (B) Type II PPB arising within a Type I PPB. Type II PPB is defined as malignant mesenchymal cells expanding the septa forming solid nodules. Note the lower portion and lower right showing residual Type I areas with thin septa. Note the areas with dense subepithelial mesenchyme (cambium layers). (C) Type III PPBs are purely solid and composed of high-grade primitive cells with anaplasia. (H&E; original magnification x200 (A), ×200 (B), ×400(C)).

lung or due to a rupture of the air-filled cyst in the chest (pneumothorax).[1–3,41,46] Very often these benign-appearing cysts are mistaken for the unrelated entities congenital pulmonary airway malformation (CPAM) or lobar emphysema.[47] In later stages of tumorigenesis, the mesenchymal cells expand and overgrow the epithelial cysts, forming an overtly malignant cystic and solid (Type II) or purely solid sarcoma (Type III).[1–3,41] These children are older than those with Type I PPB; the median ages at diagnosis for Type II and Type III PPB are 36 and 41 months, respectively. A child with Type II or III PPB typically presents with weight loss, fever, shortness of breath, and a large lung mass.[2,3] These pathologic types, Type I (cystic), Type II (cystic/solid), and Type III (solid), reflect the observed biological progression within individual patients and in age-segregated subgroups.[1–3,41] Children presenting with Type III PPB represent approximately one-third of all PPB cases. It is unclear if all the Type III PPBs go through a cystic stage; most have not had radiographic studies prior to presentation. On the other end of the spectrum, it has recently become clear that not all Type I PPB are destined to progress to Type II and then Type III PPB.[41] Through our family study, we identified older children and adults with asymptomatic lung cysts.[41] Microscopically, these cysts have the architecture of Type I PPB but lack a primitive mesenchymal component. Whether these cysts result from "regression" of Type I PPB (so-called Type IR) or have a distinct underlying pathogenesis remains an open question.

Type I PPB is treated with resection alone or resection followed by adjuvant chemotherapy, usually (vincristine (V), dactinomycin (A), cyclophosphamide (C)) VAC or VA. The overall survival for Type I PPB is 91%.[2] Types II and III PPB are considered high-grade sarcomas and require surgical removal and intensive, adjuvant chemotherapy. The current recommendation for adjuvant chemotherapy for Types II and III PPB is (ifosfamide, vincristine, dactinomycin, and doxorubicin) IVADo. Types II and III often also require sarcoma dose radiation, commonly given when residual disease remains after second look surgery. Each of these is associated with a significant risk for acute and late toxicities, including cardiomyopathy and secondary malignant neoplasms. Despite these intensive treatments, outcomes remain poor for individuals with advanced disease. The overall survival for Type II and Type III PPBs are are only 74% and 53%, respectively.[2]

III. The PPB Family Cancer Syndrome

As early as the first reported series of PPB, it was recognized that there was a familial predisposition to development of PPB; two of the first 11 cases reported were siblings,[1] and reports of PPB and other rare tumor entities in probands and close relatives followed suit.[5,13,18,29,39] Recognition of this unusual lung tumor as PPB by pathologists, followed by encouragement of children and extended families to submit details about their cases and family histories to the International PPB Registry (the Registry), provided a unique opportunity to learn about the genetic basis of the disease.[48] By 2005, there were three families in the Registry who had more than one child with PPB. An additional group of eight families had one member with PPB and one or more relatives with lung cysts, cystic nephroma, Sertoli–Leydig tumor, embryonal rhabdomyosarcoma, and thyroid nodules. The pattern of disease in these families appeared to follow an autosomal dominant pattern with reduced penetrance. A power analysis performed on the four most informative pedigrees suggested that if this was a monogenic disease, our likelihood of finding the gene was ~95%.

It took ~18 months to enroll these families into the family study and locate and recruit each of the affected family members. We hosted a meeting for families in St. Louis where 24 families from 17 different states attended and donated blood for the study. Our study team took trips to Minnesota, New York, Connecticut, and Vermont where we drew blood from family members in hotel rooms, porches, church basements, and even a gazebo. By the end of the second year, we had enrolled 66 families, of whom 11 had more than one affected family member. Our approach was to perform a linkage study using DNA extracted from the blood of the individuals in the four most informative families with multiple affected family members, using single-nucleotide polymorphisms (SNPs) as markers. Our core facility at Washington University used Affymetrix 6.0 SNP arrays (Genome-Wide Human SNP Array 6.0, Affymetrix, Santa Clara, CA) to generate the marker data for each family member. Dr. Brian

Suarez, our team's statistical geneticist, and his colleague, Tony Hinrichs, prepared a set of ~4117 marker SNPs across the 22 autosomes based on pairwise measurements of linkage disequilibrium (LD) and estimates of heterozygosity.[6] Data from each of these markers was added to a pedigree file containing information about the individual's sex, family relationships, and affected status (disease-positive, married-in, or unknown). We imported the data into the free software program easyLINKAGE-Plus.[49] Using the GeneHunter algorithm to analyze the data, a peak logarithm of odds (LOD) score of 3.71 was identified on chromosome 14q. The analysis was repeated using all SNPs from chromosome 14q to create haplotypes. Then using the embedded HaploPainter program, pedigrees were drawn with colored presentation of SNP markers and haplotypes providing a visual depiction of recombination events.[50] With the marker SNPs bordering recombination events visualized in HaploPainter, we were able to narrow down the chromosomal region linked to disease in the four families to 7 Mb. There were ~70 characterized genes in this 7 Mb interval on chromosome 14q. Each gene was queried in PubMed for an association with lung and/or development. There was one manuscript describing DICER1 and lung development by Harris *et al.*[51] It showed that conditional DICER1 ablation in lung epithelium early in mouse development led to lung cysts.[51] The pathologists on the team, who recognized lung cysts as the earliest feature of PPB, knew we had struck gold. Subsequent *DICER1* sequencing has shown that about 70% of individuals with PPB have germline mutations in *DICER1*.[52] The most immediate benefit of this finding has been in the development of a diagnostic test to determine which individuals in a family are at increased risk for development of tumors.[53] In addition, the identification of DICER1 as the genetic basis of this childhood cancer predisposition syndrome also has shed light on the suspected relationship between organ development and cancer.

IV. DICER1 and the miRNA Defect Responsible for Tumorigenesis

DICER1 encodes an RNase III-family endonuclease that cleaves precursor miRNAs (pre-miRNAs) into mature, active miRNAs.[54,55] Mature miRNAs regulate the expression of genes at the post-transcriptional stage.

miRNAs have been shown to play a prominent role in development, cell cycle control, and oncogenesis.[56–58] Sequencing studies of PPB have revealed a unique pattern of biallelic mutations in the DICER1 gene.[25,59–61] Generally, one allele bears a nonsense or frameshift mutation (often germline) predicted to cause full loss-of-function (LOF), while the other allele carries a missense mutation in the RNase IIIb domain of DICER1. The RNase IIIb missense mutations in DICER1 syndrome tumors always affect one of five "hotspot" codons for key amino acids in the metal-binding catalytic cleft: E1705, D1709, G1809, D1810, and E1813.[25,59–61] Substitutions at these positions do not result in complete loss of catalytic function in miRNA processing. Rather, they cause a selective disruption that results in failure to cleave mature 5p miRNAs from the 5′ arm of pre-miRNA hairpin.[25,59–62] Mature miRNAs are cleaved from the 3p end normally.[59] Depletion of 5p miRNAs, including the let-7 family, alters expression of numerous downstream target mRNAs, including several genes critical for embryogenesis or tumor suppression.[59–62] The pleiotropic nature of DICER1 syndromic disease likely reflects the diverse array of protein-coding genes regulated by miRNAs in different tissues at different developmental timepoints. Biallelic LOF mutations have not been identified in PPB, suggesting that retention of some 3p miRNA processing is usually required for tumor survival. Most but not all malignant tumors in DICER1 syndrome have now been shown to follow this biallelic, LOF-plus-RNase IIIb-hotspot rule. Pineoblastomas appear to be an exception. Loss of function mutations with subsequent loss of heterozygosity appear to be common in pineoblastomas.[35]

V. Molecular Diagnostics of DICER1 Syndrome

We recently completed a genotype–phenotype analysis of 124 consecutive children with PPB.[52] Over 72% of children with PPB (90/124) had germline DICER1 LOF mutations; the overwhelming majority of mutations (90%) truncate the DICER1 open reading frame proximal to the RNase IIIb domain (ClinVar accession numbers SCV000195560–SCV000195643). About 84% of these germline mutations were inherited from a parent. Targeted, next-generation sequencing (NGS) revealed that 8% of patients with PPB (10/124) have mosaicism for DICER1 mutations, with mutant allele frequencies that ranged from 1.1% to 17.2% of reads in DNA from blood, saliva, fibroblasts, or other normal tissue cells. Another 10% of children (12/124), all with solitary PPB, had mutations detectable only in the tumor tissue. We were able to identify DICER1 mutations in blood, tumor, or both in 115 of 124 children with PPB. The remaining nine children had normal DICER1 coding sequence, copy number, and exome sequencing on germline samples. We believe that these cases are also likely DICER1 related, but with incomplete evaluations (i.e., tumor-specific mutations but tumor unavailable, mosaicism below our level of detection, or deep intronic mutations not currently detected by our assay). It is not clear if the same percentages of germline, mosaic, and tumor-specific mutations can be extrapolated to other DICER1 syndrome cancers. In a study of 36 ovarian SLCT, 22 (61%) women had germline DICER1 mutations, 3/36 (8%) were mosaic, and 11/36 (31%) tumors had mutations limited to tumor tissue.[63]

One of the most interesting results of the sequencing study was identification of a distinct and severe phenotype associated with mosaicism for RNase IIIb hotspot mutations.[52] As mentioned above, the typical child with PPB has a germline LOF mutation and a tumor-specific RNase IIIb hotspot mutation. In five children, however, we found that multiple primary tumors shared the same RNase IIIb hotspot mutation, and this hotspot mutation was also detectable in non-neoplastic tissues (e.g., blood, saliva, normal lung, lymph node, normal fallopian tube) but each cancer had unique loss of function mutation or allele loss. These children are clinically distinguished from those who bear germline or mosaic LOF mutations by earlier onset of neoplastic disease, often before 1 year of age, and multiple, successive foci of disease involving more than one syndromic organ site (Figure 2). We hypothesize that the extreme phenotypes of this patient group may relate to the order in which RNase IIIb domain hotspot mutations and LOF mutations were acquired during development. In most DICER1 cancers, DICER1 is not a classical tumor suppressor gene for which "two hits" — loss of function in both alleles — are required to enable tumorigenesis. Neither is it haploinsufficient in the usual sense, i.e., that cells with only one expressed allele make wild-type protein, but not in sufficient quantity to fulfill its function. Based on this unique genetic group, we hypothesize that substitutions of key amino

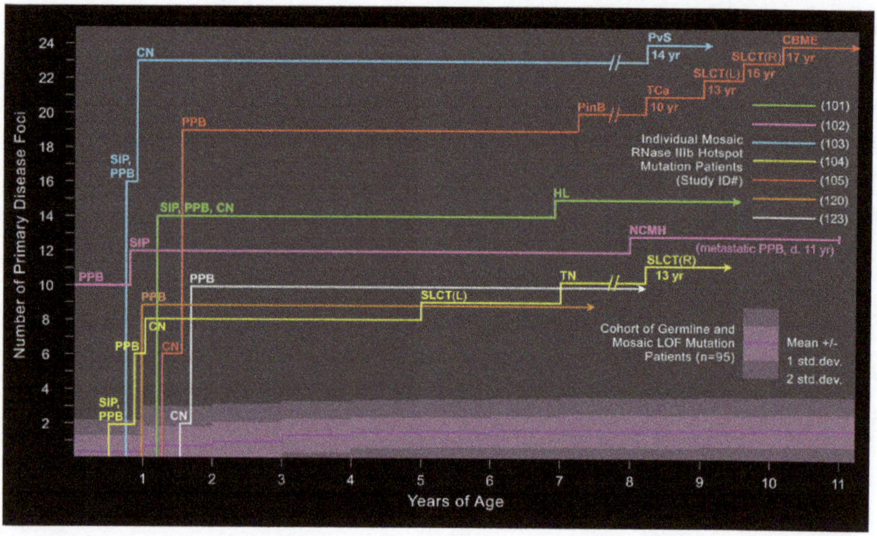

Figure 2. Numbers and types of disease foci in DICER1 syndrome patients with mosaic RNase IIIb domain hotspot mutations. For each of the seven mosaic hotspot children identified in this study, an individual timeline indicates the numbers of discrete foci of neoplastic disease and their histopathological types, graphed with respect to patient age at diagnosis. Across the lower portion of the chart, a single aggregate timeline (dark violet) represents the mean number of disease foci for all PPB/DICER1 syndrome patients with predisposing loss of function (LOF) mutations identified in this study, graphed with respect to patient age at diagnosis. The shaded areas (in lighter violet) surrounding the timeline for LOF mutation patients indicates 1 and 2 standard deviations above and below the mean. The range of foci number among all LOF mutation patients was 0 to 6 in all years of age represented (not shown). Abbreviations: CN cystic nephroma; CBME ciliary body medulloepithelioma (eye); NCMH nasal chondromesenchymal hamartoma; PPB pleuropulmonary blastoma; PinB pineoblastoma; PvS pelvic sarcoma; SIP small intestinal polyp(s); SLCT Sertoli-Leydig cell tumor (ovary); TCa thyroid carcinoma; TN thyroid nodule(s). *Source*: Taken from reference 52.

acids in the RNase IIIb domain (i.e., hotspots) causes tumor suppression to falter when it is not masked by expression of wild-type DICER1 protein. Unmasking of an RNase IIIb hotspot mutation may arise through any form of LOF mutation in the wild-type allele, including allele loss. The two mutational events, RNase IIIb missense and LOF, may occur in any order and both are generally required to precipitate the initiation of tumorigenesis. However, by simple probability, RNase IIIb hotspots are a minute mutational target (36 potential nucleotide substitutions) compared with the spectrum of potential LOF mutations in the 1922 codon full DICER1 gene. Thus, in a developing embryo or child with a germline (or mosaic) DICER1 LOF mutation, "second hits" occurring in a somatic cell will almost always be another LOF mutation, usually resulting in cell death or limited proliferation at most. Rarely, a second hit will be an RNase IIIb hotspot mutation, which allows for continued cell viability and growth, though at the cost of skewed miRNA processing that may ultimately promote tumorigenesis in the surviving clade of cells. However, the low likelihood of incurring an RNase IIIb hotspot missense mutation in somatic cells means that months, years, or a lifetime may elapse before one occurs. Further, the developmental context in which a second, hotspot mutation occurs may be important. There are apparently windows of risk for transformation, perhaps coinciding with certain periods of organ/tissue development when an "oncofetal" gene program is normally active and subject to miRNA modulation, i.e., lung and kidney in the embryo, uterine cervix and ovaries in pubertal females. A low probability of RNase IIIb hotspot mutations as second hits during windows of risk may underlie the low penetrance and variable expression of familial LOF mutations in DICER1 syndrome. We also speculate that wider tissue/organ distribution of aberrant miRNA processing in children with mosaic hotspot mutations may produce unique syndromic features such as juvenile-type small intestinal polyps, or generalized somatic overgrowth noted by Klein *et al.*[64]

Sequencing of both germline DNA and tumor DNA provide more information than germline testing alone in the 30–40% of individuals who will be negative for coding sequence abnormalities in the germline. Some deletions and intronic mutations are not detectable by standard coding region sequencing. Identifying detectable pathogenic mutations in tumor tissue can help determine if more in-depth germline testing for deletions or aberrant splice patterns from intronic mutations creating new splice sites is indicated.[65] Tumor testing may also resolve whether or not a germline variant of unknown significance (VUS) is pathogenic (i.e., is it the coexistent loss of function mutation in the tumor).

Classifying an individual as having predisposing germline or mosaic mutations rather than tumor-specific mutations is important. Results from proband testing will determine if other family members need to be tested and screened. It is also important in guiding education and subsequent surveillance. Individuals who have tumor-limited mutations may not need screening for additional DICER1 conditions, whereas children mosaic for RNase IIIb mutations may require a much more intensive surveillance program given the apparent higher risk for development of multiple tumors.

VI. Current Mutation-based Surveillance in DICER1 Syndrome

Early detection and removal of Type I PPB before malignant progression is one of the main benefits of mutation testing; however, the value of germline DICER1 mutation testing relies on a previous diagnosis of DICER1 syndrome condition in a family member leading to knowing to test — so it does not benefit "first in family" cases. For children in whom a DICER1 mutation is identified, current surveillance guidelines developed at the first International DICER1 Symposium, and affirmed at a recent American Association of Cancer Research (AACR) Childhood Cancer Predisposition Workshop, recommend surveillance with computed tomography (CT) scans for DICER1 carriers between 3 and 6 months of age and between 2.5 and 3.5 years of age.[66] Chest X-rays are recommended at 4–6-month intervals until age 8 and an annual chest X-ray until age 12 based on previous data from the PPB Registry indicating the peak risk for development of PPB.[46] Although these recommendations may lead to early diagnosis in some children, there are several limitations of an "imaging-only" approach. First, serial CT scans involve significant cumulative doses of diagnostic radiation and may increase secondary malignant neoplasm risk.[67] MRI has been considered as a screening modality but lacks sensitivity for pulmonary cysts.[46,47] Second, sedation is often required for CT and MR scans in young children. Finally, with a desire to limit the cumulative radiation dose from receiving a number of CTs, longer intervals between scans are advocated, sometimes allowing significant tumor progression between the widely spaced scans.

In addition to pulmonary screening, family education and other targeted imaging tests are currently suggested. To facilitate early detection of cystic nephroma, renal sarcoma, and Wilms tumor, abdominal ultrasound is recommended every 6 months until age 8 and annually until age 12. Cystic nephroma is benign; however, early resection may be nephron sparing. Renal sarcoma and Wilms tumor are both mostly curable with lowest burden of therapy when found at an early stage.

To facilitate the early detection of ovarian and pelvic tumors in girls, transabdominal pelvic ultrasound is recommended every 6–12 months. The optimal age range of screening is not determined; however, 95% of cases have occurred by 40 years of age.[66] Women with a history of SLCT are at particular risk for metachronous disease and may require more frequent surveillance. The likelihood of contralateral tumor development is at least 5–10% and must be carefully distinguished from recurrent disease as, if found as stage Ia, metachronous disease features a much more favorable prognosis.

Thyroid nodules are common, especially in women with DICER1 mutations. Thyroid cancer may also be seen and has generally been associated with a favorable prognosis, but rare examples of poorly differentiated cancers are now being recognized (unpublished data). Current recommendations include physical examination of the thyroid gland yearly and baseline thyroid ultrasound by 8 years of age and then every 3 years thereafter with more frequent ultrasound with anticipated chemotherapy or radiation therapy.

To screen for intraocular tumors, annual dilated eye exam is recommended from at least ages 3–10 years. Red reflex and extraocular movements should be carefully examined at each clinical opportunity and urgent brain MRI should be considered for any child presenting with symptoms of intracranial pathology. For all DICER1 tumors, family education regarding potential signs and symptoms is critical.

VII. Future Directions

While much progress has been made in the last 10 years, we have a long way to go in optimizing early detection and treating progressive disease. Molecular diagnostics will continue to play a central role. Whether it be development of disease biomarkers such as ctDNA assays designed to detect RNase IIIb mutants or development of new therapies based on the miRNA deficiencies, molecular diagnostics will continue to play a critical role in the diagnosis and management of PPB and other DICER1 syndrome disorders.

Acknowledgments

The work presented in this chapter is the product of many dedicated individuals, only a few of whom are named here. We would like to extend our sincere thanks to Pepper Dehner and Jack Priest, Yoav Messinger, Gretchen Williams, Anne Harris, Ann Mason, Mercedes Wilhelm, Jessica Havens, Paul Goodfellow, Jennifer Ivanovich, Chris Gurnett, and all of the pathologists and pediatric oncologists who covered our clinical duties when we needed time to recruit families, analyze data, and write grants. Philanthropic and granting agencies supporting this work over the years include Hope Street Kids, National Institutes of Health, National Cancer Institute Grant No. R01-CA143167, The Parson's Foundation, St Baldrick's Foundation, Pine Tree Apple Tennis Classic, Hyundai Hope on Wheels, Diana and Stephen Goldberg, and the Randy Shaver Cancer Research and Community Fund. We would also like to think Avi Rosenberg for helpful suggestions and editing.

References

1. Manivel JC, Priest JR, Watterson J, Steiner M, Woods WG, Wick MR, Dehner LP. Pleuropulmonary blastoma. The so-called pulmonary blastoma of childhood. *Cancer.* 1988; **62**(8): 1516–26. Epub 1988/10/15. PubMed PMID: 3048630.
2. Messinger YH, Stewart DR, Priest JR, Williams GM, Harris AK, Schultz KA, Yang J, Doros L, Rosenberg PS, Hill DA, Dehner LP. Pleuropulmonary blastoma: A report on 350 central pathology-confirmed pleuropulmonary blastoma cases by the International Pleuropulmonary Blastoma Registry. *Cancer* 2015; **121**(2): 276–85. Epub 2014/09/12. doi: 10.1002/cncr.29032. PubMed PMID: 25209242; PMCID: PMC4293209.
3. Priest JR, McDermott MB, Bhatia S, Watterson J, Manivel JC, Dehner LP. Pleuropulmonary blastoma: A clinicopathologic study of 50 cases. *Cancer* 1997; **80**(1): 147–61.
4. Dehner LP. Pleuropulmonary blastoma is THE pulmonary blastoma of childhood. *Semin Diagn Pathol* 1994; **11**(2): 144–51. Epub 1994/05/01. PubMed PMID: 7809508.
5. Priest JR, Watterson J, Strong L, Huff V, Woods WG, Byrd RL, Friend SH, Newsham I, Amylon MD, Pappo A, Mahoney DH, Langston C, Heyn R, Kohut G, Freyer DR,

Bostrom B, Richardson MS, Barredo J, Dehner LP. Pleuropulmonary blastoma: A marker for familial disease. *J Pediatr* 1996;**128**(2): 220–4. Epub 1996/02/01. PubMed PMID: 8636815.

6. Hill DA, Ivanovich J, Priest JR, Gurnett CA, Dehner LP, Desruisseau D, Jarzembowski JA, Wikenheiser-Brokamp KA, Suarez BK, Whelan AJ, Williams G, Bracamontes D, Messinger Y, Goodfellow PJ. DICER1 mutations in familial pleuropulmonary blastoma. *Science* 2009;**325**(5943): 965. Epub 2009/06/27. doi: 10.1126/science.1174334. PubMed PMID: 19556464; PMCID: 3098036.
7. Slade I., Bacchelli C, Davies H, Murray A, Abbaszadeh F, Hanks S, Barfoot R, Burke A, Chisholm J, Hewitt M, *et al.* DICER1 syndrome: Clarifying the diagnosis, clinical features and management implications of a pleiotropic tumour predisposition syndrome. *J Med Genet* 2011;**48**: 273–8.
8. Rio Frio T, Bahubeshi A, Kanellopoulou C, Hamel N, Niedziela M, Sabbaghian N, Pouchet C, Gilbert L, O'Brien PK, Serfas K, Broderick P, Houlston RS, Lesueur F, Bonora E, Muljo S, Schimke RN, Bouron-Dal Soglio D, Arseneau J, Schultz KA, Priest JR, Nguyen VH, Harach HR, Livingston DM, Foulkes WD, Tischkowitz M. DICER1 mutations in familial multinodular goiter with and without ovarian Sertoli-Leydig cell tumors. *J Am Med Assoc* 2011;**305**(1): 68–77. Epub 2011/01/06. doi: 305/1/68 [pii] 10.1001/jama.2010.1910. PubMed PMID: 21205968.
9. Foulkes WD, Bahubeshi A, Hamel N, Pasini B, Asioli S, Baynam G, Choong CS, Charles A, Frieder RP, Dishop MK, *et al.* Extending the phenotypes associated with DICER1 mutations. *Hum Mutat* 2011; **32** 1381–4.
10. Doros L, Schultz, KA, Stewart DR, Bauer AJ, Williams G, Rossi CT, Carr A, Yang J, Dehner LP, *et al.* DICER1-related disorders. In *GeneReviews*™ [Internet], R.A. Pagon, ed. Seattle (WA): University of Washington, 2014 April 24. PMID: 24761742.
11. Schultze-Florey RE, Graf N, Vorwerk P, Koscielniak E, Schneider DT, Kratz CP. DICER1 syndrome: A new cancer syndrome. *Klin Padiatr* 2013;**225**: 177–178.
12. Schultz KA, Yang J, Doros L, Williams GM, Harris A, Stewart DR, Messinger Y, Field A, Dehner LP, Hill DA. DICER1-pleuropulmonary blastoma familial tumor predisposition syndrome: A unique constellation of neoplastic conditions. *Pathol Case Reviews* 2014;**19**: 90–100.
13. Delahunt B, Thomson KJ, Ferguson AF, Neale TJ, Meffan PJ, Nacey JN. Familial cystic nephroma and pleuropulmonary blastoma. *Cancer* 1993;**71**: 1338–1342.
14. Bahubeshi A, Bal N, Frio TR, Hamel N, Pouchet C, Yilmaz A, Bouron-Dal Soglio D, Williams GM, Tischkowitz M, Priest JR, Foulkes WD. Germline DICER1 mutations and familial cystic nephroma. *J Med Genet* 2010;**47**(12): 863–6. Epub 2010/11/03. doi: jmg.2010.081216 [pii]10.1136/jmg.2010.081216. PubMed PMID: 21036787.
15. Boman F, Hill DA, Williams GM, Chauvenet A, Fournet JC, Soglio DB, Messinger Y, Priest JR. Familial association of pleuropulmonary blastoma with cystic nephroma and other renal tumors: A report from the International Pleuropulmonary Blastoma Registry. *J Pediatr* 2006;**149**(6): 850–4. Epub 2006/12/02. doi: 10.1016/j.jpeds.2006.08.068. PubMed PMID: 17137906.

16. Doros LA, Rossi CT, Yang J, Field A, Williams GM, Messinger Y, Cajaiba MM, Perlman EJ, K AS, Cathro HP, Legallo RD, LaFortune KA, Chikwava KR, Faria P, Geller JI, Dome JS, Mullen EA, Gratias EJ, Dehner LP, Hill DA. DICER1 mutations in childhood cystic nephroma and its relationship to DICER1-renal sarcoma. *Mod Pathol* 2014;**27**(9): 1267–80. Epub 2014/02/01. doi: 10.1038/modpathol.2013.242. PubMed PMID: 24481001; PMCID: PMC4117822.
17. Bhardwaj AK, Sharma PD, Mittal A, Sharma A. Bilateral cystic nephroma with pleuropulmonary blastoma. [Internet] *BMJ Case Report*, 2011.
18. Bouron-Dal Soglio D, Harvey I, Yazbeck S, Rypens F, Oligny LL, Fournet, JC. An association of pleuropulmonary blastoma and cystic nephroma: Possible genetic associatin. *Pediatr Dev Pathol* 2006;**9**: 61–64.
19. Bal N, Kayaselcuk F, Polat A, Bolat F, Yilmaz Z, Tuncer I. Familial cystic nephroma in two siblings with pleuropulmonary blastoma. *Pathol Oncol Res* 2005;**11**: 53–56.
20. Bahubeshi A, Bal N, Rio Frio T, Hamel N, Pouchet C, Yilmaz A, *et al.* Germline DICER1 mutations and familial cystic nephroma. *J Med Genet.* 2010;**47**(12):863–6.
21. Lopez-Andreu JA, Ferris J, Esquembre C, Verdeguer A, Gomez J, Castel V. Familial cystic nephroma and pleuropulmonary blastoma. *Cancer* 1993;**72**: 2792–2793.
22. Kousari YM, Khanna G, Hill DA, Dehner LP. Case 211: Pleuropulmonary blastoma in association with cystic nephroma-DICER1 syndrome. *Radiology* 2014;**273**(2): 622–5
23. Wu MK, Sabbaghian N, Xu B, Addidou-Kalucki S, Bernard C, Zou D, Reeve AE, Eccles MR, Cole C, Choong CS, *et al.* Biallelic DICER1 mutations occur in Wilms tumours. *J Pathol* 2013;**230**: 154–64.
24. Shaheen, IS, Fitzpatrick M, Brownlee K, Bhuskute N, Elliott M, Powis M, Ahmad N, Tyerman K. Bilateral progressive cystic nephroma in a 9-month-old male infant requiring renal replacement therapy. *Pediatr Nephrol* 2010;**25**: 1755–58.
25. Heravi-Moussavi A, Anglesio MS, Cheng SW, Senz J, Yang W, Prentice L, Fejes AP, Chow C, Tone A, Kalloger SE, Hamel N, Roth A, Ha G, Wan AN, Maines-Bandiera S, Salamanca C, Pasini B, Clarke BA, Lee AF, Lee CH, Zhao C, Young RH, Aparicio SA, Sorensen PH, Woo MM, Boyd N, Jones SJ, Hirst M, Marra MA, Gilks B, Shah SP, Foulkes WD, Morin GB, Huntsman DG. Recurrent somatic DICER1 mutations in nonepithelial ovarian cancers. *N Engl J Med* 2012;**366**(3): 234–42. Epub 2011/12/23. doi: 10.1056/NEJMoa1102903. PubMed PMID: 22187960.
26. Schultz KA, Harris A, Messinger Y, Sencer S, Baldinger S, Dehner LP, Hill DA. Ovarian tumors related to intronic mutations in DICER1: A report from the international ovarian and testicular stromal tumor registry. *Fam Cancer* 2016;**15**(1): 105–10. doi: 10.1007/s10689–015-9831-y. PubMed PMID: 26289771; PMCID: PMC4699873.
27. Schultz KA, Pacheco MC, Yang J, Williams GM, Messinger Y, Hill DA, Dehner LP, Priest JR. Ovarian sex cord-stromal tumors, pleuropulmonary blastoma and DICER1 mutations: A report from the International Pleuropulmonary Blastoma Registry. *Gynecol Oncol* 2011;**122**(2): 246–50. Epub 2011/04/20. doi: 10.1016/j.ygyno.2011.03.024. PubMed PMID: 21501861; PMCID: PMC3138876.

28. Witkowski L, Mattina J, Schonberger S, Murray MJ, Huntsman DG, Reis-Filho JS, McCluggage WG, Nicholson, JC, Coleman N, Calaminus G, *et al.* DICER1 hotspot mutations in non-epithelial gonadal tumours. *Br J Cancer* 2013;**109**: 2744–50.
29. McDermott MB, Ponder TB, Dehner LP. Nasal chondromesenchymal hamartoma: An upper respiratory tract analogue of the chest wall mesenchymal hamartoma. *Am J Surg Pathol* 1998;**22**: 425–33.
30. Priest JR, Williams GM, Mize WA, Dehner LP, McDermott MB. Nasal chondromesenchymal hamartoma in children with pleuropulmonary blastoma — A report from the International Pleuropulmonary Blastoma Registry. *Int J Pediatr Otorhinolaryngol* 2010;**74**(11): 1240–4. Epub 2010/09/09. doi: S0165-5876(10)00366-6 [pii] 10.1016/j.ijporl.2010.07.022. PubMed PMID: 20822816.
31. Stewart DR, Messinger Y, Williams GM, Yang J, Field A, Schultz KA, Harney LA, Doros LA, Dehner LP, Hill DA. Nasal chondromesenchymal hamartomas arise secondary to germline and somatic mutations of DICER1 in the pleuropulmonary blastoma tumor predisposition disorder. *Hum Genet* 2014;**133**(11): 1443–50. doi: 10.1007/s00439-014-1474-9. PubMed PMID: 25118636; PMCID: PMC4185226.
32. Dehner LP, Jarzembowski JA, Hill DA. Embryonal rhabdomyosarcoma of the uterine cervix: A report of 14 cases and a discussion of its unusual clinicopathological associations. *Mod Pathol* 2012;**25**(4): 602–14.
33. Doros L, Yang J, Dehner L, Rossi C. Skiver K, Jarzembowski JA, Messinger Y, Schultz KA, Williams G, Andre N, *et al.* DICER1 mutations in embryonal rhabdomyosarcomas from children with and without familial PPB-tumor predisposition syndrome. *Pediatr Blood Cancer* 2012;**59**: 558–560
34. Priest JR, Williams GM, Manera R, Jenkinson H, Brundler MA, Davis S, Murray TG, Galliani CA, Dehner LP. Ciliary body medulloepithelioma: Four cases associated with pleuropulmonary blastoma — A report from the International Pleuropulmonary Blastoma Registry. *Br J Ophthalmol* 2011;**95**(7):1001–5. Epub 2010/12/16. doi: 10.1136/bjo.2010.189779. PubMed PMID: 21156700.
35. de Kock L, Sabbaghian N, Druker H, Weber E, Hamel N, Miller S, Choong CS, Gottardo NG, Kees UR, Rednam SP, van Hest LP, Jongmans MC, Jhangiani S, Lupski JR, Zacharin M, Bouron-Dal Soglio D, Huang A, Priest JR, Perry A, Mueller S, Albrecht S, Malkin D, Grundy RG, Foulkes WD. Germ-line and somatic DICER1 mutations in pineoblastoma. *Acta Neuropathol* 2014;**128**(4): 583–95. Epub 2014/07/16. doi: 10.1007/s00401-014-1318-7. PubMed PMID: 25022261; PMCID: PMC4381868.
36. Sabbaghian N, Hamel N, Srivastava A, Albrecht S, Priest JR, Foulkes WD. Germline DICER1 mutation and associated loss of heterozygosity in a pineoblastoma. *J Med Genet* 2012;**49**(7): 417–9. Epub 2012/06/22. doi: 10.1136/jmedgenet-2012-100898. PubMed PMID: 22717647.
37. de Kock L, Sabbaghian N, Plourde F, Srivastava A, Weber E, Bouron-Dal Soglio D, Hamel N, Choi JH, Park SH, Deal CL, Kelsey MM, Dishop MK, Esbenshade A,

Kuttesch JF, Jacques TS, Perry A, Leichter H, Maeder P, Brundler MA, Warner J, Neal J, Zacharin M, Korbonits M, Cole T, Traunecker H, McLean TW, Rotondo F, Lepage P, Albrecht S, Horvath E, Kovacs K, Priest JR, Foulkes WD. Pituitary blastoma: A pathognomonic feature of germ-line DICER1 mutations. *Acta Neuropathol* 2014;**128**(1):111–22. Epub 2014/05/21. doi: 10.1007/s00401-014-1285-z. PubMed PMID: 24839956; PMCID: PMC4129448.

38. Sahakitrungruang T, Srichomthong C, Pornkunwilai S, Amornfa J, Shuangshoti S, Kulawonganunchai S, Suphapeetiporn K, Shotelersuk V. Germline and somatic DICER1 mutations in a pituitary blastoma causing infantile-onset Cushing's disease. *J Clin Endocrinol Metab*. 2014;**99**(8):E1487–92. doi: 10.1210/jc.2014-1016. PubMed PMID: 24823459.
39. Rome A, Gentet JC, Coze C, Andre N. Pediatric thyroid cancer arising as a fourth cancer in a child with pleuropulmonary blastoma. *Pediatr Blood Cancer*. 2008;**50**(5):1081. Epub 2007/11/15. doi: 10.1002/pbc.21431. PubMed PMID: 18000857.
40. de Kock L, Sabbaghian N, Soglio DB, Guillerman RP, Park BK, Chami R, Deal CL, Priest JR, Foulkes WD. Exploring the association Between DICER1 mutations and differentiated thyroid carcinoma. *J Clin Endocrinol Metab* 2014;**99**(6):E1072–7. Epub 2014/03/13. doi: 10.1210/jc.2013–4206. PubMed PMID: 24617712.
41. Hill DA, Jarzembowski JA, Priest JR, Williams G, Schoettler P, Dehner LP. Type I pleuropulmonary blastoma: Pathology and biology study of 51 cases from the international pleuropulmonary blastoma registry. *Am J Surg Pathol* 2008;**32**(2):282–95.
42. Akeson AL, Wetzel B, Thompson FY, *et al.* Embryonic vasculogenesis by endothelial precursor cells derived from lung mesenchyme. *Dev Dyn* 2000; **217**(1):11–23.
43. Rudnick D. Developmental capacities of the chick lung in chorioallantoic grafts. *J.Exp.Zool* 1933;**66:**125–154.
44. Shannon JM, Hyatt BA. Epithelial-mesenchymal interactions in the developing lung. *Annu Rev Physiol* 2004; **66**:625–645.
45. Colvin JS, White AC, Pratt SJ, Ornitz DM. Lung hypoplasia and neonatal death in Fgf9-null mice identify this gene as an essential regulator of lung mesenchyme. *Development* 2001;**128**(11):2095–2106.
46. Priest JR, Hill DA, Williams GM, Moertel CL, Messinger Y, Finkelstein MJ, Dehner LP. Type I pleuropulmonary blastoma: A report from the International Pleuropulmonary Blastoma Registry. *J Clin Oncol* 2006;**24**(27):4492–8. Epub 2006/09/20. doi: 10.1200/JCO.2005.05.3595. PubMed PMID: 16983119.
47. Priest JR, Williams GM, Hill DA, Dehner LP, Jaffe A. Pulmonary cysts in early childhood and the risk of malignancy. *Pediatr Pulmonol* 2009;**44**(1):14–30. Epub 2008/12/09. doi: 10.1002/ppul.20917. PubMed PMID: 19061226.
48. www.ppbregistry.org
49. Hoffmann K. Lindner TH. easyLINKAGE-Plus — Automated linkage analyses using large-scale SNP data. *Bioinformatics* 2015;**21**(17): 3565–3567, https://doi.org/10.1093/bioinformatics/bti571.

50. Thiele, H. and Nurnberg, P. HaploPainter: A tool for drawing pedigrees with complex haplotypes. *Bioinformatics* 2005;(**21**):1730–1732.
51. Harris KS, Zhang Z, McManus MT, Harfe BD, Sun X. Dicer function is essential for lung epithelium morphogenesis. *Proc Natl Acad Sci USA* 2006; **103**(7):2208–2213.
52. Brenneman M, Field A, Yang J, *et al.* Temporal order of RNase IIIb and loss-of-function mutations during development determines phenotype in pleuropulmonary blastoma/DICER1 syndrome: A unique variant of the two-hit tumor suppression model [version 2; referees: 2 approved]. F1000Research 2018, 4:214
53. Schultz KA, Harris A, Williams GM, Baldinger S, Doros L, Valusek P, Frazier AL, Dehner LP, Messinger Y, Hill DA. Judicious DICER1 testing and surveillance imaging facilitates early diagnosis and cure of pleuropulmonary blastoma. *Pediatr Blood Cancer* 2014;**61**(9):1695–7. Epub 2014/05/14. doi: 10.1002/pbc.25092. PubMed PMID: 24821309; PMCID: PMC4139105.
54. Bernstein E, Caudy AA, Hammond SM, Hannon GJ. Role for a bidentate ribonuclease in the initiation step of RNA interference. *Nature* 2001;**409**(6818):363–6.
55. Denli AM, Tops BBJ, Plasterk RHA, Ketting RF, Hannon GJ. Processing of primary microRNAs by the Microprocessor complex. *Nature* 2004;**432**(7014):231–5.
56. Bernstein E, Kim SY, Carmell MA, Murchison EP, Alcorn H, Li MZ, *et al.* Dicer is essential for mouse development. *Nat Genet* 2003;**35**(3):215–7.
57. Kumar MS, Lu J, Mercer KL, Golub TR, Jacks T. Impaired microRNA processing enhances cellular transformation and tumorigenesis. *Nat Genet* 2007;**39**(5):673–7.
58. Murchison EP, Partridge JF, Tam OH, Cheloufi S, Hannon GJ. Characterization of Dicer-deficient murine embryonic stem cells. Proc *Natl Acad Sci USA* 2005;**102**(34):12135–40.
59. Pugh TJ, Yu W, Yang J, Field AL, Ambrogio L, Carter SL, Cibulskis, Giannikopoulos P, Kiezun A, Kim J, McKenna A, Nickerson E, Getz G, Hoffher S, Messinger YH, Dehner LP, Roberts CWM, Rodriguez- Galindo CR, Williams GM, Rossi CT, Meyerson M*, Hill, DA*. Exome sequencing of pleuropulmonary blastoma reveals frequent biallelic loss of TP53 and two-hits in DICER1 resulting in retention of 5p-derived miRNA hairpin loop sequences. *Oncogene* 2014;**33**(45):5295–302. PMID: 24909177 NIHMSID: 586233.
60. Seki M, Yoshida K, Shiraishi Y, Shimamura T, Sato Y, Nishimura R, *et al.* Biallelic DICER1 mutations in sporadic pleuropulmonary blastoma. *Cancer Res* 2014;**74**:2742–9.
61. Anglesio MS, Wang Y, Yang W, Senz J, Wan A, Heravi-Moussavi A, *et al.* Cancer-associated somatic DICER1 hotspot mutations cause defective miRNA processing and reverse-strand expression bias to predominantly mature 3p strands through loss of 5p strand cleavage. *J Pathol* 2013;**229**:400–409.
62. Gurtan AM, Lu V, Bhutkar A, and Sharp PA. *In vivo* structure-function analysis of human Dicer reveals directional processing of precursor miRNAs. *RNA* 2012;**18**:1116–1122.
63. Schultz KAP, Harris AK, Finch M, Dehner LP, Brown JB, Gershenson DM, Young RH, Field A, Yu W, Turner J, Cost NG, Schneider DT, Stewart DR, Frazier AL, Messinger Y, Hill DA. DICER1-related Sertoli–Leydig cell tumor and gynandroblas-

toma: Clinical and genetic findings from the International Ovarian and Testicular Stromal Tumor Registry. *Gynecol Oncol* 2017; **147**(3): 521–7. doi: 10.1016/j.ygyno.2017.09.034. Epub 2017 Oct 14.

64. Klein S, Lee H, Ghahremani S, Kempert P, Ischander M, Teitell MA, Nelson SF, Martinez-Agosto JA. Expanding the phenotype of mutations in DICER1: Mosaic missense mutations in the RNase IIIb domain of DICER1 cause GLOW syndrome. *J Med Genet* 2014; **51**(5): 294–302. Epub 2014/03/29. doi: jmedgenet-2013-101943 [pii] 10.1136/jmedgenet-2013-101943. PubMed PMID: 24676357; PMCID: PMC4429769.
65. Verrier F, Dubois d'Enghien C, Gauthier-Villars M, Bonadona V, Faure-Conter C, Dijoud F, Stoppa-Lyonnet D, Houdayer C, Golmard L. Multiple DICER1-related lesions associated with a germline deep intronic mutation. *Pediatr Blood Cancer* (2018). doi: 10.1002/pbc.27005. [Epub ahead of print] PMID: 29469200.
66. Schultz KAP, Williams GM, Kamihara J, Stewart DR, Harris AK, Bauer AJ, Turner J, Shah R, Schneider K, Schneider KW, Carr AG, Harney LA, Baldinger S, Frazier AL, Orbach D, Schneider DT, Malkin D, Dehner LP, Messinger YH, Hill DA. *DICER1* and Associated Conditions: Identification of At-risk Individuals and Recommended Surveillance Strategies. *Clin Cancer Res* 2018. doi: 10.1158/1078-0432.CCR-17-3089. [Epub ahead of print] Review.
67. Berrington de Gonzalez A, Salotti JA, McHugh K, Little MP, Harbron RW, Lee C, Ntowe E, Braganza MZ, Parker L, Rajaraman P, Stiller C, Stewart DR, Craft AW, Pearce MS. Relationship between paediatric CT scans and subsequent risk of leukaemia and brain tumours: assessment of the impact of underlying conditions. *Br J Cancer* 2016;**114**(4): 388–94. doi: 10.1038/bjc.2015.415. PubMed PMID: 26882064; PMCID: PMC4815765

15 Precision Medicine: Challenges in Changing the Paradigm for Medical Care in the 21st Century

Timothy J. O'Leary

Office of Research and Development
Veterans Health Administration
Washington DC 20420
And
Department of Pathology
University of Maryland School of Medicine
Baltimore, MD 20201

Contents

I. Introduction

While there are potentially many ways to date the birth of genomic medicine, the development of dye-terminator sequencing by Frederick Sanger in 1977 made possible for the first time the systematic exploration of the genome. At the time of Sanger's work, it typically took upwards of a year to completely sequence even a relatively small gene, however, and a complete understanding of the human genome seemed unattainable. Technical innovations both in sequencing technology and in informatics have led to a dramatic change in the past 40 years, and current sequencing technologies make it possible to sequence a human genome in only a few hours. While this represents remarkable progress, the effective use of this information for the understanding, prevention, and treatment of human diseases requires that we overcome some formidable conceptual and technical obstacles. For neoplastic diseases, many of these obstacles are well understood, and immense progress has been made in addressing them. An understanding of genomic changes in cancer has led to the development of new therapies and the repurposing of old treatments in a manner that achieves better results with less toxicity. Increasingly, cancer drugs are paired with companion diagnostics that are used to determine the potential for treatment effectiveness, and current investigations offer the hope that genetic characterization of a tumor will definitively determine the treatment to be employed.

Genomic medicine has not had a similar impact on the treatment of non-neoplastic diseases, however. Although there are occasional successes with the use of pharmacogenomics testing, such as the use of genetic testing to reduce the risk of agranulocytosis associated with clozapine treatment,[1–4] clinical use of genomic testing in non-neoplastic disease has more typically yielded results that are inconsistent, as has been seen in clinical trials or meta-analyses to determine optimal warfarin dosing.[5–20]

The failure of pharmacogenetics testing for warfarin to "catch on" illustrates a number of challenges facing any genetic testing strategy associated with chronic disease in the adult population. These challenges include modest predictive value, a clinical outcome strongly dependent on non-genetic factors (such as consumption of Vitamin K-rich vegetables), the use of alternative testing strategies that better account for non-genetic factors of clinical importance (such as home-use prothrombin time testing), and the development of better therapies which obviate the need for genetic testing (such as newer anticoagulants that are more effective than warfarin in preventing stroke in patients with atrial fibrillation — the most common indication for chronic anticoagulation — and have a smaller risk of bleeding). One or another of these challenges has significantly dimmed enthusiasm for genetic testing in a variety of clinical situations which beg for better initial treatment, and for which there are genetic testing strategies available, including depression[21,22] and bipolar disorder.[23]

There are, perhaps no diseases that have been better studied from a genetic perspective than type II diabetes and coronary artery disease, each of which, illustrate two additional challenges facing the use of genomic testing in routine adult medical practice. The first is that, as additional data accumulate, genetic risk prediction models, for these diseases change. Were these changes only small refinements in the models, there would be no problem, but in fact, for many diseases, up to 25% of patients are reclassified from "low risk" to "high risk" or vice versa.[24] A second problem, of course, is that even were the classification model stable, we do not currently have effective ways to use the information. Although a case can be made (without the use of genetic information) that in older males long-term use of statins to lower low-density-lipoprotein cholesterol (LDL-C) is useful, this recommendation is controversial and imposes both a financial burden and a small but real medical risk; for diabetes, there is no similar pharmacologic prevention. For both of these conditions — as well as for many other conditions for which gene–environment interactions play a role in pathogenesis — lifestyle changes are required. While lifestyle interventions for promoting weight reduction and regular exercise are available and effective, they require substantial psychosocial support that is not currently available for most patients,[25–28] and most individuals fail to make such changes on their own.

A final barrier to routine use of genetic information in routine medical practice is information overload. A PubMed search of papers relating to human genetics or genomics found nearly 10,000 papers published in January 2015 alone. Nearly 200 of these papers related in some way to a clinical trial. Online databases, which undergo less curation than the published literature, also suffer from inconsistencies that limit their practical utility for busy clinicians. A recent study of comparison of variant classification using the Emory, Human Gene Mutation, and ClinVar databases showed discordant classifications in half the variants found in more than one database. Primary care physicians are simply unable to absorb this information into clinical practice. This inability may not be a bad thing; many genetic associations found in the literature are not replicated, and there is reason to believe that most scientific reports are in some way either false or misleading.[29–39] When added to the fact that use of exome sequencing and whole genome sequencing gives rise to incidental findings of uncertain significance,[40–47] it is not hard to understand why skepticism regarding the practical utility of these testing approaches has developed.

Although these are substantial challenges to overcome, the situation is not hopeless. This chapter will outline some of the approaches that are currently being undertaken to assure the validity of genomic data and genomic predictors. We will further explore some of the work in computational science and systems biology to develop models for effective use of genomic and lifestyle information in understanding and treating human disease. Finally, we will outline some approaches to therapeutic development that may migrate into the realm of precision medicine and eventually facilitate the development of truly individualized personal therapeutics — drugs that are indeed created for each specific patient, based on their genome, environment, and disease.

II. Getting High-Quality Data — The Use of Large Genetic Cohorts

Although large genetic and genomic studies have provided tantalizing clues regarding genetic predisposition to disease and likelihood of response to specific treatments, moving from clue to implementation is difficult. The development of large genomic cohorts is helping to remove two of the barriers — that of uncertainty about the *presence* of an effect, and that of certainty regarding the *magnitude* of an effect. Almost all current genomic association studies are designed in a manner that provides reasonable assurance of small type I error — deciding that an effect is present when it is not. Very few current genomic studies provide a small probability of type II error — failing to find an effect that is present. Thus, it should not be surprising that genomic association studies often fail to replicate the findings of earlier studies — they are not adequately powered to do so. It should be even less surprising, therefore, that studies that are not adequately powered to be sure that an effect is found when present, are not adequately powered to precisely estimate the magnitude of that effect. The only remedy for this situation is the design of studies that are statistically powered (and hence very, very large) to detect very small effect sizes with a high level of certainty, and larger effect sizes with a high level of precision. Over the last few years, significant effort has gone into developing remedies for this deficiency, in the form of large genomic cohorts. Current genomic cohorts fall into two groups — case–control studies based upon single diseases or groups of diseases, and cross-sectional cohorts that sample the entire population, from whom specific longitudinal or case–control cohorts can later be sampled. In this section, we will illustrate these approaches with a few examples, pointing out some of the strengths and weaknesses of each.

III. Disease-Specific Cohorts and Consortia

Many condition-specific cohorts and consortia have been developed, dealing with diseases that range from common to extremely rare (http://consortiapedia.fastercures.org/consortia/#W). Most cohorts established to date consist of a variety of smaller cohorts that have been assembled by individual investigators, sometimes based upon sponsorship of the cohort by a large organization. The Wellcome Trust Case–Control Consortium, the Psychiatric Genomics Consortium, and the charge Consortium provide three examples of this approach.

a. The Wellcome Trust Case Control Consortia (WTCCC)

The publication of a genome-wide association study of 14,000 cases of seven common diseases (2,000 subjects each with bipolar disorder, hypertension, coronary atherosclerosis, rheumatoid arthritis, Crohn's disease, and diabetes types I and II) and 3,000 shared controls in 2007 illustrates both the power and some limitations in this approach.[48] This study showed that adequate power for discovery (with acceptable type I error rates) is achieved by case–control studies involving a few thousand subjects, but that many of the associations uncovered have modest effect sizes which cannot be adequately estimated and which may result in non-replication in other studies. More importantly, however, the study demonstrated the ability to assemble large groups of subjects and of cooperating investigators. As a result, the consortium expanded its efforts to consider not only gene polymorphism, but copy number variation, expand its scope to a variety of additional conditions, and conduct genotyping on both an Affymetrix platform (used in the original studies) and the Illumina platform.

The WTCCC makes data from its studies available to other qualified investigators after a short embargo period to allow initial analysis and publication. Data sharing of this sort has become a feature of virtually all other large genomic efforts undertaken worldwide.

b. Psychiatric Genomics Consortium

The Psychiatric Genomics Consortium (PGC) (htttps://pgc.unc.edu), which was initiated in 2007, represents a mega-collaboration of over 800 investigators throughout the world who share samples and data to in order to analyze genome-wide genomic data for psychiatric disorders. Nearly a million samples are involved in these analyses, which have addressed such diverse disorders as schizophrenia, bipolar disorder, depression, autism, attention-deficit hyperactivity disorder, substance abuse, and Tourette syndrome. Although analyses conducted under the auspices of the consortium initially focused on genome-wide association studies, more recent efforts have involved analysis of copy number variation and sequence analysis. PGC investigators have made data available for other investigators through a data repository maintained by the National Institute of Mental Health, which currently maintains data related to schizophrenia, bipolar disorder, and major depressive disorder (https://www.nimhgenetics.org/available_data/data_biosamples/pgc_public.php), as well as through other download sites (https://www.med.unc.edu/pgc/downloads). Consortium members also make available a variety of tools for data visualization, such as ricopili (https://www.broadinstitute.org/mpg/ricopili/), a tool for visualizing specific regions of genomic interest.

The importance and success, as well as some limitations, of the PGC may be found in the analyses they have published. For example, the PGC GWAS of nearly 7,500 subjects with bipolar disorder and over 9000 controls identified over 30 SNPs associated with the disease (http://www.ncbi.nlm.nih.gov/pmc/articles/PMC3637176/), and suggested that calcium channel disorders may be related to disease pathogenesis. Analysis of data from over 50,000 subjects, with and without schizophrenia, implicated a micro RNA, MIR137, a regulator of neuronal development, in schizophrenia etiology (http://www.ncbi.nlm.nih.gov/pmc/articles/PMC3303194/), among others. However, the strength of association identified in these studies is high from a statistical perspective, but the relative risk associated with the individual variants is generally less than 1.2, and was no more than 1.4 in either study. This highlights one of the major limitations in the use of single-disease mega-cohort GWAS studies from the perspective of molecular diagnostics — while they may provide clues for further investigation, they do not yield results having immediate diagnostic or clinical utility. These studies also may suffer both in power for discovery and utility for further investigations, from heterogeneity in subjects provided by individual investigators who may not have used identical diagnostic criteria for subject identification. This characteristic also limits the use of the data set for exploring the genomics of specific endophenotypes, such as the level of functional disability. On the other hand, assembly of a mega-cohort by combination of data or samples from smaller existing cohorts may be accomplished much more quickly.

c. The Cohorts for Heart and Aging Research in Genomic Epidemiology (CHARGE) Consortium

The CHARGE consortium consists of ten large longitudinal cardiovascular disease cohorts that have each collected genetic specimens and data, and who have been working together for most of a decade.[49,50] Each of the underlying studies (see table below) has developed a very carefully phenotyped population using relatively well-standardized criteria for measurement of cardiovascular function and disease susceptibility, creating opportunities for discovery and replication that are not generally available in the "cohort of cohorts" approach.

Reykjavik Age, Gene, Environment Susceptibility Study
Atherosclerosis Risk in Communities Study
Cardiovascular Health Study
Framingham Heart Study
Rotterdam Study
Coronary Artery Disk Development in Young Adults Cohort
Family Heart Study
Health, aging, and Body Composition Study
Jackson Heart Study
The Multi-Ethnic Study of Atherosclerosis

Additional strengths of the CHARGE consortium include substantial diversity, not only ethnic but environmental, which enables more effective exploration of gene–environment interactions and development of risk models that generalize to a broad population. The completeness of cardiovascular phenotype information has made possible some amazing studies, ranging from genomic associations with hypertension[51,52] and coronary atherosclerosis[53,54] to studies of serum electrolyte levels,[55] sleep[56] and overall longevity.[57,58]

The phenotypic strengths of this study highlight two of the challenges facing all genetic association studies — that of adequately defining the phenotype, and that of defining the reference population. By and large, disease is not static, but association studies are performed at a point in time. The difficulty is that the probability of clinically manifesting disease is not constant — a 20 year old athlete is not likely to demonstrate type II diabetes, but a 70 year old obese former athlete is quite likely to suffer from this disease. In a sense, one can say that an individual's phenotype is not adequately defined unless the time course of every disease that will ever be present in that individual is mapped out; this is not possible before death. Given that this is true, the difficulty in defining a reference population is manifest; the ideal population probably consists of healthy centenarians!

IV. Population-based Cohorts

In contrast to the condition-specific cohort, the population-based cohort is constructed from a broad population of individuals, with and without disease, each of whom agrees to provide access to genetic material and health information for research. Studies of individual health conditions are performed by sampling this larger cohort. New studies can thus be completed without the effort of assembling a new cohort, which may take several years. The price for this flexibility is an inability to collect the level of detailed phenotypic information that is frequently available for each and every condition in a disease-specific cohort. Over time, a number of large cohorts have been developed as shown in the list below; we will focus on six of these cohorts — DeCode, the United Kingdom Biobank, the Kaiser Permanente Research Program on Genes, Environment, and Health, the China Kadoori Biobank, Vanderbilt's BioVu, and the US Department of Veterans Affairs Million Veteran Program, to explore some of the strengths and weaknesses of this approach.

Europe

DeCODE/Icelandic Biobank
UK Biobank
Banco Nacional de AND (Spain)
GenomEUtwin
Finnish Biobank
Swedish Biobank
German biobank, KORA
Estonian Biobank
Family-based collections (Nordic)
Generation Scotland
EPIC, European (Cancer)
Danubian Biobank Consortium
GATiB Genome Austria Tissue Bank
Biobank Hungary

North America

Canadian Consortium
dBGap, NIH
National Children's Study
Psychiatric Genome Consortium
Marshfield Clinic
Kaiser Permanente
BioVu (Vanderbilt)
Howard University African Diaspora
Million Veteran Program
Mayo Clinic
Geisinger/Regeneron
Joint Pathology Center Biobank (Department of Defense)

Asia

Kadoori Biobank
Korea Biobank Project
Biobank Japan
Taiwan Biobank

Australia

Genetic Repositories Australia
Australasian Biospecimen Network

Middle East

National Laboratory for the Genetics of Israeli Populations

a. DeCODE

In the late 1990s, the company DeCODE demonstrated the potential value of a large genetic biobank when it proposed to collect blood specimens for genotyping, together with genealogical data and electronic health records from the entire population of Iceland, which was over 250,000 individuals. Although the company suffered from a variety of legal and financial setbacks that required significant redefinition of the project, the scientific contributions from this effort have been enormous; over 200 papers have been published covering a range of conditions from schizophrenia to diabetes, cardiovascular disease, stroke, and various cancers.

Although the scientific contributions attributable to the DeCODE biobank are unquestionable, the greatest significance of this project is that it showed beyond the shadow of a doubt the feasibility of combining, on a large scale, medical data and genomic specimens to be used for research purposes. Unlike most subsequent efforts, the DeCODE dataset allows the use of traditional linkage analysis, based on genealogical information, for gene discovery. In general, this allows for greater discovery power for a data set of any given size. The relative genetic homogeneity of the Icelandic population is both a blessing and a curse, increasing the likelihood of discovering associations that are present in that population, but precluding discovery of associations that may be common elsewhere, but not in Iceland. Although DeCODE scientists have collaborated with other researchers, there are significant restrictions on the use of this dataset by outside investigators.

b. The UK Biobank

The UK Biobank recruited 500,000 participants between ages 40 and 69 from 2006 to 2010, with the support of both governmental and private organizations in the United Kingdom.[59–62] Individuals who enrolled in the biobank project provided not only blood specimens, but also urine and saliva samples, together with a broad consent for use of these materials and information in future research. Recruitment was carried out by a network of 22 regional centers, which carried out a variety of health assessments as a part of the enrollment process, including health and lifestyle questionnaires, a detailed medical history interview, measurements of blood pressure, height, weight, body composition, and grip strength, as well as tests of memory, pulmonary function, and bone density, hearing, vision and ocular health, and an electrocardiogram. More recently, the biobank has invited participants to wear a wrist activity monitor, has performed repeat and extended measurements on about 5% of participants. and is currently undertaking an MRI body and DEXA bone scanning enhancement. Many participants have also completed detailed web-based questionnaires on their diet, cognitive function, and work history. All data, including genetic, biochemistry, and imaging data, are being made available for research as they become ready. More recently, information from the National Health Service medical system is being linked to the dataset. Genotyping is performed using a customized Affymetrix array, and information is made broadly available for research purposes immediately following curation. Results of genetic analyses are available for research purposes only, and are not returned to participants or used for medical care.

The UK Biobank project is notable for the quality of phenotypic information obtained at the time of initial enrollment, which exceeds that which is universally obtainable from medical records information alone; limitations include both a fairly heterogeneous medical records system in the National Health Service, as well as the upper age limit of the biobank population, which excludes what may be an ideal reference set for a healthy genome — individuals who live into their 90s or hundreds with little or no evidence of disease. In addition, the UK population, while considerably more diverse than that of Iceland, lacks the genetic and cultural diversity found in the United States and other "melting pot" populations. In spite of these limitations, however, the UK Biobank at present is probably the most valuable genomic information source in the world, and it has provided a template used by other biobanks in planning their efforts. As a result, over 1000 researchers are now using UK Biobank resources for more than 200 research projects. Over 100 papers have been published that relate to the use of this resource, ranging from factors influencing mate selection to studies on cardiovascular disease and cancer. Almost half were published in 2015 alone, demonstrating the rapidly growing importance of this major resource.

c. Kaiser Permanente Research Program on Genes, Environment, and Health

The Research Program on Genes, Environment, and Health (RPGEH) is an effort by the California Kaiser Permanente organization to capitalize on the scientific expertise of its staff and the size (6 million members) and stability of the Kaiser Permanente patient population for improving care. Kaiser intends to build a database that includes genomic information obtained from saliva and/or blood on upwards of 500,000 participants who also provide access to health records and other health information. Analyses are performed using an Affymetrix array that is similar to that employed by the UK Biobank, and are linked with health records and are performed for research purposes only; they are not returned to participants or used to guide medical treatment. Over 200,000 individuals have provided specimens to date; a subset of approximately 100,000 of these individuals in the Genetic Epidemiology Research and Adult Health and Aging (GERA) cohort have undergone extensive genetic characterization which has led to publications on the role of telomere length in longevity, genomic susceptibility to macular degeneration, and other work.[63–67]

Like the UK Biobank, the Kaiser effort builds upon a strong healthcare system, but has the advantage of more extensive electronic health records. Also like the UK Biobank, Kaiser is committed to data sharing. The Kaiser Biobank has somewhat greater ethnic diversity, but does suffer from a greater likelihood that participating individuals will at some time in the future seek care from healthcare systems that are not participating in the effort.

d. China Kadoori Biobank

The China Kadoori Biobank (formerly the Kadoori Study of Chronic Diseases in China) consists of approximately 510,000 Chinese adults who were recruited from 10 different regions of China, all of whom completed questionnaires, underwent a series of physical measurements, and donated blood for future research; approximately 25,000 participants are periodically resurveyed.[68–70] Baseline data is currently available to Chinese investigators, and will be available more generally in the near future; the baseline questionnaire is available online (http://www.ckbiobank.org/site/binaries/content/assets/resources/pdf/qs_baseline-final-from10june2004.pdf). SNP genotyping of approximately 100,000 participants is planned, with data to be released first to Chinese investigators, and then more broadly. While this study has contributed to the understanding of disease epidemiology in China,[68,71–81] only a single genetic study, on phospholipase A2 loss of function variant, has been published to date.[82]

When genetic results become available, the Kadoori Biobank will likely contribute enormously to understanding the role of genetic variation in the pathogenesis of disease, particularly because it reflects an ethnic group that is, relative to the global population, substantially underrepresented in other biobank efforts.

e. Vanderbilt's BioVU

BioVU is a biobank consisting of over 215,000 specimens originating in the Vanderbilt University's healthcare system, which are linked to the electronic health records of participating subjects.[83–90] BioVU's approach to recruitment is unique. While most biobanking efforts use an informed consent mechanism that requires participants to explicitly "opt in" to participate, BioVU originally employed an "opt out" strategy that enrolled patients who were being treated unless they specifically indicated that they did not wish to enroll; more recently the study has switched to an "opt in" approach. A second unique feature of BioVU is its reliance on the use of residual specimens that were originally sent to clinical laboratories for diagnostic purposes. This strategy enables efficient ongoing collection of specimens for longitudinal study — a capability that is rare in large banking efforts. BioVU investigators have made substantial contributions to a variety of problem areas using the resource, including pharmacogenomics,[91–97] cardiovascular disease,[98–100] and reproductive health.[101–103]

Studies using the BioVU resource require the participation of a Vanderbilt University faculty member; collaborative efforts with outside investigators are frequent, and Vanderbilt is a lead institution for the eMERGE consortium (above).

f. The Department of Veterans Affairs Million Veteran Program

The Million Veteran Program (MVP), which enrolled its first subjects in 2011, represents over 450,000 Veterans of the United States uniformed services (Army, Navy, Air Force, Marines, Coast Guard, and Commissioned Corps of the Public Health Service and National Oceanographic and Atmospheric Administration), who receive treatment in the Veterans Health Administration (104). Each participant in this study completes baseline and lifestyle questionnaires (http://www.jclinepi.com/cms/attachment/2046226040/2057250640/mmc1.pdf), provides a blood specimen for genetic analysis, gives permission for investigators to perform virtually any type of genetic or health records research using this information and information from their VA electronic health record, and agrees to allow recontact. Initial genetic characterization of the blood specimens is carried out using a specialized Affymetrix array that is similar to the exome array used by Kaiser and the UK Biobank, but is enriched for polymorphisms that are more frequent in the African American population or that have been identified in association with psychiatric disease by the Psychiatric Genomics Consortium. Approximately 400,000 specimens have been genotyped to date, and approximately 100,000 specimens are undergoing exome sequencing. Like most other biobanks discussed in this chapter, MVP makes de-identified data, rather than specimens, available to investigators; since this data has not been analyzed in a clinical laboratory, it is not returned to participants. In order to prevent investigators from reidentifying individual subjects by comparing this data against that of other databases, data analysis is carried out in a secure high-performance computing environment that allows analysis to be performed without download of individual patient records or genomic results. Since the computing systems for data analysis are currently under development, data is available to a small number of "alpha test" and "beta test" investigators in the VA, but the expectation is that it will eventually be available to government and academic investigators throughout the United States.

Although the MVP cohort has relatively good representation of African American and Hispanic men, the Veteran population served by VA is 90% male. Underrepresentation of women represents one of its primary limitations.

g. The NIH precision medicine initiative

Recently, the NIH has proposed creating a cohort that, like that planned by the Department of Veterans Affairs, will consist of a million or more participants who agree to provide genomic information, access to health records, and permission for broad use in research (https://www.nih.gov/precision-medicine-initiative-cohort-program). It is proposed that at least some results be returned to participants, and that data obtained as wait part of the initiative will be available to investigators worldwide.

h. The eMERGE network

The eMERGE network is not a genomic cohort, but rather a consortium assembled to develop methods for using electronic health records (EHR) for genomic research. Although it may seem that electronic health records are ideal for assembling large amounts of data on gene–disease associations, the reality is that these data are seldom collected and maintained in a manner that is suitable for research. To put it simply, EHR data are messy. While data collected prospectively for research purposes is highly structured and collected in the same way for every participant, data obtained in the course of health care delivery is, by comparison, rather haphazard. There are many reasons for this. Patients do not all seek medical care on a regular basis, so the intervals between data collections can vary from patient to patient. Patients do not all see physicians with the same background and training, and physicians whose training varies may use different criteria to make a diagnosis. For example, neurologists are likely to be particularly discriminating when making a diagnosis of Alzheimer's disease, while family practitioners may apply the eponym for almost any dementia. Patients present at different times with different complaints — this triggers different use of laboratory and radiologic examination. Finally, patients may see different physicians, each of whom uses a different electronic health record system; data interchange may be in the form of non-computable free-text, or image data. For these and other reasons, extraction of high-quality medical information for research requires great care, and is, in many cases, institution-specific.

The eMERGE consortium is devoted to developing efficient methods for in-depth phenotyping, and maintains a repository of well-validated approaches for use with EHRs, PheKB (https://www.phekb.org/phenotypes). To date, 33 phenotypes have been made public, ranging from Abdominal Aortic Aneurism to Warfarin Dose Response. Although individual institutions need to program these phenotyping algorithms specifically for their own EHR systems, the use of uniform phenotyping tools created by eMERGE will allow more efficient exploitation of genomic databases such as those described above.

V. Genomic Cohorts — Moving Forward

The difficulty in understanding human gene–disease associations should not be underestimated. There are, in principle, many more combinations of bases available than there are human beings; it is statistically impossible to completely solve this problem. Fortunately, the single-nucleotide polymorphisms that are responsible for human variation are much fewer in number — only about 10–15 million. Nevertheless, this is still larger than the number of people whose health information and genetic material are available in genomic biobanks.

The large cohorts described above do have considerable power to identify gene–environment–disease associations, as noted by the Precision Medicine Initiative Working Group assembled by NIH to advise the initiative (http://www.nih.gov/sites/default/files/research-training/initiatives/pmi/pmi-working-group-report-20150917-2.pdf), as well as by the UK Biobank (http://www.ukbiobank.ac.uk/wp-content/uploads/2011/11/UK-Biobank-Protocol.pdf). For common alleles and diseases (allele in 10% of the cohort, disease present in 25,000 subjects), there is an 80% likelihood of identifying an odds ratio of at least 1.16. The NIH working group concluded that a cohort of a million individuals will be able to identify disease–gene associations for hundreds of health outcomes. For those of European ancestry, the cohorts described above provide more than sufficient statistical power not only to identify gene–disease associations, but to replicate the findings.

African populations and indigenous populations in remote Asia, the Americas and in Australia, New Zealand, and the Pacific Islands are highly underrepresented. Although it is likely that most disease–gene associations are similar to those of other populations, it will nevertheless require the development of additional genomic cohorts to assure that advances from genomic medicine will be available worldwide.

VI. Systems Biology — Putting Genomic Data into Context

Although, as we have seen, a great deal of information can be gleaned from the analysis of gene–environment–disease associations using large datasets such as those described above, it is statistically unlikely that simple association analyses will be sufficient to understand the interplay of gene polymorphisms in metabolic pathways that may involve even as few as five or six of these gene products. The reason for this is simple mathematics. Suppose we consider a pathway involving six genes, each of which has ten sites at which a non-lethal sequence variation can occur, and suppose we further specify that within any given gene, only one of these variations can occur at a time. There are in this example a million different combinations of sequence variation that can, in principle, occur. Even in a database of 2 million individuals, there is insufficient statistical power to describe such a system.

Fortunately, substantial progress is being made in three realms of investigation that should, with time, obviate this difficulty — problem simplification, protein modeling, and network modeling. In the next few paragraphs, we will describe the progress that has been made in each of these areas.

a. Problem simplification — Transcriptomics and proteomics

When a physician evaluates a patient that has appeared with a complaint, he/she first attempts to isolate the problem to an organ system or systems.

Through a small number of physical, laboratory, and radiologic examinations he/she will come to the conclusion that the major issue to address, for example, is that of cardiac arrhythmia. For genetic cardiac disease, this allows a focus on ion channel abnormalities and the ability to discard, in all probability, whatever may be going on with the patient's tumor suppressor genes.

One of the reasons that the acquired genetic alterations in cancer have been so effectively characterized is that gene expression arrays were used to compare normal and tumor tissue. Genes that were not active in either the tumors or in normal surrounding tissue could then be mentally discarded, and focus could be given to those genes in which expression differed significantly between normal and tumor tissue. The result is that the cancer biologist has learned to focus on a few hundred genes rather than the 30,000 or so that constitute the human genome.

Similar techniques can be used for medical diseases, although the availability of biospecimens is hampered by the fact that, unlike cancer, large specimens are not generally obtained as a part of routine medical care. Nevertheless, many body sites are available for biopsy, and all for autopsy; furthermore, gene expression patterns tend to be relatively similar across closely related species. If we apply this thinking to the liver, for example, we find that only about half of human proteins are expressed at any time during development, and that half of them are "housekeeping" genes that function similarly throughout the organism (http://www.proteinatlas.org/humanproteome/liver). Only 436 of these proteins are elevated in comparison with other tissue types. Discovery power is considerably enhanced when association analysis is limited to the transcriptome most likely to be associated with that disease process, rather than the entire human genome. Even greater power is obtained when it is further limited to those mRNAs or proteins that differ significantly between healthy normal and diseased tissue. The ability to conduct these experiments efficiently has been facilitated by recent advances that have enabled proteomic analysis of archival tissue to yield essentially the same results as those previously obtained using fresh or frozen specimens; this makes the vast tissue archives of hospital pathology departments a treasure trove for further investigation.[105–112] Whether expression analysis is carried out using measures of mRNA expression or of proteins, however, it enables a significant reduction in the complexity of the analysis required to establish the presence and nature of gene–disease associations. This also reduces the size of the system that must be developed (and solved) if the normal function of an organ is to be described mathematically (see Section VI.c).

b. Protein modeling

During most of the 20th century, protein structure determination was based upon the ability to crystallize proteins and then infer their structure using X-ray diffraction techniques. These methods, while effective, required months or years of work for a single protein, and for each sequence variant. For this reason, only a small number of protein sequence variants were ever explored using this approach, which could not be employed for non-crystallizable proteins such as those found in cell membranes. Other chemical techniques, including infrared and Raman spectroscopy, and nuclear magnetic resonance spectroscopy, were employed to obtain some understanding of the structure of these non-crystallizable proteins, but were not capable of providing the fine detail available by crystallography. As computing systems and computational techniques have become more powerful, however, it has become possible to infer the structure of many proteins from the information obtained about the forces governing the interaction of atoms within the protein with each other and with solvent.[113–122] More recently, it is becoming possible to model protein–protein interactions, such as those involved in macromolecular assembly, using similar computational techniques,[123–126] which are also being employed to understand the interactions of small molecules with large biomolecules.[127–132] As such computational methods are improved, it will be relatively straightforward to predict the effect of changing a nucleic acid (and thus protein) sequence on these interactions. Programs are currently available to perform docking simulations [https://en.wikipedia.org/wiki/Docking_(molecular)]; while they are not yet sufficiently robust for clinical use, it is clear that in silico computation of protein–protein, protein–ligand, and protein–drug interactions will provide a robust platform upon which to develop a systems-biology understanding of the genome.

c. Network modeling

The second requirement for truly understanding the complexity of genomic polymorphism is a significant improvement of our understanding of biological pathways. Considerable effort has gone into mapping biological pathways over the last 60 or 70 years. For the most part, the yield from this effort is found in the kind of biological pathway map seen in biochemistry and cell biology textbooks. These maps provide great insight into the effects of gene mutations, but this insight is often limited to the cases in which either a protein is constitutively activated or totally inactivated. In a few simple cases, such as drug metabolism via cytochrome oxidase pathways, the effects of partial gain or loss of function are somewhat understood, but the impact of these changes is not usually so large as to enable decisions on drug dose alterations. In order to make this information truly useful, it must be considered in the context of body mass and composition, diet, absorption kinetics, and other factors that vary significantly among individuals. To obtain such an understanding requires that we go beyond the "flat map" of biological pathways and create an understanding of biology that is founded in well-validated mathematical models such as those that form the basis of most contemporary chemistry and all physics.

Although mathematical modeling of biological phenomena is not new, most modeling efforts have focused on the application of known physical and chemical principals to isolated systems, such as nerve conduction or cardiovascular flow. Systems biologists are currently developing models of more complex systems, such as gene regulatory networks, based upon such frameworks coupled with differential equations, stochastic equations, or Boolean networks. These models can be validated (or disproven) in laboratory models and used as building blocks for a complex model of normal (and abnormal) human physiology.

It seems unlikely that functional alterations associated with genetic polymorphisms will be treatable simply by directing a small molecule to bind to a single altered protein. Rather, a biological therapy for a complex multigene disease will require interventions at several points along the disease pathway; without predictive models to guide this therapy, it is likely that most attempted interventions will be ineffective. On the other hand, a detailed and accurate mathematical understanding of the molecular abnormalities underlying an individual patient's disease processes will enable effective personalized treatment.

While these models are relatively complex by comparison with those that have historically dominated biological modeling, they are not particularly large when compared, for example, with weather prediction. I believe it is likely that the development of such models is simply a matter of time and patience.

VII. Rational Drug Design — Toward True Personalized Precision Medicine

In a sense, both the term "personalized medicine" and the term "precision medicine" are misleading. The limited armamentarium of laboratory targets, together with the even more limited arsenal of therapeutic drugs, assures that while treatment decisions are based upon analysis of more genetically homogeneous groups, they are nevertheless statistically rational guesses based upon group outcomes, rather than therapies that are truly targeted to an individual's disease. If, ultimately, a mathematical model can be developed that accurately describes an individual's disease, then it becomes possible to think about the development of therapies that are truly directed towards a single specific individual. To accomplish this, however, requires a significant expansion of the therapeutic options available for treatment. This cannot practically be accomplished through a traditional pharmaceutical development model, but rather requires that it be possible to design, and synthesize, new drugs on a patient-by-patient basis. While it is likely to take decades to implement such a goal, the scientific efforts required for success are already underway — rational drug design, combinatorial chemistry, and high-throughput screening.

In its most complete form, rational drug design would refer to the design of a drug that will bind a specific biological target to known effect, and for which the absorption, metabolism, and transformation could be predicted. More typically the term is used to refer to the design of drugs based upon analysis of the interaction of the drug with a protein target — ligand binding. One of the first drugs to result from targeting a specific protein function was fluoxetine (Prozac), which was intended to selectively inhibit serotonin reuptake in the brain.[133,134] Rather than using a computational approach as outlined above, a series of compounds structurally related to diphenhydramine were synthesized and tested in a model system; fluoxetine proved to be the most selective and potent of the reuptake inhibitors.

The renin inhibitor aliskiren was developed for treating hypertension through renin inhibition, and designed based on a combination of molecular modeling and X-ray crystallography methods.[135] Although peptide inhibitors of renin had previously been investigated, they were therapeutically unsatisfactory; hence the desire to develop a non-peptide inhibitor, for which there were no predicate molecules. Although the drug was generally safe and effective in clinical trials,[136–146] and received approval by the Food and Drug Administration, it remains a second- or third-line treatment option since most patients tolerate thiazide diuretics, angiotensin-converting enzyme inhibitors, and calcium channel blockers well, achieve therapeutic results similar to those seen with aliskiren, and incur lower costs. Nevertheless, the success of aliskiren demonstrates that the concept of rational drug design based upon molecular modeling of protein–ligand docking is, in principle, a feasible goal. Needless to say, however, significant technology development will be required before this type of approach can be employed on a patient-by-patient basis.

Although this example demonstrates the potential feasibility of designing molecules on a truly personalized basis, it does not illustrate the solution to the problem of actually synthesizing them. Those who have taken a synthetic organic chemistry course in college will recognize that the synthesis of a new molecular entity can be difficult to impossible in practice. Hence, to be practical, an approach that uses molecular models of ligand-docking to assess potential therapeutic efficacy must be coupled with actual synthesis of potential ligands to assure success. Combinatorial chemistry provides one such approach (https://en.wikipedia.org/wiki/Combinatorial_chemistry); the generation of random primers for nucleic acid sequencing is an example that is familiar to molecular biologists. Basically, combinatorial chemistry is an approach to chemical synthesis that leads quickly to a large library of different entities, for essentially the same cost as the synthesis of any individual chemical entity using traditional methods.[147] While combinatorial approaches are most firmly established in the solid-phase synthesis of peptides, nucleic acids, and closely related molecules, an approach known as "click chemistry" can be used to generate a wide variety of molecular shapes, based upon molecular substrates having a limited number of reactive functional groups.[148] Click chemistry is similar to peptide or primer synthesis in that each reaction is highly reliable and the conformation of the final product is predictable. The result of such synthesis may thus be tested either computationally, experimentally, or both. Although not all possibly useful ligands can be synthesized using such highly automated approaches, a very significant number can, and pharmaceutical companies currently screen hundreds of thousands of compounds produced by combinatorial methods each year for potential therapeutic activity.

Although an ideal approach to truly personalized drug development would be based on computation alone, it is likely that it will be many decades before this is possible. High-throughput screening of compounds — a technique currently utilized in drug development — may provide a part of the pathway toward personalized drug development.[149] Typically, screening is performed by adding a ligand to a cell-based assay system in a microtiter plate, and each well receives a different compound synthesized as part of the combinatorial chemistry process. The effects of each compound are then assessed by an appropriate technique, and the most desirable compounds selected for further use; similar approaches can also be used to screen combinations of compounds.[150] Recently, a microfluidic approach has been developed for some screening applications that can perform up to 1 million assays per hour;[151] while this approach is not generalizable to every desirable screening application, it is an efficient method for identifying materials that can be produced using biological production systems, such as yeast or bacteria.

An alternative approach to screening and repurposing previously approved therapeutic agents is based on comparing disease-associated gene expression profiles with drug-induced profile alterations in cellular model system; this approach identified tricyclic antidepressants as possible therapeutics for small cell lung cancer and, if successful, could provide another avenue for improving individualized therapy without requiring the synthesis of new molecular entities.[152] It is not unreasonable to expect that further technological improvements will bring these techniques into the realm of practicality for individualized therapy, just as improvements in genomic sequencing have brought whole-genome sequencing into today's reality for molecular diagnosis.

VIII. Conclusion

In this chapter, we have shown that, although "precision medicine" has not yet found its way into routine practice outside the realm of cancer, there is an effective pathway for implementing personalized treatment at a level that is not generally anticipated in today's scientific community. The development of biobanks and disease-specific consortia provides a base of several million patients by which genetic alterations can be associated with specific diseases — a process that can be made more efficient through the use of tissue-specific transcript analysis and proteomics, both of which can be effectively employed in formalin-fixed, paraffin-embedded tissue archives that are ubiquitous in medical institutions. The large populations available for research also make possible adequate statistical power for replicating these studies. The current availability of whole genome sequencing at a reasonable speed and cost makes it possible to use this information for diagnostic purposes almost as soon as it becomes validated.

Nevertheless, the ability to rapidly sequence the genomes of millions of people to establish gene–disease associations is insufficient to enable significant advances in truly personalized medicine. While they will enable one to more effectively use therapeutic agents that are currently "on the shelf," more robust, personalized drug discovery must become possible. Achieving this goal requires further developments in the realm of systems biology, to enable identification of all possible targets for personalized intervention. The goal should be nothing less than a mathematical model of human physiology at both the microscopic and macroscopic level, although significant impact can be achieved by a more quantitative description of the function of individual organs, or even of individual gene pathways. Ultimately, however, it is only by creating the ability to design and test new therapeutic agents on a patient-by-patient basis will we truly achieve a personalized and precise medical practice. The seeds for success have already been sowed by major advances in the mathematical analysis of drug–protein interactions, the ability to rapidly design large families of potential therapies through combinatorial chemistry, and the ability to test these therapies in cellular systems using high-throughput screening techniques.

This goal, while ambitious, is not unachievable. In less than a decade, sequencing the genome has gone from an enterprise involving hundreds of scientists and many years to a fully automated process that can be carried out in a few hours. Many of the concepts that have gone into this transformation also contribute to the future transformation proposed in this chapter, which can be achieved within our children's lifetimes, if not our own.

References

1. Arranz MJ, Rivera M, Munro JC. Pharmacogenetics of response to antipsychotics in patients with schizophrenia. *CNS Drugs* 2011; **25**: 933–969.
2. Chowdhury NI, Remington G, Kennedy JL. Genetics of antipsychotic-induced side effects and agranulocytosis. *Curr Psychiatry Rep* 2011; **13**: 156–165.
3. Goldstein JI, Jarskog LF, Hilliard C, Alfirevic A, Duncan L, Fourches D, Huang H, Lek M, Neale BM, Ripke S, Shianna K, Szatkiewicz JP, Tropsha A, van den Oord EJ, Cascorbi I, Dettling M, Gazit E, Goff DC, Holden AL, Kelly DL, Malhotra AK, Nielsen J, Pirmohamed M, Rujescu D, Werge T, Levy DL, Josiassen RC, Kennedy JL, Lieberman JA, Daly MJ, Sullivan PF. Clozapine-induced agranulocytosis is associated with rare HLA-DQB1 and HLA-B alleles. *Nat Commun* 2014; **5**: 4757.
4. Opgen-Rhein C, Dettling M. Clozapine-induced agranulocytosis and its genetic determinants. *Pharmacogenomics* 2008; **9**: 1101–1111.
5. Anderson JL, Horne BD, Stevens SM, Woller SC, Samuelson KM, Mansfield JW, Robinson M, Barton S, Brunisholz K, Mower CP, Huntinghouse JA, Rollo JS, Siler D, Bair TL, Knight S, Muhlestein JB, Carlquist JF. A randomized and clinical effectiveness trial comparing two pharmacogenetic algorithms and standard care for individualizing warfarin dosing (CoumaGen-II). *Circulation* 2012; **125**: 1997–2005.
6. Baker WL, Chamberlin KW. New oral anticoagulants vs. warfarin treatment: No need for pharmacogenomics? *Clin Pharmacol Ther* 2014; **96**: 17–19.
7. Borgman MP, Pendleton RC, McMillin GA, Reynolds KK, Vazquez S, Freeman A, Wilson A, Valdes R, Jr., Linder MW. Prospective pilot trial of PerMIT versus standard anticoagulation service management of patients initiating oral anticoagulation. *Thromb Haemost* 2012; **108**: 561–569.
8. Burmester JK, Berg RL, Yale SH, Rottscheit CM, Glurich IE, Schmelzer JR, Caldwell MD. A randomized controlled trial of genotype-based Coumadin initiation. *Genet Med* 2011; **13**: 509–518.

9. Caraco Y, Blotnick S, Muszkat M. CYP2C9 genotype-guided warfarin prescribing enhances the efficacy and safety of anticoagulation: A prospective randomized controlled study. *Clin Pharmacol Ther* 2008; **83**: 460–470.
10. Cavallari LH, Nutescu EA. Warfarin pharmacogenetics: To genotype or not to genotype, that is the question. *Clin Pharmacol Ther* 2014; **96**: 22–24.
11. Eriksson N, Wadelius M. Prediction of warfarin dose: Why, when and how? *Pharmacogenomics* 2012; **13**: 429–440.
12. Ferder NS, Eby CS, Deych E, Harris JK, Ridker PM, Milligan PE, Goldhaber SZ, King CR, Giri T, McLeod HL, Glynn RJ, Gage BF. Ability of VKORC1 and CYP2C9 to predict therapeutic warfarin dose during the initial weeks of therapy. *J Thromb Haemost* 2010; **8**: 95–100.
13. Fung E, Patsopoulos NA, Belknap SM, O'Rourke DJ, Robb JF, Anderson JL, Shworak NW, Moore JH. Effect of genetic variants, especially CYP2C9 and VKORC1, on the pharmacology of warfarin. *Semin Thromb Hemost* 2012; **38**: 893–904.
14. Jonas DE, Evans JP, McLeod HL, Brode S, Lange LA, Young ML, Shilliday BB, Bardsley MM, Swinton-Jenkins NJ, Weck KE. Impact of genotype-guided dosing on anticoagulation visits for adults starting warfarin: A randomized controlled trial. *Pharmacogenomics* 2013; **14**: 1593–1603.
15. McMillin GA, Melis R, Wilson A, Strong MB, Wanner NA, Vinik RG, Peters CL, Pendleton RC. Gene-based warfarin dosing compared with standard of care practices in an orthopedic surgery population: A prospective, parallel cohort study. *Ther Drug Monit* 2010; **32**: 338–345.
16. Pirmohamed M, Burnside G, Eriksson N, Jorgensen AL, Toh CH, Nicholson T, Kesteven P, Christersson C, Wahlstrom B, Stafberg C, Zhang JE, Leathart JB, Kohnke H, Maitland-van der Zee AH, Williamson PR, Daly AK, Avery P, Kamali F, Wadelius M. A randomized trial of genotype-guided dosing of warfarin. *N Engl J Med* 2013; **369**: 2294–2303.
17. Scott SA, Lubitz SA. Warfarin pharmacogenetic trials: Is there a future for pharmacogenetic-guided dosing? *Pharmacogenomics* 2014; **15**: 719–722.
18. Stergiopoulos K, Brown DL. Genotype-guided vs clinical dosing of warfarin and its analogues: Meta-analysis of randomized clinical trials. *JAMA Intern Med* 2014; **174**: 1330–1338.
19. van Schie RM, Wadelius MI, Kamali F, Daly AK, Manolopoulos VG, de BA, Barallon R, Verhoef TI, Kirchheiner J, Haschke-Becher E, Briz M, Rosendaal FR, Redekop WK, Pirmohamed M, Maitland van der Zee AH. Genotype-guided dosing of coumarin derivatives: The European pharmacogenetics of anticoagulant therapy (EU-PACT) trial design. *Pharmacogenomics* 2009; **10**: 1687–1695.
20. Xu H, Xie X, Wang B, Chen Y, Meng T, Ma S, Wang F. Meta-analysis of efficacy and safety of genotype-guided pharmacogenetic dosing of warfarin. *Int J Cardiol* 2014; **177**: 654–657.
21. Hall-Flavin DK, Winner JG, Allen JD, Carhart JM, Proctor B, Snyder KA, Drews MS, Eisterhold LL, Geske J, Mrazek DA. Utility of integrated pharmacogenomic

testing to support the treatment of major depressive disorder in a psychiatric outpatient setting. *Pharmacogenet Genomics* 2013; **23**: 535–548
22. Hall-Flavin DK, Winner JG, Allen JD, Jordan JJ, Nesheim RS, Snyder KA, Drews MS, Eisterhold LL, Biernacka JM, Mrazek DA. Using a pharmacogenomic algorithm to guide the treatment of depression. *Transl Psychiatry* 2012; **2**: e172.
23. Salloum NC, McCarthy MJ, Leckband SG, Kelsoe JR. Towards the clinical implementation of pharmacogenetics in bipolar disorder. *BMC Med* 2014; **12**: 90.
24. Krier J, Barfield R, Green RC, Kraft P. Reclassification of genetic-based risk predictions as GWAS data accumulate. *Genome Med* 2016; **8**: 20.
25. Unick JL, Neiberg RH, Hogan PE, Cheskin LJ, Dutton GR, Jeffery R, Nelson JA, Pi-Sunyer X, West DS, Wing RR. Weight change in the first 2 months of a lifestyle intervention predicts weight changes 8 years later. *Obesity* (Silver Spring) 2015; **23**: 1353–1356.
26. Unick JL, Leahey T, Kent K, Wing RR. Examination of whether early weight loss predicts 1-year weight loss among those enrolled in an Internet-based weight loss program. *Int J Obes* (Lond) 2015; **39**: 1558–1560.
27. Unick JL, Beavers D, Jakicic JM, Kitabchi AE, Knowler WC, Wadden TA, Wing RR. Effectiveness of lifestyle interventions for individuals with severe obesity and type 2 diabetes: Results from the Look AHEAD trial. *Diabetes Care* 2011; **34**: 2152–2157.
28. Unick JL, Jakicic JM, Marcus BH. Contribution of behavior intervention components to 24-month weight loss. *Med Sci Sports Exerc* 2010; **42**: 745–753.
29. Bruns SB, Ioannidis JP. p-Curve and p-Hacking in Observational Research. *PLoS One* 2016; **11**: e0149144.
30. Ioannidis JP. An epidemic of false claims. Competition and conflicts of interest distort too many medical findings. *Sci Am* 2011; **304**: 16.
31. Ioannidis JP. To replicate or not to replicate: The case of pharmacogenetic studies: Have pharmacogenomics failed, or do they just need larger-scale evidence and more replication? *Circ Cardiovasc Gene*t 2013; **6**: 413–418.
32. Ioannidis JP. How to make more published research true. *PLoS Med* 2014; **11**: e1001747.
33. Ioannidis JP. Discussion: Why "An estimate of the science-wise false discovery rate and application to the top medical literature" is false. *Biostatistics* 2014; **15**: 28–36.
34. Manrai AK, Patel CJ, Gehlenborg N, Tatonetti NP, Ioannidis JP, Kohane IS. Methods to enhance the reproducibility of precision medicine. *Pac Symp Biocomput* 2016; **21**: 180–182.
35. Ioannidis JP, Ntzani EE, Trikalinos TA, Contopoulos-Ioannidis DG. Replication validity of genetic association studies. *Nat Genet* 2001; **29**: 306–309.
36. Ioannidis JP, Trikalinos TA, Ntzani EE, Contopoulos-Ioannidis DG. Genetic associations in large versus small studies: An empirical assessment. *Lancet* 2003; **361**: 567–571.
37. Ioannidis JP. Genetic associations: False or true? *Trends Mol Med* 2003; **9**: 135–138.

38. Ioannidis JP. Contradicted and initially stronger effects in highly cited clinical research. *JAMA* 2005; **294**: 218–228.
39. Ioannidis JP. Why most published research findings are false. *PLoS Med* 2005; **2**: e124.
40. Points to consider for informed consent for genome/exome sequencing. *Genet Med* 2013; **15**: 748–749.
41. ACMG policy statement: Updated recommendations regarding analysis and reporting of secondary findings in clinical genome-scale sequencing. *Genet Med* 2015; **17**: 68–69.
42. Allyse M, Michie M. Not-so-incidental findings: The ACMG recommendations on the reporting of incidental findings in clinical whole genome and whole exome sequencing. *Trends Biotechnol* 2013; **31**: 439–441.
43. Bennette CS, Gallego CJ, Burke W, Jarvik GP, Veenstra DL. The cost-effectiveness of returning incidental findings from next-generation genomic sequencing. *Genet Med* 2015; **17**: 587–595.
44. Blackburn HL, Schroeder B, Turner C, Shriver CD, Ellsworth DL, Ellsworth RE. Management of incidental findings in the era of next-generation sequencing. *curr genomics* 2015; **16**: 159–174.
45. Hofmann B. Incidental findings of uncertain significance: To know or not to know — that is not the question. *BMC Med Ethics* 2016; **17**: 13.
46. Park JY, Clark P, Londin E, Sponziello M, Kricka LJ, Fortina P. Clinical exome performance for reporting secondary genetic findings. *Clin Chem* 2015; **61**: 213–220.
47. Rosenblatt DS. Who's on first in exome and whole genome sequencing? Is it the patient or the incidental findings? *Mol Genet Metab* 2013; **110**: 1–2.
48. Genome-wide association study of 14,000 cases of seven common diseases and 3,000 shared controls. *Nature* 2007; **447**: 661–678.
49. Psaty BM, Sitlani C. The Cohorts for Heart and Aging Research in Genomic Epidemiology (CHARGE) Consortium as a model of collaborative science. *Epidemiology* 2013; **24**: 346–348.
50. Psaty BM, O'Donnell CJ, Gudnason V, Lunetta KL, Folsom AR, Rotter JI, Uitterlinden AG, Harris TB, Witteman JC, Boerwinkle E. Cohorts for Heart and Aging Research in Genomic Epidemiology (CHARGE) Consortium: Design of prospective meta-analyses of genome-wide association studies from 5 cohorts. *Circ Cardiovasc Genet* 2009; **2**: 73–80.
51. Bis JC, Sitlani C, Irvin R, Avery CL, Smith AV, Sun F, Evans DS, Musani SK, Li X, Trompet S, Krijthe BP, Harris TB, Quibrera PM, Brody JA, Demissie S, Davis BR, Wiggins KL, Tranah GJ, Lange LA, Sotoodehnia N, Stott DJ, Franco OH, Launer LJ, Sturmer T, Taylor KD, Cupples LA, Eckfeldt JH, Smith NL, Liu Y, Wilson JG, Heckbert SR, Buckley BM, Ikram MA, Boerwinkle E, Chen YD, de Craen AJ, Uitterlinden AG, Rotter JI, Ford I, Hofman A, Sattar N, Slagboom PE, Westendorp RG, Gudnason V, Vasan RS, Lumley T, Cummings SR, Taylor HA, Jr., Post W, Jukema JW, Stricker BH, Whitsel EA, Psaty BM, Arnett D. Drug–gene interactions

of antihypertensive medications and risk of incident cardiovascular disease: A pharmacogenomics study from the CHARGE consortium. *PLoS One* 2015; **10**: e0140496.

52. Morrison AC, Bis JC, Hwang SJ, Ehret GB, Lumley T, Rice K, Muzny D, Gibbs RA, Boerwinkle E, Psaty BM, Chakravarti A, Levy D. Sequence analysis of six blood pressure candidate regions in 4, 178 individuals: The Cohorts for Heart and Aging Research in Genomic Epidemiology (CHARGE) targeted sequencing study. *PLoS One* 2014; **9**: e109155.
53. Bis JC, White CC, Franceschini N, Brody J, Zhang X, Muzny D, Santibanez J, Gibbs R, Liu X, Lin H, Boerwinkle E, Psaty BM, North KE, Cupples LA, O'Donnell CJ. Sequencing of 2 subclinical atherosclerosis candidate regions in 3669 individuals: Cohorts for Heart and Aging Research in Genomic Epidemiology (CHARGE) Consortium Targeted Sequencing Study. Circ *Cardiovasc Genet* 2014; **7**: 359–364.
54. Bis JC, Kavousi M, Franceschini N, Isaacs A, Abecasis GR, Schminke U, Post WS, Smith AV, Cupples LA, Markus HS, Schmidt R, Huffman JE, Lehtimaki T, Baumert J, Munzel T, Heckbert SR, Dehghan A, North K, Oostra B, Bevan S, Stoegerer EM, Hayward C, Raitakari O, Meisinger C, Schillert A, Sanna S, Volzke H, Cheng YC, Thorsson B, Fox CS, Rice K, Rivadeneira F, Nambi V, Halperin E, Petrovic KE, Peltonen L, Wichmann HE, Schnabel RB, Dorr M, Parsa A, Aspelund T, Demissie S, Kathiresan S, Reilly MP, Taylor K, Uitterlinden A, Couper DJ, Sitzer M, Kahonen M, Illig T, Wild PS, Orru M, Ludemann J, Shuldiner AR, Eiriksdottir G, White CC, Rotter JI, Hofman A, Seissler J, Zeller T, Usala G, Ernst F, Launer LJ, D'Agostino RB, Sr., O'Leary DH, Ballantyne C, Thiery J, Ziegler A, Lakatta EG, Chilukoti RK, Harris TB, Wolf PA, Psaty BM, Polak JF, Li X, Rathmann W, Uda M, Boerwinkle E, Klopp N, Schmidt H, Wilson JF, Viikari J, Koenig W, Blankenberg S, Newman AB, Witteman J, Heiss G, Duijn C, Scuteri A, Homuth G, Mitchell BD, Gudnason V, O'Donnell CJ. Meta-analysis of genome-wide association studies from the CHARGE consortium identifies common variants associated with carotid intima media thickness and plaque. *Nat Gene*t 2011; **43**: 940–947.
55. Meyer TE, Verwoert GC, Hwang SJ, Glazer NL, Smith AV, van Rooij FJ, Ehret GB, Boerwinkle E, Felix JF, Leak TS, Harris TB, Yang Q, Dehghan A, Aspelund T, Katz R, Homuth G, Kocher T, Rettig R, Ried JS, Gieger C, Prucha H, Pfeufer A, Meitinger T, Coresh J, Hofman A, Sarnak MJ, Chen YD, Uitterlinden AG, Chakravarti A, Psaty BM, van Duijn CM, Kao WH, Witteman JC, Gudnason V, Siscovick DS, Fox CS, Kottgen A. Genome-wide association studies of serum magnesium, potassium, and sodium concentrations identify six Loci influencing serum magnesium levels. *PLoS Gene*t 2010; 6.
56. Gottlieb DJ, Hek K, Chen TH, Watson NF, Eiriksdottir G, Byrne EM, Cornelis M, Warby SC, Bandinelli S, Cherkas L, Evans DS, Grabe HJ, Lahti J, Li M, Lehtimaki T, Lumley T, Marciante KD, Perusse L, Psaty BM, Robbins J, Tranah GJ, Vink JM, Wilk JB, Stafford JM, Bellis C, Biffar R, Bouchard C, Cade B, Curhan GC, Eriksson JG, Ewert R, Ferrucci L, Fulop T, Gehrman PR, Goodloe R, Harris TB,

Heath AC, Hernandez D, Hofman A, Hottenga JJ, Hunter DJ, Jensen MK, Johnson AD, Kahonen M, Kao L, Kraft P, Larkin EK, Lauderdale DS, Luik AI, Medici M, Montgomery GW, Palotie A, Patel SR, Pistis G, Porcu E, Quaye L, Raitakari O, Redline S, Rimm EB, Rotter JI, Smith AV, Spector TD, Teumer A, Uitterlinden AG, Vohl MC, Widen E, Willemsen G, Young T, Zhang X, Liu Y, Blangero J, Boomsma DI, Gudnason V, Hu F, Mangino M, Martin NG, O'Connor GT, Stone KL, Tanaka T, Viikari J, Gharib SA, Punjabi NM, Raikkonen K, Volzke H, Mignot E, Tiemeier H. Novel loci associated with usual sleep duration: The CHARGE Consortium Genome-Wide Association Study. *Mol Psychiatry* 2015; **20**: 1232–1239.

57. Broer L, Buchman AS, Deelen J, Evans DS, Faul JD, Lunetta KL, Sebastiani P, Smith JA, Smith AV, Tanaka T, Yu L, Arnold AM, Aspelund T, Benjamin EJ, De Jager PL, Eirkisdottir G, Evans DA, Garcia ME, Hofman A, Kaplan RC, Kardia SL, Kiel DP, Oostra BA, Orwoll ES, Parimi N, Psaty BM, Rivadeneira F, Rotter JI, Seshadri S, Singleton A, Tiemeier H, Uitterlinden AG, Zhao W, Bandinelli S, Bennett DA, Ferrucci L, Gudnason V, Harris TB, Karasik D, Launer LJ, Perls TT, Slagboom PE, Tranah GJ, Weir DR, Newman AB, van Duijn CM, Murabito JM. GWAS of longevity in CHARGE consortium confirms APOE and FOXO3 candidacy. *J Gerontol A Biol Sci Med Sci* 2015; **70**: 110–118.
58. Newman AB, Walter S, Lunetta KL, Garcia ME, Slagboom PE, Christensen K, Arnold AM, Aspelund T, Aulchenko YS, Benjamin EJ, Christiansen L, D'Agostino RB, Sr., Fitzpatrick AL, Franceschini N, Glazer NL, Gudnason V, Hofman A, Kaplan R, Karasik D, Kelly-Hayes M, Kiel DP, Launer LJ, Marciante KD, Massaro JM, Miljkovic I, Nalls MA, Hernandez D, Psaty BM, Rivadeneira F, Rotter J, Seshadri S, Smith AV, Taylor KD, Tiemeier H, Uh HW, Uitterlinden AG, Vaupel JW, Walston J, Westendorp RG, Harris TB, Lumley T, van Duijn CM, Murabito JM. A meta-analysis of four genome-wide association studies of survival to age 90 years or older: The Cohorts for Heart and Aging Research in Genomic Epidemiology Consortium. *J Gerontol A Biol Sci Med Sci* 2010; **65**: 478–487.
59. Sudlow C, Gallacher J, Allen N, Beral V, Burton P, Danesh J, Downey P, Elliott P, Green J, Landray M, Liu B, Matthews P, Ong G, Pell J, Silman A, Young A, Sprosen T, Peakman T, Collins R. UK biobank: An open access resource for identifying the causes of a wide range of complex diseases of middle and old age. *PLoS Med* 2015; **12**: e1001779.
60. Allen NE, Sudlow C, Peakman T, Collins R. UK biobank data: Come and get it. *Sci Transl Med* 2014; **6**: 224ed4.
61. Collins R. What makes UK Biobank special? *Lancet* 2012; **379**: 1173–1174.
62. Zielhuis GA. Biobanking for epidemiology. *Public Health* 2012; **126**: 214–216.
63. Shen L, Hoffmann TJ, Melles RB, Sakoda LC, Kvale MN, Banda Y, Schaefer C, Risch N, Jorgenson E. Differences in the genetic susceptibility to age-related macular degeneration clinical subtypes. *Invest Ophthalmol Vis Sci* 2015; **56**: 4290–4299.
64. Kvale MN, Hesselson S, Hoffmann TJ, Cao Y, Chan D, Connell S, Croen LA, Dispensa BP, Eshragh J, Finn A, Gollub J, Iribarren C, Jorgenson E, Kushi LH, Lao R,

Lu Y, Ludwig D, Mathauda GK, McGuire WB, Mei G, Miles S, Mittman M, Patil M, Quesenberry CP, Jr., Ranatunga D, Rowell S, Sadler M, Sakoda LC, Shapero M, Shen L, Shenoy T, Smethurst D, Somkin CP, Van Den Eeden SK, Walter L, Wan E, Webster T, Whitmer RA, Wong S, Zau C, Zhan Y, Schaefer C, Kwok PY, Risch N. Genotyping Informatics and Quality Control for 100,000 Subjects in the Genetic Epidemiology Research on Adult Health and Aging (GERA) Cohort. *Genetics* 2015; **200**: 1051–1060.

65. Lapham K, Kvale MN, Lin J, Connell S, Croen LA, Dispensa BP, Fang L, Hesselson S, Hoffmann TJ, Iribarren C, Jorgenson E, Kushi LH, Ludwig D, Matsuguchi T, McGuire WB, Miles S, Quesenberry CP, Jr., Rowell S, Sadler M, Sakoda LC, Smethurst D, Somkin CP, Van Den Eeden SK, Walter L, Whitmer RA, Kwok PY, Risch N, Schaefer C, Blackburn EH. Automated Assay of Telomere Length Measurement and Informatics for 100,000 Subjects in the Genetic Epidemiology Research on Adult Health and Aging (GERA) Cohort. *Genetics* 2015; **200**: 1061–1072
66. Banda Y, Kvale MN, Hoffmann TJ, Hesselson SE, Ranatunga D, Tang H, Sabatti C, Croen LA, Dispensa BP, Henderson M, Iribarren C, Jorgenson E, Kushi LH, Ludwig D, Olberg D, Quesenberry CP, Jr., Rowell S, Sadler M, Sakoda LC, Sciortino S, Shen L, Smethurst D, Somkin CP, Van Den Eeden SK, Walter L, Whitmer RA, Kwok PY, Schaefer C, Risch N. Characterizing Race/Ethnicity and Genetic Ancestry for 100,000 Subjects in the Genetic Epidemiology Research on Adult Health and Aging (GERA) Cohort. *Genetics* 2015; **200**: 1285–1295.
67. Wen CC, Yee SW, Liang X, Hoffmann TJ, Kvale MN, Banda Y, Jorgenson E, Schaefer C, Risch N, Giacomini KM. Genome-wide association study identifies ABCG2 (BCRP) as an allopurinol transporter and a determinant of drug response. *Clin Pharmacol Ther* 2015; **97**: 518–525.
68. Chen Z, Chen J, Collins R, Guo Y, Peto R, Wu F, Li L. China Kadoorie Biobank of 0.5 million people: Survey methods, baseline characteristics and long-term follow-up. *Int J Epidemiol* 2011; **40**: 1652–1666.
69. Chen Z, Lee L, Chen J, Collins R, Wu F, Guo Y, Linksted P, Peto R. Cohort profile: The Kadoorie study of chronic disease in China (KSCDC). *Int J Epidemiol* 2005; **34**: 1243–1249.
70. Li L, Guo Y, Chen Z, Chen J, Peto R. Epidemiology and the control of disease in China, with emphasis on the Chinese Biobank study. *Public Health* 2012; **126**: 210–213.
71. Chen Z, Peto R, Zhou M, Iona A, Smith M, Yang L, Guo Y, Chen Y, Bian Z, Lancaster G, Sherliker P, Pang S, Wang H, Su H, Wu M, Wu X, Chen J, Collins R, Li L. Contrasting male and female trends in tobacco-attributed mortality in China: Evidence from successive nationwide prospective cohort studies. *Lancet* 2015; **386**: 1447–1456.
72. Du H, Li L, Whitlock G, Bennett D, Guo Y, Bian Z, Chen J, Sherliker P, Huang Y, Zhang N, Zheng X, Li Z, Hu R, Collins R, Peto R, Chen Z. Patterns and socio-demo-

graphic correlates of domain-specific physical activities and their associations with adiposity in the China Kadoorie Biobank study. *BMC Public Health* 2014; **14**: 826.

73. Du H, Bennett D, Li L, Whitlock G, Guo Y, Collins R, Chen J, Bian Z, Hong LS, Feng S, Chen X, Chen L, Zhou R, Mao E, Peto R, Chen Z. Physical activity and sedentary leisure time and their associations with BMI, waist circumference, and percentage body fat in 0.5 million adults: The China Kadoorie Biobank study. *Am J Clin Nutr* 2013; **97**: 487–496.
74. Kurmi OP, Li L, Wang J, Millwood IY, Chen J, Collins R, Guo Y, Bian Z, Li J, Chen B, Xie K, Jia W, Gao Y, Peto R, Chen Z. COPD and its association with smoking in the Mainland China: A cross-sectional analysis of 0.5 million men and women from ten diverse areas. *Int J Chron Obstruct Pulmon Dis* 2015; **10**: 655–665.
75. Kurmi OP, Li L, Smith M, Augustyn M, Chen J, Collins R, Guo Y, Han Y, Qin J, Xu G, Wang J, Bian Z, Zhou G, Davis K, Peto R, Chen Z, Li L, Chen Z, Chen J, Collins R, Peto R, Chen Z, Lancaster G, Yang X, Williams A, Smith M, Yang L, Chang Y, Millwood I, Chen Y, Lewington S, Sansome S, Walters R, Kurmi O, Guo Y, Bian Z, Hou C, Tan Y, Wang Z, Cai X, Zhou H, Chen X, Pang Z, Li S, Wang S, Lv S, Zhao Z, Liu S, Pang Z, Yang L, He H, Yu B, Wang S, Wang H, Chen C, Zheng X, Hu X, Zhou M, Wu M, Tao R, Wang Y, Hu Y, Ma L, Zhou R, Tang Z, Chen N, Huang Y, Li M, Gan Z, Meng J, Qin J, Wu X, Zhang N, Luo G, Que X, Chen X, Ge P, Ren X, Dong C, Zhang H, Mao E, Li Z, Zhou G, Feng S, Gao Y, He T, Jiang L, Sun H, Yu M, Su D, Lu F, Qian Y, Shi K, Han Y, Chen L, Li G, Liu H, Yin L, Xiong Y, Tan Z, Jia W. Regional variations in the prevalence and misdiagnosis of air flow obstruction in China: Baseline results from a prospective cohort of the China Kadoorie Biobank (CKB). *BMJ Open Respir Res* 2014; **1**: e000025.
76. Lewington S, Li L, Sherliker P, Guo Y, Millwood I, Bian Z, Whitlock G, Yang L, Collins R, Chen J, Wu X, Wang S, Hu Y, Jiang L, Yang L, Lacey B, Peto R, Chen Z. Seasonal variation in blood pressure and its relationship with outdoor temperature in 10 diverse regions of China: The China Kadoorie Biobank. *J Hypertens* 2012; **30**: 1383–1391.
77. Lewington S, Li L, Murugasen S, Hong LS, Yang L, Guo Y, Bian Z, Collins R, Chen J, He H, Wu M, He T, Ren X, Meng J, Peto R, Chen Z. Temporal trends of main reproductive characteristics in ten urban and rural regions of China: The China Kadoorie biobank study of 300,000 women. *Int J Epidemiol* 2014; **43**: 1252–1262.
78. Mezuk B, Chen Y, Yu C, Guo Y, Bian Z, Collins R, Chen J, Pang Z, Wang H, Peto R, Que X, Zhang H, Tan Z, Kendler KS, Li L, Chen Z. Depression, anxiety, and prevalent diabetes in the Chinese population: Findings from the China Kadoorie Biobank of 0.5 million people. *J Psychosom Res* 2013; **75**: 511–517.
79. Smith M, Li L, Augustyn M, Kurmi O, Chen J, Collins R, Guo Y, Han Y, Qin J, Xu G, Wang J, Bian Z, Zhou G, Peto R, Chen Z. Prevalence and correlates of airflow obstruction in approximately 317,000 never-smokers in China. *Eur Respir J* 2014; **44**: 66–77.

80. Yang L, Zhou M, Sherliker P, Cai Y, Peto R, Wang L, Millwood I, Smith M, Hu Y, Yang G, Chen Z. Alcohol drinking and overall and cause-specific mortality in China: Nationally representative prospective study of 220,000 men with 15 years of follow-up. *Int J Epidemiol* 2012; **41**: 1101–1113.
81. Yang L, Li L, Lewington S, Guo Y, Sherliker P, Bian Z, Collins R, Peto R, Liu Y, Yang R, Zhang Y, Li G, Liu S, Chen Z. Outdoor temperature, blood pressure, and cardiovascular disease mortality among 23,000 individuals with diagnosed cardiovascular diseases from China. *Eur Heart J* 2015; **36**: 1178–1185.
82. Millwood IY, Bennett DA, Walters RG, Clarke R, Waterworth D, Johnson T, Chen Y, Yang L, Guo Y, Bian Z, Hacker A, Yeo A, Parish S, Hill MR, Chissoe S, Peto R, Cardon L, Collins R, Li L, Chen Z. Lipoprotein-associated phospholipase A2 loss-of-function variant and risk of vascular diseases in 90,000 Chinese Adults. *J Am Coll Cardiol* 2016; **67**: 230–231.
83. Denny JC, Ritchie MD, Basford MA, Pulley JM, Bastarache L, Brown-Gentry K, Wang D, Masys DR, Roden DM, Crawford DC. PheWAS: Demonstrating the feasibility of a phenome-wide scan to discover gene-disease associations. *Bioinformatics* 2010; **26**: 1205–1210.
84. Denny JC, Ritchie MD, Crawford DC, Schildcrout JS, Ramirez AH, Pulley JM, Basford MA, Masys DR, Haines JL, Roden DM. Identification of genomic predictors of atrioventricular conduction: Using electronic medical records as a tool for genome science. *Circulation* 2010; **122**: 2016–2021.
85. Dumitrescu L, Ritchie MD, Brown-Gentry K, Pulley JM, Basford M, Denny JC, Oksenberg JR, Roden DM, Haines JL, Crawford DC. Assessing the accuracy of observer-reported ancestry in a biorepository linked to electronic medical records. *Genet Med* 2010; **12**: 648–650.
86. Pulley J, Clayton E, Bernard GR, Roden DM, Masys DR. Principles of human subjects protections applied in an opt-out, de-identified biobank. *Clin Transl Sci* 2010; **3**: 42–48.
87. Ritchie MD, Denny JC, Crawford DC, Ramirez AH, Weiner JB, Pulley JM, Basford MA, Brown-Gentry K, Balser JR, Masys DR, Haines JL, Roden DM: Robust replication of genotype-phenotype associations across multiple diseases in an electronic medical record. *Am J Hum Genet* 2010; **86**: 560–572.
88. Brothers KB, Morrison DR, Clayton EW. Two large-scale surveys on community attitudes toward an opt-out biobank. *Am J Med Genet A* 2011; 155A: 2982–2990.
89. Langanke M, Brothers KB, Erdmann P, Weinert J, Krafczyk-Korth J, Dorr M, Hoffmann W, Kroemer HK, Assel H. Comparing different scientific approaches to personalized medicine: research ethics and privacy protection. *Per Med* 2011; **8**: 437–444.
90. Pendergrass S, Dudek SM, Roden DM, Crawford DC, Ritchie MD. Visual integration of results from a large DNA biobank (BioVU) using synthesis-view. *Pac Symp Biocomput* 2011;265–275.
91. Birdwell KA, Grady B, Choi L, Xu H, Bian A, Denny JC, Jiang M, Vranic G, Basford M, Cowan JD, Richardson DM, Robinson MP, Ikizler TA, Ritchie MD,

Stein CM, Haas DW. The use of a DNA biobank linked to electronic medical records to characterize pharmacogenomic predictors of tacrolimus dose requirement in kidney transplant recipients. *Pharmacogenet Genomics* 2012; **22**: 32–42.

92. Delaney JT, Ramirez AH, Bowton E, Pulley JM, Basford MA, Schildcrout JS, Shi Y, Zink R, Oetjens M, Xu H, Cleator JH, Jahangir E, Ritchie MD, Masys DR, Roden DM, Crawford DC, Denny JC. Predicting clopidogrel response using DNA samples linked to an electronic health record. *Clin Pharmacol Ther* 2012; **91**: 257–263.
93. Oetjens MT, Denny JC, Ritchie MD, Gillani NB, Richardson DM, Restrepo NA, Pulley JM, Dilks HH, Basford MA, Bowton E, Masys DR, Wilke RA, Roden DM, Crawford DC. Assessment of a pharmacogenomic marker panel in a polypharmacy population identified from electronic medical records. *Pharmacogenomics* 2013; **14**: 735–744.
94. Oetjens M, Bush WS, Birdwell KA, Dilks HH, Bowton EA, Denny JC, Wilke RA, Roden DM, Crawford DC. Utilization of an EMR-biorepository to identify the genetic predictors of calcineurin-inhibitor toxicity in heart transplant recipients. *Pac Symp Biocomput* 2014;253–264.
95. Wei WQ, Feng Q, Weeke P, Bush W, Waitara MS, Iwuchukwu OF, Roden DM, Wilke RA, Stein CM, Denny JC. Creation and validation of an EMR-based algorithm for identifying major adverse cardiac events while on statins. *AMIA Jt Summits Transl Sci Proc* 2014; **2014**: 112–119.
96. Dahlin A, Denny J, Roden DM, Brilliant MH, Ingram C, Kitchner TE, Linneman JG, Shaffer CM, Weeke P, Xu H, Kubo M, Tamari M, Clemmer GL, Ziniti J, McGeachie MJ, Tantisira KG, Weiss ST, Wu AC. CMTR1 is associated with increased asthma exacerbations in patients taking inhaled corticosteroids. *Immun Inflamm Dis* 2015; **3**: 350–359.
97. Laper SM, Restrepo NA, Crawford DC. The challenges in using electronic health records for pharmacogenomics and precision medicine research. *Pac Symp Biocomput* 2016; **21**: 369–380.
98. White CC, Feng Q, Cupples LA, Gainer JV, Dawson EP, Wilke RA, Brown NJ. CYP4A11 variant is associated with high-density lipoprotein cholesterol in women. *Pharmacogenomics J* 2013; **13**: 44–51.
99. Friedman EA, Texeira L, Delaney J, Weeke PE, Lynch DR, Jr., Kasasbeh E, Song Y, Harrell FE, Jr., Denny JC, Hamm HE, Roden DM, Cleator JH. Evaluation of the F2R IVS-14A/T PAR1 polymorphism with subsequent cardiovascular events and bleeding in patients who have undergone percutaneous coronary intervention. *J Thromb Thrombolysis* 2016; **41**(4): 656–662.
100. Weeke P, Denny JC, Basterache L, Shaffer C, Bowton E, Ingram C, Darbar D, Roden DM. Examining rare and low-frequency genetic variants previously associated with lone or familial forms of atrial fibrillation in an electronic medical record system: A cautionary note. *Circ Cardiovasc Genet* 2015; **8**: 58–63

101. Edwards TL, Michels KA, Hartmann KE, Velez Edwards DR. BET1L and TNRC6B associate with uterine fibroid risk among European Americans. *Hum Genet* 2013; **132**: 943–953.
102. Malinowski J, Farber-Eger E, Crawford DC. Development of a data-mining algorithm to identify ages at reproductive milestones in electronic medical records. *Pac Symp Biocomput* 2014; 376–387.
103. Aissani B, Zhang K, Wiener H. Evaluation of GWAS candidate susceptibility loci for uterine leiomyoma in the multi-ethnic NIEHS uterine fibroid study. *Front Genet* 2015; **6**: 241.
104. Gaziano JM, Concato J, Brophy M, Fiore L, Pyarajan S, Breeling J, Whitbourne S, Deen J, Shannon C, Humphries D, Guarino P, Aslan M, Anderson D, LaFleur R, Hammond T, Schaa K, Moser J, Huang G, Muralidhar S, Przygodzki R, O'Leary TJ. Million Veteran Program: A mega-biobank to study genetic influences on health and disease. *J Clin Epidemiol* 2016; **70**: 214–223.
105. Fowler CB, Cunningham RE, O'Leary TJ, Mason JT. 'Tissue surrogates' as a model for archival formalin-fixed paraffin-embedded tissues. *Lab Invest* 2007; **87**: 836–846.
106. Fowler CB, Cunningham RE, Waybright TJ, Blonder J, Veenstra TD, O'Leary TJ, Mason JT. Elevated hydrostatic pressure promotes protein recovery from formalin-fixed, paraffin-embedded tissue surrogates. *Lab Invest* 2008; **88**: 185–195.
107. Fowler CB, Chesnick IE, Moore CD, O'Leary TJ, Mason JT. Elevated pressure improves the extraction and identification of proteins recovered from formalin-fixed, paraffin-embedded tissue surrogates. *PLoS One* 2010; **5**: e14253.
108. Fowler CB, O'Leary TJ, Mason JT. Protein mass spectrometry applications on FFPE tissue sections. *Methods Mol Biol* 2011; **724**: 281–295.
109. Fowler CB, Waybright TJ, Veenstra TD, O'Leary TJ, Mason JT. Pressure-assisted protein extraction: A novel method for recovering proteins from archival tissue for proteomic analysis. *J Proteome Res* 2012; **11**: 2602–2608.
110. Shi SR, Taylor CR, Fowler CB, Mason JT. Complete solubilization of formalin-fixed, paraffin-embedded tissue may improve proteomic studies. *Proteomics Clin Appl* 2013; **7**: 264–272.
111. Fowler CB, O'Leary TJ, Mason JT. Toward improving the proteomic analysis of formalin-fixed, paraffin-embedded tissue. *Expert Rev Proteomics* 2013; **10**: 389–400.
112. Fowler CB, O'Leary TJ, Mason JT. Improving the proteomic analysis of archival tissue by using pressure-assisted protein extraction: A mechanistic approach. *J Proteomics Bioinform* 2014; **7**: 151–157.
113. Berlin K, O'Leary DP, Fushman D. Fast approximations of the rotational diffusion tensor and their application to structural assembly of molecular complexes. *Proteins* 2011; **79**: 2268–2281.

114. Berlin K, O'Leary DP, Fushman D. Improvement and analysis of computational methods for prediction of residual dipolar couplings. *J Magn Reson* 2009; **201**: 25–33.
115. Dunlavy DM, O'Leary DP, Klimov D, Thirumalai D. HOPE: A homotopy optimization method for protein structure prediction. *J Comput Biol* 2005; **12**: 1275–1288.
116. Zotenko E, O'Leary DP, Przytycka TM. Secondary structure spatial conformation footprint: A novel method for fast protein structure comparison and classification. *BMC Struct Biol* 2006; **6**: 12.
117. Zotenko E, Dogan RI, Wilbur WJ, O'Leary DP, Przytycka TM. Structural footprinting in protein structure comparison: the impact of structural fragments. *BMC Struct Biol* 2007; **7**: 53.
118. Cecchini M, Krivov SV, Spichty M, Karplus M. Calculation of free-energy differences by confinement simulations. Application to peptide conformers. *J Phys Chem B* 2009; **113**: 9728–9740.
119. Karplus M. 2 His Nobel Lecture and his speculative commentary on "What does the Future Hold"? *J Biomol Struct Dyn* 2015; 33 (Suppl 1): 1.
120. Karplus M. Behind the folding funnel diagram. *Nat Chem Biol* 2011; **7**: 401–404.
121. Karplus M. Development of multiscale models for complex chemical systems: From H+H(2) to biomolecules (Nobel Lecture). *Angew Chem Int Ed Engl* 2014; **53**: 9992–10005.
122. Ovchinnikov V, Cecchini M, Karplus M. A simplified confinement method for calculating absolute free energies and free energy and entropy differences. *J Phys Chem B* 2013; **117**: 750–762.
123. Fenollar-Ferrer C, Stockner T, Schwarz TC, Pal A, Gotovina J, Hofmaier T, Jayaraman K, Adhikary S, Kudlacek O, Mehdipour AR, Tavoulari S, Rudnick G, Singh SK, Konrat R, Sitte HH, Forrest LR. Structure and regulatory interactions of the cytoplasmic terminal domains of serotonin transporter. *Biochemistry* 2014; **53**: 5444–5460.
124. Yadav S, Gupta S, Selvaraj C, Doharey PK, Verma A, Singh SK, Saxena JK. In silico and *in vitro* studies on the protein-protein interactions between Brugia malayi immunomodulatory protein calreticulin and human C1q. *PLoS One* 2014; **9**: e106413.
125. Spiga E, Degiacomi MT, Dal PM. New strategies for integrative dynamic modeling of macromolecular assembly. *Adv Protein Chem Struct Biol* 2014; **96**: 77–111.
126. Sivakamavalli J, Selvaraj C, Singh SK, Vaseeharan B. Modelling of macromolecular proteins in prophenoloxidase cascade through experimental and computational approaches. *Biotechnol Appl Biochem* 2016; **63**(6): 779–788.
127. Fukunishi Y, Nakamura H. Definition of drug-likeness for compound affinity. *J Chem Inf Model* 2011; **51**: 1012–1016.
128. Fukunishi Y, Nakamura H. Statistical estimation of the protein-ligand binding free energy based on direct protein-ligand interaction obtained by molecular dynamics simulation. *Pharmaceuticals (Basel)* 2012; **5**: 1064–1079.

129. Fukunishi Y, Nakamura H. Integration of ligand-based drug screening with structure-based drug screening by combining maximum volume overlapping score with ligand docking. *Pharmaceuticals (Basel)* 2012; **5**: 1332–1345.
130. Fukunishi Y, Nakamura H. Improved estimation of protein-ligand binding free energy by using the ligand-entropy and mobility of water molecules. *Pharmaceuticals (Basel)* 2013; **6**: 604–622.
131. Nguyen H, Tran T, Fukunishi Y, Higo J, Nakamura H, Le L. Computational study of drug binding affinity to influenza a neuraminidase using smooth reaction path generation (SRPG) method. *J Chem Inf Model* 2015; **55**: 1936–1943.
132. Ballante F, Marshall GR. An automated strategy for binding-pose selection and docking assessment in structure-based drug design. *J Chem Inf Model* 2016; **56**: 54–72.
133. Wong DT, Horng JS, Bymaster FP, Hauser KL, Molloy BB. A selective inhibitor of serotonin uptake: Lilly 110140, 3-(p-trifluoromethylphenoxy)-N-methyl-3-phenylpropylamine. *Life Sci* 1974; **15**: 471–479.
134. Wong DT, Bymaster FP, Engleman EA. Prozac (fluoxetine, Lilly 110140), the first selective serotonin uptake inhibitor and an antidepressant drug: Twenty years since its first publication. *Life Sci* 1995; **57**: 411–441.
135. Rahuel J, Rasetti V, Maibaum J, Rueger H, Goschke R, Cohen N-C, Stutz S, Cumin F, Fuhrer W, Wood JM, Grutter MG. Structure-based drug design: the discovery of novel nonpeptide orally active inhibitors of human renin. *Chemistry and Biology* 2000; **7**: 493–504.
136. Gradman AH, Schmieder RE, Lins RL, Nussberger J, Chiang Y, Bedigian MP. Aliskiren, a novel orally effective renin inhibitor, provides dose-dependent antihypertensive efficacy and placebo-like tolerability in hypertensive patients. *Circulation* 2005; **111**: 1012–1018.
137. Gradman AH, Vivas Y. New drugs for hypertension: What do they offer? *Curr Hypertens Rep* 2006; **8**: 425–432.
138. Gradman AH, Traub D. The efficacy of aliskiren, a direct renin inhibitor, in the treatment of hypertension. *Rev Cardiovasc Med* 2007; 8 (Suppl 2): S22–S30.
139. Nussberger J, Gradman AH, Schmieder RE, Lins RL, Chiang Y, Prescott MF. Plasma renin and the antihypertensive effect of the orally active renin inhibitor aliskiren in clinical hypertension. *Int J Clin Pract* 2007; **61**: 1461–1468.
140. Gradman AH, Kad R. Renin inhibition in hypertension. *J Am Coll Cardiol* 2008; **51**: 519–528.
141. Gradman AH, Pinto R, Kad R. Current concepts: Renin inhibition in the treatment of hypertension. *Curr Opin Pharmacol* 2008; **8**: 120–126.
142. Sever PS, Gradman AH, Azizi M. Managing cardiovascular and renal risk: The potential of direct renin inhibition. *J Renin Angiotensin Aldosterone Syst* 2009; **10**: 65–76.
143. Pinto R, Gradman AH. Direct renin inhibition: An update. *Curr Hypertens Rep* 2009; **11**: 456–462.

144. Stanton AV, Gradman AH, Schmieder RE, Nussberger J, Sarangapani R, Prescott MF. Aliskiren monotherapy does not cause paradoxical blood pressure rises: Meta-analysis of data from 8 clinical trials. *Hypertension* 2010; **55**: 54–60.
145. Gradman AH, Weir MR, Wright M, Bush CA, Keefe DL. Efficacy, safety and tolerability of aliskiren, a direct renin inhibitor, in women with hypertension: A pooled analysis of eight studies. *J Hum Hypertens* 2010; **24**: 721–729.
146. Sica D, Gradman AH, Lederballe O, Kolloch RE, Zhang J, Keefe DL. Long-term safety and tolerability of the oral direct renin inhibitor aliskiren with optional add-on hydrochlorothiazide in patients with hypertension: A randomized, open-label, parallel-group, multicentre, dose-escalation study with an extension phase. *Clin Drug Investig* 2011; **31**: 825–837.
147. Combinatorial chemistry. *Nat Biotechnol* 2000; 18 (Suppl): IT50–IT52.
148. Kolb HC, Finn MG, Sharpless KB. Click chemistry: Diverse chemical function from a few good reactions. *Angew Chem Int Ed Engl* 2001; **40**: 2004–2021.
149. Yu P, Liu B, Kodadek T. A high-throughput assay for assessing the cell permeability of combinatorial libraries. *Nat Biotechnol* 2005; **23**: 746–751.
150. Tan X, Hu L, Luquette LJ, III, Gao G, Liu Y, Qu H, Xi R, Lu ZJ, Park PJ, Elledge SJ. Systematic identification of synergistic drug pairs targeting HIV. *Nat Biotechnol* 2012; **30**: 1125–1130.
151. Agresti JJ, Antipov E, Abate AR, Ahn K, Rowat AC, Baret JC, Marquez M, Klibanov AM, Griffiths AD, Weitz DA. Ultrahigh-throughput screening in drop-based microfluidics for directed evolution. *Proc Natl Acad Sci U S A* 2010; **107**: 4004–4009.
152. Jahchan NS, Dudley JT, Mazur PK, Flores N, Yang D, Palmerton A, Zmoos AF, Vaka D, Tran KQ, Zhou M, Krasinska K, Riess JW, Neal JW, Khatri P, Park KS, Butte AJ, Sage J. A drug repositioning approach identifies tricyclic antidepressants as inhibitors of small cell lung cancer and other neuroendocrine tumors. *Cancer Discov* 2013; **3**: 1364–1377.

Index

www.ingramcontent.com/pod-product-compliance
Ingram Content Group UK Ltd.
Pitfield, Milton Keynes, MK11 3LW, UK
UKHW020351150926
8624IPUK00004B/107

9 789813 276222